ATF Federal Explosives Law and Regulations

Plus

ATF National Firearms Act Handbook

Table of Contents

Organized Crime Control Act of 1970, Title XI

Regulation of Explosives

Public Law 91-452, Approved October 15, 1970 (as Amended)

[**NOTE:** Any reference to the Internal Revenue Code of 1954 refers to the Internal Revenue Code of 1986 (§2, Public Law 99-514, 100 Stat. 2085, October 22, 1986.)]

Purpose

SEC. 1101. The Congress hereby declares that the purpose of this title is to protect interstate and foreign commerce against interference and interruption by reducing the hazard to persons and property arising from misuse and unsafe or insecure storage of explosive materials. It is not the purpose of this title to place any undue or unnecessary Federal restrictions or burdens on law-abiding citizens with respect to the acquisition, possession, storage, or use of explosive materials for industrial, mining, agricultural, or other lawful purposes, or to provide for the imposition by Federal regulations of any procedures or requirements other than those reasonably necessary to implement and effectuate the provisions of this title.

Sec. 1102. Title 18, United States Code, is amended by adding after chapter 39 the following chapter:

Chapter 40. Importation, Manufacture, Distribution and Storage of Explosive Materials

Editor's Note: The sections of law set out herein were added by Public Law 91-452, Title XI, §1102(a), Oct. 15, 1970, 84 Stat. 952-959, and remain un-changed unless otherwise footnoted.

Sec.

§ 841. Definitions

As used in this chapter

(a) "Person" means any individual, corporation, company, association, firm, partnership, society, or joint stock company.

(b) "Interstate or foreign commerce" means commerce between any place in a State and any place outside of that State, or within any possession of the United States (not including the Canal Zone) or the District of Columbia, and commerce between places within the same State but through any place outside of that State. "State" includes the District of Columbia, the Commonwealth of Puerto Rico, and the possessions of the United States (not including the Canal Zone).

(c) "Explosive materials" means explosives, blasting agents, and detonators.

(d) Except for the purposes of subsections (d), (e), (f), (g), (h), (i), and (j) of section 844 of this title, **"explosives"** means any chemical compound mixture, or device, the primary or common purpose of which is to function by explosion; the term includes, but is not limited to, dynamite and other high explosives, black powder, pellet powder, initiating explosives, detonators, safety fuses, squibs, detonating cord, igniter cord, and igniters. The Attorney General shall publish and revise at least annually in the Federal Register a list of these and any additional explosives which he determines to be within the coverage of this chapter. For the purposes of subsections (d), (e), (f), (g), (h), and (i) of section 844 of this title, the term "explosive" is defined in subsection (j) of such section 844.

(e) "Blasting agent" means any material or mixture, consisting of fuel and oxidizer, intended for blasting, not otherwise defined as an explosive: Provided, That the finished product, as mixed for use or shipment, cannot be detonated by means of a numbered 8 test blasting cap when unconfined.

(f) "Detonator" means any device containing a detonating charge that is used for initiating detonation in an explosive; the term includes, but is not limited to, electric blasting caps of instantaneous and delay types, blasting caps for use with safety fuses and detonating-cord delay connectors.

(g) "Importer" means any person engaged in the business of importing or bringing explosive materials into the United States for purposes of sale or distribution.

(h) "Manufacturer" means any person engaged in the business of manufacturing explosive materials for purposes of sale or distribution or for his own use.

(i) "Dealer" means any person engaged in the business of distributing explosive materials at wholesale or retail.

(j) "Permittee" means any user of explosives for a lawful purpose, who has obtained either a user permit or a limited permit under the provisions of this chapter.

(k) "Attorney General" means the Attorney General of the United States.

(l) "Crime punishable by imprisonment for a term exceeding one year" shall not mean (1) any Federal or State offenses pertaining to antitrust violations, unfair trade practices, restraints of trade, or other similar offenses relating to the regulation of business practices as the Attorney General may be regulation designate, or (2) any State offense (other than one involving a firearm or explosive) classified by the laws of the State as a misdemeanor and punishable by a term of imprisonment of two years or less.

(m) "Licensee" means any importer, manufacturer, or dealer licensed under the provisions of this chapter.

(n) "Distribute" means sell, issue, give, transfer, or otherwise dispose of.

(o) "Convention on the Marking of Plastic Explosives" means the Convention on the Marking of Plastic Explosives for the Purpose of Detection, Done at Montreal on 1 March 1991.

(p) "Detection agent" means any one of the substances specified in this subsection when introduced into a plastic explosive or formulated in such explosive as a part of the manufacturing process in such a manner as to achieve homogeneous distribution in the finished explosive, including—

(1) Ethylene glycol dinitrate (EGDN), C[2]H[4](NO[3])[2], molecular weight 152, when the minimum concentration in the finished explosive is 0.2 percent by mass;

(2) 2,3-Dimethyl-2,3-dinitrobutane (DMNB), C[6]H[12](NO[2])[2], molecular weight 176, when the minimum concentration in the finished explosive is 0.1 percent by mass;

(3) Para-Mononitrotoluene (p-MNT), C[7]H[7]NO[2], molecular weight 137, when the minimum concentration in the finished explosive is 0.5 percent by mass;

(4) Ortho-Mononitrotoluene (o-MNT), C[7]H[7]NO[2], molecular weight 137, when the minimum concentration in the finished explosive is 0.5 percent by mass; and

(5) any other substance in the concentration specified by the Attorney General, after consultation with the Secretary of State and the Secretary of Defense, that has been added to the table in part 2 of the Technical Annex to the Convention on the Marking of Plastic Explosives.

(q) "Plastic explosive" means an explosive material in flexible or elastic sheet form formulated with one or more high explosives which in their pure form has a vapor pressure less than 10^{-}<4> Pa at a temperature of 25 degrees C., is formulated with a binder material, and is as a mixture malleable or flexible at normal room temperature.

(r) "Alien" means any person who is not a citizen or national of the United States.

(s) "Responsible person" means an individual who has the power to direct the management and policies of the applicant pertaining to explosive materials.

(Added Pub. L. 91-452, title XI, Sec. 1102(a), Oct. 15, 1970, 84 Stat. 952; amended Pub. L. 104-132, title VI, Sec. 602, Apr. 24, 1996, 110 Stat. 1288; Pub. L. 107-296, title XI, Sec. 1112(e)(1), (3), 1122(a), Nov. 25, 2002, 116 Stat. 2276, 2280.)

§ 842. Unlawful acts

(a) It shall be unlawful for any person—

(1) to engage in the business of importing, manufacturing, or dealing in explosive materials without a license issued under this chapter;

(2) knowingly to withhold information or to make any false or fictitious oral or written statement or to furnish or exhibit any false, fictitious, or misrepresented identification, intended or likely to deceive for the purpose of obtaining explosive materials, or a license, permit, exemption, or relief from disability under the provisions of this chapter;

(3) other than a licensee or permittee knowingly–

(A) to transport, ship, cause to be transported, or receive any explosive materials; or

(B) to distribute explosive materials to any person other than a licensee or permittee; or

(4) who is a holder of a limited permit—

(A) to transport, ship, cause to be transported, or receive in interstate or foreign commerce any explosive materials; or

(B) to receive explosive materials from a licensee or permittee, whose premises are located outside the State of residence of the limited permit holder, or on more than 6 separate occasions, during the period of the permit, to receive explosive materials from 1 or more licensees or permittees whose premises are located within the State of residence of the limited permit holder.

(b) It shall be unlawful for any licensee or permittee to knowingly distribute any explosive materials to any person other than—

(1) a licensee;

(2) a holder of a user permit; or

(3) a holder of a limited permit who is a resident of the State here distribution is made and in which the premises of the ansferor are located.

) It shall be unlawful for any licensee to distribute explosive aterials to any person who the licensee has reason to believe tends to transport such explosive materials into a State where the archase, possession, or use of explosive materials is prohibited or hich does not permit its residents to transport or ship explosive aterials into it or to receive explosive materials in it.

) It shall be unlawful for any person knowingly to distribute xplosive materials to any individual who:

(1) is under twenty-one years of age;

(2) has been convicted in any court of a crime punishable by aprisonment for a term exceeding one year;

(3) is under indictment for a crime punishable by imprisonent for a term exceeding one year;

(4) is a fugitive from justice;

(5) is an unlawful user of or addicted to any controlled abstance (as defined in section 102 of the Controlled Substances ct);

(6) has been adjudicated a mental defective or who has been ommitted to a mental institution;

(7) is an alien, other than an alien who—

(A) is lawfully admitted for permanent residence (as efined in section 101 (a)(20) of the Immigration and Nationality ct);

(B) is in lawful nonimmigrant status, is a refugee dmitted under section 207 of the Immigration and Nationality ct or is in asylum status under section 208 of the Immigration ad Nationality Act and—

(i) is a foreign law enforcement officer of a iendly foreign government, as determined by the Secretary in onsultation with the Secretary of State, entering the United States a official law enforcement business, and the shipping, transportg, possession, or receipt of explosive materials is in furtherance this official law enforcement business; or

(ii) is a person having the power to direct or cause e direction of the management and policies of a corporation, artnership, or association licensed pursuant to section 843(a), and e shipping, transporting, possession, or receipt of explosive aterials is in furtherance of such power;

(C) is a member of a North Atlantic Treaty Organization NATO) or other friendly foreign military force, as determined by e Attorney General in consultation with the Secretary of efense, who is present in the United States under military orders r training or other military purpose authorized by the United ates and the shipping, transporting, possession, or receipt of xplosive materials is in furtherance of the authorized military urpose; or

(D) is lawfully present in the United States in cooperation with the Director of Central Intelligence, and the shipment, transportation, receipt, or possession of the explosive materials is in furtherance of such cooperation;

(8) has been discharged from the armed forces under dishonorable conditions;

(9) having been a citizen of the United States, has renounced the citizenship of that person.

(e) It shall be unlawful for any licensee knowingly to distribute any explosive materials to any person in any State where the purchase, possession, or use by such person of such explosive materials would be in violation of any State law or any published ordinance applicable at the place of distribution.

(f) It shall be unlawful for any licensee or permittee willfully to manufacture, import, purchase, distribute, or receive explosive materials without making such records as the Attorney General may by regulation require, including, but not limited to, a statement of intended use, the name, date, place of birth, social security number or taxpayer identification number, and place of residence of any natural person to whom explosive materials are distributed. If explosive materials are distributed to a corporation or other business entity, such records shall include the identity and principal and local places of business and the name, date, place of birth, and place of residence of the natural person acting as agent of the corporation or other business entity in arranging the distribution.

(g) It shall be unlawful for any licensee or permittee knowingly to make any false entry in any record which he is required to keep pursuant to this section or regulations promulgated under section 847 of this title.

(h) It shall be unlawful for any person to receive, possess, transport, ship, conceal, store, barter, sell, dispose of, or pledge or accept as security for a loan, any stolen explosive materials which are moving as, which are part of, which constitute, or which have been shipped or transported in, interstate or foreign commerce, either before or after such materials were stolen, knowing or having reasonable cause to believe that the explosive materials were stolen.

(i) It shall be unlawful for any person—

(1) who is under indictment for, or who has been convicted in any court of, a crime punishable by imprisonment for a term exceeding one year;

(2) who is a fugitive from justice;

(3) who is an unlawful user of or addicted to any controlled substance (as defined in section 102 of the Controlled Substances Act);

(4) who has been adjudicated as a mental defective or who has been committed to a mental institution;

(5) who is an alien, other than an alien who—

(A) is lawfully admitted for permanent residence (as that term is defined in section 101(a)(20) of the Immigration and Nationality Act);

(B) is in lawful nonimmigrant status, is a refugee admitted under section 207 of the Immigration and Nationality Act or is in asylum status under section 208 of the Immigration and Nationality Act and--

(i) is a foreign law enforcement officer of a friendly foreign government, as determined by the Secretary in consultation with the Secretary of State, entering the United States on official law enforcement business, and the shipping, transporting, possession, or receipt of explosive materials is in furtherance of this official law enforcement business; or

(ii) is a person having the power to direct or cause the direction of the management and policies of a corporation, partnership, or association licensed pursuant to section 843(a), and the shipping, transporting, possession, or receipt of explosive materials is in furtherance of such power;

(C) is a member of a North Atlantic Treaty Organization (NATO) or other friendly foreign military force, as determined by the Attorney General in consultation with the Secretary of Defense, who is present in the United States under military orders for training or other military purpose authorized by the United States and the shipping, transporting, possession, or receipt of explosive materials is in furtherance of the authorized military purpose; or

(D) is lawfully present in the United States in cooperation with the Director of Central Intelligence, and the shipment, transportation, receipt, or possession of the explosive materials is in furtherance of such cooperation;

(6) who has been discharged from the armed forces under dishonorable conditions;

(7) who, having been a citizen of the United States, has renounced the citizenship of that person; to ship or transport any explosive in or affecting interstate or foreign commerce or to receive or possess any explosive which has been shipped or transported in or affecting interstate or foreign commerce.

(j) It shall be unlawful for any person to store any explosive material in a manner not in conformity with regulations promulgated by the Attorney General. In promulgating such regulations, the Attorney General shall take into consideration the class, type, and quantity of explosive materials to be stored, as well as the standards of safety and security recognized in the explosives industry.

(k) It shall be unlawful for any person who has knowledge of the theft or loss of any explosive materials from his stock, to fail to report such theft or loss within twenty-four hours of discovery thereof, to the Attorney General and to appropriate local authorities.

(l) It shall be unlawful for any person to manufacture any plastic explosive that does not contain a detection agent.

(m) (1) It shall be unlawful for any person to import or bring i the United States, or export from the United States, any plastic explosive that does not contain a detection agent.

(2) This subsection does not apply to the importation or bringing into the United States, or the exportation from the Uni States, of any plastic explosive that was imported or brought int or manufactured in the United States prior to the date of enactment of this subsection [enacted April 24, 1996] by or on behal of any agency of the United States performing military or polic functions (including any military reserve component) or by or c behalf of the National Guard of any State, not later than 15 yea after the date of entry into force of the Convention on the Mark of Plastic Explosives, with respect to the United States.

(n) (1) It shall be unlawful for any person to ship, transport, transfer, receive, or possess any plastic explosive that does not contain a detection agent.

(2) This subsection does not apply to—

(A) the shipment, transportation, transfer, receipt, or possession of any plastic explosive that was imported or brougł into, or manufactured in the United States prior to the date of enactment of this subsection [enacted April 24, 1996] by any person during the period beginning on that date and ending 3 years after that date of enactment; or

(B) the shipment, transportation, transfer, receipt, or possession of any plastic explosive that was imported or brougł into, or manufactured in the United States prior to the date of enactment of this subsection [enacted April 24, 1996] by or on behalf of any agency of the United States performing a military police function (including any military reserve component) or t or on behalf of the National Guard of any State, not later than 1 years after the date of entry into force of the Convention on the Marking of Plastic Explosives, with respect to the United State

(o) It shall be unlawful for any person, other than an agency o the United States (including any military reserve component) o the National Guard of any State, possessing any plastic explosi on the date of enactment of this subsection [enacted April 24, 1996], to fail to report to the Attorney General within 120 days after such date of enactment the quantity of such explosives possessed, the manufacturer or importer, any marks of identific tion on such explosives, and such other information as the Attorney General may prescribe by regulation.

(p) Distribution of information relating to explosives, destruct devices, and weapons of mass destruction.

(1) Definitions. In this subsection—

(A) the term "destructive device" has the same meani as in section 921(a)(4);

(B) the term "explosive" has the same meaning as in section 844(j); and

(C) the term "weapon of mass destruction" has the sa meaning as in section 2332a(c)(2).

(2) Prohibition. It shall be unlawful for any person—

(A) to teach or demonstrate the making or use of an xplosive, a destructive device, or a weapon of mass destruction, to distribute by any means information pertaining to, in whole in part, the manufacture or use of an explosive, destructive evice, or weapon of mass destruction, with the intent that the aching, demonstration, or information be used for, or in rtherance of, an activity that constitutes a Federal crime of olence; or

(B) to teach or demonstrate to any person the making or se of an explosive, a destructive device, or a weapon of mass estruction, or to distribute to any person, by any means, informa-on pertaining to, in whole or in part, the manufacture or use of n explosive, destructive device, or weapon of mass destruction, nowing that such person intends to use the teaching, demonstra-on, or information for, or in furtherance of, an activity that onstitutes a Federal crime of violence.

dded Pub. L. 91-452, title XI, Sec. 1102(a), Oct. 15, 1970, 84 Stat. 3; amended Pub. L. 100-690, title VI, Sec. 6474(c), (d), Nov. 18, 1988, 2 Stat. 4380; Pub. L. 101-647, title XXXV, Sec. 3521, Nov. 29, 1990, 4 Stat. 4923; Pub. L. 103-322, title XI, Sec. 110508, 10516, Sept. 13, 94, 108 Stat. 2018, 2020; Pub. L. 104-132, title VI, Sec. 603, title VII, c. 707, Apr. 24, 1996, 110 Stat. 1289, 1296; Pub. L. 106-54, Sec. 2(a), ug. 17, 1999, 113 Stat. 398; Pub. L. 107-296, title XI, Sec. 1112(e)(3), 22(b), 1123, Nov. 25, 2002, 116 Stat. 2276, 2280, 2283.)

843. Licenses and user permits

) An application for a user permit or limited permit or a cense to import, manufacture, or deal in explosive materials all be in such form and contain such information as the ttorney General shall by regulation prescribe, including the mes of and appropriate identifying information regarding all nployees who will be authorized by the applicant to possess plosive materials, as well as fingerprints and a photograph of ch responsible person. Each applicant for a license or permit all pay a fee to be charged as set by the Attorney General, said e not to exceed $ 50 for a limited permit and $ 200 for any her license or permit. Each license or user permit shall be valid r not longer than 3 years from the date of issuance and each nited permit shall be valid for not longer than 1 year from the te of issuance. Each license or permit shall be renewable upon e same conditions and subject to the same restrictions as the iginal license or permit, and upon payment of a renewal fee not exceed one-half of the original fee.

) Upon the filing of a proper application and payment of the escribed fee, and subject to the provisions of this chapter and her applicable laws, the Attorney General shall issue to such plicant the appropriate license or permit if--

(1) the applicant (or, if the applicant is a corporation, rtnership, or association, each responsible person with respect the applicant) is not a person described in section 842(i);

(2) the applicant has not willfully violated any of the ovisions of this chapter or regulations issued hereunder;

(3) the applicant has in a State premises from which he conducts or intends to conduct business;

(4) **(A)** the Secretary verifies by inspection or, if the application is for an original limited permit or the first or second renewal of such a permit, by such other means as the Secretary determines appropriate, that the applicant has a place of storage for explosive materials which meets such standards of public safety and security against theft as the Attorney General by regulations shall prescribe; and

(B) subparagraph (A) shall not apply to an applicant for the renewal of a limited permit if the Secretary has verified, by inspection within the preceding 3 years, the matters described in subparagraph (A) with respect to the applicant; and

(5) the applicant has demonstrated and certified in writing that he is familiar with all published State laws and local ordinances relating to explosive materials for the location in which he intends to do business;

(6) none of the employees of the applicant who will be authorized by the applicant to possess explosive materials is any person described in section 842(i); and

(7) in the case of a limited permit, the applicant has certified in writing that the applicant will not receive explosive materials on more than 6 separate occasions during the 12-month period for which the limited permit is valid.

(c) The Attorney General shall approve or deny an application within a period of 90 days for licenses and permits beginning on the date such application is received by the Attorney General.

(d) The Attorney General may revoke any license or permit issued under this section if in the opinion of the Attorney General the holder thereof has violated any provision of this chapter or any rule or regulation prescribed by the Attorney General under this chapter, or has become ineligible to acquire explosive materials under section 842(d). The Attorney General's action under this subsection may be reviewed only as provided in subsection (e)(2) of this section.

(e) **(1)** Any person whose application is denied or whose license or permit is revoked shall receive a written notice from the Attorney General stating the specific grounds upon which such denial or revocation is based. Any notice of a revocation of a license or permit shall be given to the holder of such license or permit prior to or concurrently with the effective date of the revocation.

(2) If the Attorney General denies an application for, or revokes a license, or permit, he shall, upon request by the aggrieved party, promptly hold a hearing to review his denial or revocation. In the case of a revocation, the Attorney General may upon a request of the holder stay the effective date of the revocation. A hearing under this section shall be at a location convenient to the aggrieved party. The Attorney General shall give written notice of his decision to the aggrieved party within a reasonable time after the hearing. The aggrieved party may, within sixty days after receipt of the Secretary's written decision, file a petition with the United States court of appeals for the district in which he resides or has his principal place of business for a judicial review of such denial or revocation, pursuant to sections 701-706 of title 5, United States Code.

(f) Licensees and holders of user permits shall make available for inspection at all reasonable times their records kept pursuant to this chapter or the regulations issued hereunder, and licensees and permittees shall submit to the Attorney General such reports and information with respect to such records and the contents thereof as he shall by regulations prescribe. The Attorney General may enter during business hours the premises (including places of storage) of any licensee or holder of a user permit, for the purpose of inspecting or examining (1) any records or documents required to be kept by such licensee or permittee, under the provisions of this chapter or regulations issued hereunder, and (2) any explosive materials kept or stored by such licensee or permittee at such premises. Upon the request of any State or any political subdivision thereof, the Attorney General may make available to such State or any political subdivision thereof, any information which he may obtain by reason of the provisions of this chapter with respect to the identification of persons within such State or political subdivision thereof, who have purchased or received explosive materials, together with a description of such explosive materials. The Secretary may inspect the places of storage for explosive materials of an applicant for a limited permit or, at the time of renewal of such permit, a holder of a limited permit, only as provided in subsection (b)(4).

(g) Licenses and user permits issued under the provisions of subsection (b) of this section shall be kept posted and kept available for inspection on the premises covered by the license and permit.

(h) **(1)** If the Secretary receives, from an employer, the name and other identifying information of a responsible person or an employee who will be authorized by the employer to possess explosive materials in the course of employment with the employer, the Secretary shall determine whether the responsible person or employee is one of the persons described in any paragraph of section 842(i). In making the determination, the Secretary may take into account a letter or document issued under paragraph (2).

(2) **(A)** If the Secretary determines that the responsible person or the employee is not one of the persons described in any paragraph of section 842(i), the Secretary shall notify the employer in writing or electronically of the determination and issue, to the responsible person or employee, a letter of clearance, which confirms the determination.

(B) If the Secretary determines that the responsible person or employee is one of the persons described in any paragraph of section 842(i), the Secretary shall notify the employer in writing or electronically of the determination and issue to the responsible person or the employee, as the case may be, a document that—

(i) confirms the determination;

(ii) explains the grounds for the determination;

(iii) provides information on how the disability may be relieved; and

(iv) explains how the determination may be appealed.

(i) Furnishing of samples.

(1) In general. Licensed manufacturers and licensed importers and persons who manufacture or import explosive materials or ammonium nitrate shall, when required by letter issued by the Secretary, furnish—

(A) samples of such explosive materials or ammonium nitrate;

(B) information on chemical composition of those products; and

(C) any other information that the Secretary determines is relevant to the identification of the explosive materials or to identification of the ammonium nitrate.

(2) Reimbursement. The Secretary shall, by regulation, authorize reimbursement of the fair market value of samples furnished pursuant to this subsection, as well as the reasonable costs of shipment.

(Added Pub. L. 91-452, title XI, Sec. 1102(a), Oct. 15, 1970, 84 Stat. 955; amended Pub. L. 107-296, title XI, Sec. 1112(e)(3), 1122(c)-(h), 1124, Nov. 25, 2002, 116 Stat. 2276, 2281, 2282, 2285.)

§ 844. Penalties

(a) Any person who—

(1) violates any of subsections (a) through (i) or (l) through (o) of section 842 shall be fined under this title, imprisoned for not more than 10 years, or both; and

(2) violates subsection (p)(2) of section 842, shall be fined under this title, imprisoned not more than 20 years, or both.

(b) Any person who violates any other provision of section 842 of this chapter shall be fined under this title or imprisoned not more than one year, or both.

(c) **(1)** Any explosive materials involved or used or intended to be used in any violation of the provisions of this chapter or any other rule or regulation promulgated thereunder or any violation of any criminal law of the United States shall be subject to seizure and forfeiture, and all provisions of the Internal Revenue Code 1986 relating to the seizure, forfeiture, and disposition of firearms, as defined in section 5845(a) of that Code, shall, so far as applicable, extend to seizures and forfeitures under the provisions of this chapter.

(2) Notwithstanding paragraph (1), in the case of the seizure of any explosive materials for any offense for which the materials would be subject to forfeiture in which it would be impracticable or unsafe to remove the materials to a place of storage or would be unsafe to store them, the seizing officer may destroy the explosive materials forthwith. Any destruction under this paragraph shall be in the presence of at least 1 credible witness. The seizing officer shall make a report of the seizure and take samples as the Attorney General may by regulation prescribe.

(3) Within 60 days after any destruction made pursuant to paragraph (2), the owner of (including any person having an interest in) the property so destroyed may make application to the Attorney General for reimbursement of the value of the property. If the claimant establishes to the satisfaction of the Attorney General that—

(A) the property has not been used or involved in a violation of law; or

(B) any unlawful involvement or use of the property was without the claimant's knowledge, consent, or willful blindness, the Attorney General shall make an allowance to the claimant not exceeding the value of the property destroyed.

(d) Whoever transports or receives, or attempts to transport or receive, in interstate or foreign commerce any explosive with the knowledge or intent that it will be used to kill, injure, or intimidate any individual or unlawfully to damage or destroy any building, vehicle, or other real or personal property, shall be imprisoned for not more than ten years, or fined under this title, or both; and if personal injury results to any person, including any public safety officer performing duties as a direct or proximate result of conduct prohibited by this subsection, shall be imprisoned for not more than twenty years or fined under this title, or both; and if death results to any person, including any public safety officer performing duties as a direct or proximate result of conduct prohibited by this subsection, shall be subject to imprisonment for any term of years, or to the death penalty or to life imprisonment.

(e) Whoever, through the use of the mail, telephone, telegraph, or other instrument of interstate or foreign commerce, or in or affecting interstate or foreign commerce, willfully makes any threat, or maliciously conveys false information knowing the same to be false, concerning an attempt or alleged attempt being made, or to be made, to kill, injure, or intimidate any individual or unlawfully to damage or destroy any building, vehicle, or other real or personal property by means of fire or an explosive shall be imprisoned for not more than 10 years or fined under this title, or both.

(f) **(1)** Whoever maliciously damages or destroys, or attempts to damage or destroy, by means of fire or an explosive, any building, vehicle, or other personal or real property in whole or in part owned or possessed by, or leased to, the United States, or any department or agency thereof, or any institution or organization receiving Federal financial assistance, shall be imprisoned for not less than 5 years and not more than 20 years, fined under this title, or both.

(2) Whoever engages in conduct prohibited by this subsection, and as a result of such conduct, directly or proximately causes personal injury or creates a substantial risk of injury to any person, including any public safety officer performing duties, shall be imprisoned for not less than 7 years and not more than 40 years, fined under this title, or both.

(3) Whoever engages in conduct prohibited by this subsection, and as a result of such conduct directly or proximately causes the death of any person, including any public safety officer performing duties, shall be subject to the death penalty, or imprisoned for not less than 20 years or for life, fined under this title, or both.

(g) **(1)** Except as provided in paragraph (2), whoever possesses an explosive in an airport that is subject to the regulatory authority of the Federal Aviation Administration, or in any building in whole or in part owned, possessed, or used by, or leased to, the United States or any department or agency thereof, except with the written consent of the agency, department, or other person responsible for the management of such building or airport, shall be imprisoned for not more than five years, or fined under this title, or both.

(2) The provisions of this subsection shall not be applicable to—

(A) the possession of ammunition (as that term is defined in regulations issued pursuant to this chapter) in an airport that is subject to the regulatory authority of the Federal Aviation Administration if such ammunition is either in checked baggage or in a closed container; or

(B) the possession of an explosive in an airport if the packaging and transportation of such explosive is exempt from, or subject to and in accordance with, regulations of the Pipeline and Hazardous Materials Safety Administration for the handling of hazardous materials pursuant to chapter 51 of title 49.

(h) Whoever—

(1) uses fire or an explosive to commit any felony which may be prosecuted in a court of the United States, or

(2) carries an explosive during the commission of any felony which may be prosecuted in a court of the United States, including a felony which provides for an enhanced punishment if committed by the use of a deadly or dangerous weapon or device shall, in addition to the punishment provided for such felony, be sentenced to imprisonment for 10 years. In the case of a second or subsequent conviction under this subsection, such person shall be sentenced to imprisonment for 20 years. Notwithstanding any other provision of law, the court shall not place on probation or suspend the sentence of any person convicted of a violation of this subsection, nor shall the term of imprisonment imposed under this subsection run concurrently with any other term of imprisonment including that imposed for the felony in which the explosive was used or carried.

(i) Whoever maliciously damages or destroys, or attempts to damage or destroy, by means of fire or an explosive, any building, vehicle, or other real or personal property used in interstate or foreign commerce or in any activity affecting interstate or foreign commerce shall be imprisoned for not less than 5 years and not more than 20 years, fined under this title, or both; and if personal injury results to any person, including any public safety officer performing duties as a direct or proximate result of conduct prohibited by this subsection, shall be imprisoned for not less than 7 years and not more than 40 years, fined under this title, or both; and if death results to any person, including any public safety officer performing duties as a direct or proximate result of conduct prohibited by this subsection, shall also be subject to imprisonment for any term of years, or to the death penalty or to life imprisonment.

(j) For the purposes of subsections (d), (e), (f), (g), (h), and (i) of this section and section 842(p), the term "explosive" means gunpowders, powders used for blasting, all forms of high explosives, blasting materials, fuses (other than electric circuit breakers), detonators, and other detonating agents, smokeless powders, other explosive or incendiary devices within the meaning of paragraph (5) of section 232 of this title, and any chemical compounds, mechanical mixture, or device that contains any oxidizing and combustible units, or other ingredients, in such proportions, quantities, or packing that ignition by fire, by friction, by concussion, by percussion, or by detonation of the compound, mixture, or device or any part thereof may cause an explosion.

(k) A person who steals any explosives materials which are moving as, or are a part of, or which have moved in, interstate or foreign commerce shall be imprisoned for not more than 10 years, fined under this title, or both.

(l) A person who steals any explosive material from a licensed importer, licensed manufacturer, or licensed dealer, or from any permittee shall be fined under this title, imprisoned not more than 10 years, or both.

(m) A person who conspires to commit an offense under subsection (h) shall be imprisoned for any term of years not exceeding 20, fined under this title, or both.

(n) Except as otherwise provided in this section, a person who conspires to commit any offense defined in this chapter shall be subject to the same penalties (other than the penalty of death) as the penalties prescribed for the offense the commission of which was the object of the conspiracy.

(o) Whoever knowingly transfers any explosive materials, knowing or having reasonable cause to believe that such explosive materials will be used to commit a crime of violence (as defined in section 924(c)(3)) or drug trafficking crime (as defined in section 924(c)(2)) shall be subject to the same penalties as may be imposed under subsection (h) for a first conviction for the use or carrying of an explosive material.

(p) Theft reporting requirement.

(1) In general. A holder of a license or permit who knows that explosive materials have been stolen from that licensee or permittee, shall report the theft to the Attorney General not later than 24 hours after the discovery of the theft.

(2) Penalty. A holder of a license or permit who does not report a theft in accordance with paragraph (1), shall be fined not more than $ 10,000, imprisoned not more than 5 years, or both.

(Added Pub. L. 91-452, title XI, Sec. 1102(a), Oct. 15, 1970, 84 Stat. 956; amended Pub. L. 97-298, Sec. 2, Oct. 12, 1982, 96 Stat. 1319; Pub. L. 98-473, title II, Sec. 1014, Oct. 12, 1984, 98 Stat. 2142; Pub. L. 99-514, Sec. 2, Oct. 22, 1986, 100 Stat. 2095; Pub. L. 100-690, title VI, Sec. 6474(a), (b), Nov. 18, 1988, 102 Stat. 4379; Pub. L. 101-647, title XXXV, Sec. 3522, Nov. 29, 1990, 104 Stat. 4924; Pub. L. 103-272, Sec. 5(e)(7), July 5, 1994, 108 Stat. 1374; Pub. L. 103-322, title VI, Sec. 60003(a)(3), title XI, Sec. 110504(b), 110509, 110515(b), 110518(b), title XXXII, Sec. 320106, 320917(a), title XXXIII, Sec. 330016(1)(H), (K), (L), (N), Sept. 13, 1994, 108 Stat. 1969, 2016, 2018, 2020, 2111, 2129, 2147, 2148; Pub. L. 104-132, title VI, Sec. 604, title VII, Sec. 701, 706, 708(a), (c)(3), 724, Apr. 24, 1996, 110 Stat. 1289, 1291, 1295-1297, 1300; Pub. L. 104-294, title VI, Sec. 603(a), Oct. 11, 1996, 110 Stat. 3503; Pub. L. 106-54, Sec. 2(b), Aug. 17, 1999, 113 Stat. 399; Pub. L. 107-296, title XI, Sec. 1112(e)(3), 1125, 1127, Nov. 25, 2002, 116 Stat. 2276, 2285.)

§ 845. Exceptions; relief from disabilities

(a) Except in the case of subsections (l), (m), (n), or (o) of section 842 and subsections (d), (e), (f), (g), (h), and (i) of section 844 of this title, this chapter shall not apply to:

(1) aspects of the transportation of explosive materials via railroad, water, highway, or air that pertain to safety, including security, and are regulated by the United States Department of Transportation or the Department of Homeland Security;

(2) the use of explosive materials in medicines and medicinal agents in the forms prescribed by the official United States Pharmacopeia, or the National Formulary;

(3) the transportation, shipment, receipt, or importation of explosive materials for delivery to any agency of the United States or to any State or political subdivision thereof;

(4) small arms ammunition and components thereof;

(5) commercially manufactured black powder in quantities not to exceed fifty pounds, percussion caps, safety and pyrotechnic fuses, quills, quick and slow matches, and friction primers, intended to be used solely for sporting, recreational, or cultural purposes in antique firearms as defined in section 921(a)(16) of title 18 of the United States Code, or in antique devices as exempted from the term "destructive device" in section 921(a)(4) of title 18 of the United States Code; and

(6) the manufacture under the regulation of the military department of the United States of explosive materials for, or thei distribution to or storage or possession by the military or naval services or other agencies of the United States; or to arsenals, navy yards, depots, or other establishments owned by, or operated by or on behalf of, the United States.

) (1) A person who is prohibited from shipping, transporting, ceiving, or possessing any explosive under section 842(i) may pply to the Secretary for relief from such prohibition.

(2) The Secretary may grant the relief requested under aragraph (1) if the Secretary determines that the circumstances garding the applicability of section 842(i), and the applicant's cord and reputation, are such that the applicant will not be kely to act in a manner dangerous to public safety and that the anting of such relief is not contrary to the public interest.

(3) A licensee or permittee who applies for relief, under is subsection, from the disabilities incurred under this chapter a result of an indictment for or conviction of a crime punishble by imprisonment for a term exceeding 1 year shall not be arred by such disability from further operations under the license permit pending final action on an application for relief filed ırsuant to this section.

) It is an affirmative defense against any proceeding involving bsections (l) through (o) of section 842 if the proponent proves a preponderance of the evidence that the plastic explosive—

(1) consisted of a small amount of plastic explosive tended for and utilized solely in lawful—

(A) research, development, or testing of new or modied explosive materials;

(B) training in explosives detection or development or sting of explosives detection equipment; or

(C) forensic science purposes; or

(2) was plastic explosive that, within 3 years after the date enactment of the Antiterrorism and Effective Death Penalty Act 1996 [enacted April 24, 1996], will be or is incorporated in a ilitary device within the territory of the United States and mains an integral part of such military device, or is intended to , or is incorporated in, and remains an integral part of a military vice that is intended to become, or has become, the property of y agency of the United States performing military or police nctions (including any military reserve component) or the ational Guard of any State, wherever such device is located.

(3) For purposes of this subsection, the term "military vice" includes, but is not restricted to, shells, bombs, projeces, mines, missiles, rockets, shaped charges, grenades, perforas, and similar devices lawfully manufactured exclusively for ilitary or police purposes.

dded Pub. L. 91-452, title XI, Sec. 1102(a), Oct. 15. 1970, 84 Stat. 8; amended Pub. L. 93-639, Sec. 101, Jan. 4, 1975, 88 Stat. 2217; Pub. 104-132, title VI, Sec. 605, Apr. 24, 1996, 110 Stat. 1289; Pub. L. 107-6, title XI, Sec. 1126, Nov. 25, 2002, 116 Stat. 2285; Pub. L. 109-59, e VII, Sec. 7127, Aug. 10, 2005, 119 Stat. 1909.)

§ 846. Additional powers of the Attorney General

(a) The Attorney General is authorized to inspect the site of any accident, or fire, in which there is reason to believe that explosive materials were involved, in order that if any such incident has been brought about by accidental means, precautions may be taken to prevent similar accidents from occurring. In order to carry out the purpose of this subsection, the Attorney General is authorized to enter into or upon any property where explosive materials have been used, are suspected of having been used, or have been found in an otherwise unauthorized location. Nothing in this chapter shall be construed as modifying or otherwise affecting in any way the investigative authority of any other Federal agency. In addition to any other investigatory authority they have with respect to violations of provisions of this, the Federal Bureau of Investigation, together with the Bureau of Alcohol, Tobacco, Firearms, and Explosives, shall have authority to conduct investigations with respect to violations of subsection (d), (e), (f), (g), (h), or (i) of section 844 of this title.

(b) The Attorney General is authorized to establish a national repository of information on incidents involving arson and the suspected criminal misuse of explosives. All Federal agencies having information concerning such incidents shall report the information to the Attorney General pursuant to such regulations as deemed necessary to carry out the provisions of this subsection. The repository shall also contain information on incidents voluntarily reported to the Attorney General by State and local authorities.

(Added Pub. L. 91-452, title XI, Sec. 1102(a), Oct. 15, 1970, 84 Stat. 959; amended Pub. L. 104-208, div. A, title I, Sec. 101(f) (title VI, Sec. 654(a)), Sept. 30, 1996, 110 Stat. 3009-314, 3009-369; Pub. L. 107-296, title XI, Sec. 1112(e)(2), (3), Nov. 25, 2002, 116 Stat. 2276.)

§ 847. Rules and regulations

The administration of this chapter shall be vested in the Attorney General. The Attorney General may prescribe such rules and regulations as he deems reasonably necessary to carry out the provisions of this chapter. The Attorney General shall give reasonable public notice, and afford to interested parties opportunity for hearing, prior to prescribing such rules and regulations.

(Added Pub. L. 91-452, title XI, Sec. 1102(a), Oct. 15, 1970, 84 Stat. 959; amended Pub. L. 107-296, title XI, Sec. 1112(e)(3), Nov. 25, 2002, 116 Stat. 2276.)

§ 848. Effect on State Law

No provision of this chapter shall be construed as indicating an intent on the part of the Congress to occupy the field in which such provision operates to the exclusion of the law of any State on the same subject matter, unless there is a direct and positive conflict between such provision and the law of the State so that the two cannot be reconciled or consistently stand together.

(Added Pub. L. 91-452, title XI, Sec. 1102(a), Oct. 15, 1970, 84 Stat. 959.)

Title 27, Code of Federal Regulations
Part 555—Commerce in Explosives

Section Contents

555.126 Explosives transaction record for distribution of explosive materials prior to May 24, 2003 and Limited Permittee Transaction Report for distribution of explosive materials on and after May 24, 2003.
555.127 Daily summary of magazine transactions.
555.128 Discontinuance of business.
555.129 Exportation.
555.130 [Reserved]

Subpart H— Exemptions

555.141 Exemptions.
555.142 Relief from disabilities (effective January 24, 2003).

Subpart I—Unlawful Acts, Penalties, Seizures and Forfeitures

555.161 Engaging in business without a license.
555.162 False statement or representation.
555.163 False entry in record.
555.164 Unlawful storage.
555.165 Failure to report theft or loss.
555.166 Seizure or forfeiture.

Subpart J—Marking of Plastic Explosives

555.180 Prohibitions relating to unmarked plastic explosives.
555.181 Reporting of plastic explosives.
555.182 Exceptions.
555.183 Importation of plastic explosives on or after April 24, 1997.
555.184 Statements of process and samples.
555.185 Criminal sanctions.
555.186 Seizure or forfeiture.

Subpart K—Storage

555.201 General.
555.202 Classes of explosive materials.
555.203 Types of magazines.
555.204 Inspection of magazines.
555.205 Movement of explosive materials.
555.206 Location of magazines.
§ 555.207 Construction of type 1 magazines.
§ 555.208 Construction of type 2 magazines.
§ 555.209 Construction of type 3 magazines.
§ 555.210 Construction of type 4 magazines.
§ 555.211 Construction of type 5 magazines.
§ 555.212 Smoking and open flames.
§ 555.213 Quantity and storage restrictions.
§ 555.214 Storage within types 1, 2, 3, and 4 magazines.
§ 555.215 Housekeeping.
§ 555.216 Repair of magazines.
§ 555.217 Lighting.
§ 555.218 Table of distances for storage of explosive materials.
§ 555.219 Table of distances for storage of low explosives.
§ 555.220 Table of separation distances of ammonium nitrate and blasting agents from explosives or blasting agents.
§ 555.221 Requirements for display fireworks, pyrotechnic compositions, and explosive materials used in assembling fireworks or articles pyrotechnic.
§ 555.222 Table of distances between fireworks process buildings and between fireworks process and fireworks nonprocess buildings.
§ 555.223 Table of distances between fireworks process buildings and other specified areas.
§ 555.224 Table of distances for the storage of display fireworks (except bulk salutes).

Authority: 18 U.S.C. 847.

Source: T.D. ATF–87, 46 FR 40384, Aug. 7, 1981, unless otherwise noted.

Editor's Note: There are several name changes for titles referenced in the regulations, 27 CFR Part 555. The title "Chief, Firearms and Explosives Licensing Center" has been changed to "Chief, National Explosives Licensing Center". Also, the title "Regional director (compliance)" has been changed to "Director, Industry Operations" within the respective ATF Field Divisions. All regulations and rulings contained within this publication were current as of 1/26/07. For more recent information, please refer to the ATF website at www.atf.gov.

Subpart A—Introduction

§ 555.1 Scope of regulations.

(a) In general. The regulations contained in this part relate to commerce in explosives and implement Title XI, Regulation of Explosives (18 U.S.C. Chapter 40; 84 Stat. 952), of the Organized Crime Control Act of 1970 (84 Stat. 922), Pub. L. 103–322 (108 Stat. 1796), Pub. L. 104–132 (110 Stat. 1214), and Pub. L. 107–296 (116 Stat. 2135).

(b) Procedural and substantive requirements. This part contains the procedural and substantive requirements relative to:

(1) The interstate or foreign commerce in explosive materials;

(2) The licensing of manufacturers and importers of, and dealers in, explosive materials;

(3) The issuance of permits;

(4) The conduct of business by licensees and operations by permittees;

(5) The storage of explosive materials;

(6) The records and reports required of licensees and permittees;

(7) Relief from disabilities under this part;

(8) Exemptions, unlawful acts, penalties, seizures, and forfeitures; and

(9) The marking of plastic explosives.

[T.D. ATF–87, 46 FR 40384, Aug. 7, 1981, as amended by T.D. ATF–363, 60 F 17449, Apr. 6, 1995; T.D. ATF–387, 62 FR 8376, Feb. 25, 1997; ATF No. 1, 68 13780, Mar. 20, 2003]

§ 555.2 Relation to other provisions of law.

The provisions in this part are in addition to, and are not in li of, any other provision of law, or regulations, respecting commerce in explosive materials. For regulations applicable to commerce in firearms and ammunition, see Part 478 of this chapter. For regulations applicable to traffic in machine guns, destructive devices, and certain other firearms, see Part 479 of this chapter. For statutes applicable to the registration and licensing of persons engaged in the business of manufacturing, importing or exporting arms, ammunition, or implements of wa see section 38 of the Arms Export Control Act (22 U.S.C. 2778 and regulations of Part 447 of this chapter and in Parts 121 through 128 of Title 22, Code of Federal Regulations. For statut applicable to nonmailable materials, see 18 U.S.C. 1716 and implementing regulations. For statutes applicable to water quali standards, see 33 U.S.C. 1341.

Subpart B—Definitions

§ 555.11 Meaning of terms.

When used in this part, terms are defined as follows in this section. Words in the plural form include the singular, and vice versa, and words indicating the masculine gender include the feminine. The terms "includes" and "including" do not exclude other things not named which are in the same general class or are otherwise within the scope of the term defined.

Act. 18 U.S.C. Chapter 40.

Adjudicated as a mental defective. (a) A determination by a court, board, commission, or other lawful authority that a person, as a result of marked subnormal intelligence, or mental illness, incompetency, condition, or disease:

(1) Is a danger to himself or to others; or

(2) Lacks the mental capacity to contract or manage his own affairs.

(b) The term will include—

(1) A finding of insanity by a court in a criminal case; and

(2) Those persons found incompetent to stand trial or found not guilty by reason of lack of mental responsibility by any court or pursuant to articles 50a and 76b of the Uniform Code of Military Justice, 10 U.S.C. 850a, 876b.

Alien. Any person who is not a citizen or national of the United States.

Ammunition. Small arms ammunition or cartridge cases, primers, bullets, or smokeless propellants designed for use in small arms, including percussion caps, and 3/32 inch and other external burning pyrotechnic hobby fuses. The term does not include black powder.

Appropriate identifying information. The term means, in relation to an individual:

(a) The full name, date of birth, place of birth, sex, race, stree address, State of residence, telephone numbers (home and work country or countries of citizenship, and position at the employe business or operations of responsible persons and employees authorized to possess explosive materials;

(b) The business name, address, and license or permit numbe with which the responsible person or employee is affiliated;

(c) If an alien, INS-issued alien number or admission numbe and

(d) Social security number, as optional information (this information is not required but is helpful in avoiding misidentif cation when a background check is conducted).

Approved storage facility. A place where explosive material are stored, consisting of one or more approved magazines, conforming to the requirements of this part and covered by a license or permit issued under this part.

Articles pyrotechnic. Pyrotechnic devices for professional use similar to consumer fireworks in chemical composition and construction but not intended for consumer use. Such articles meeting the weight limits for consumer fireworks but not labeled as such and classified by U.S. Department of Transportation regulations in 49 CFR 172.101 as UN0431 or UN0432.

Artificial barricade. An artificial mound or revetted wall of earth of a minimum thickness of three feet, or any other approved barricade that offers equivalent protection.

ATF. (a) Prior to January 24, 2003. The Bureau of Alcohol, Tobacco and Firearms, Department of the Treasury, Washington, DC.

(b) On and after January 24, 2003. The Bureau of Alcohol, Tobacco, Firearms and Explosives, Department of Justice, Washington, DC.

ATF officer. (a) Prior to January 24, 2003. An officer or employee of the Bureau of Alcohol, Tobacco and Firearms (ATF) authorized to perform any function relating to the administration or enforcement of this part.

(b) On and after January 24, 2003. An officer or employee of the Bureau of Alcohol, Tobacco, Firearms and Explosives (ATF) authorized to perform any function relating to the administration or enforcement of this part.

Authority having jurisdiction for fire safety. The fire department having jurisdiction over sites where explosives are manufactured or stored.

Barricaded. The effective screening of a magazine containing explosive materials from another magazine, a building, a railway, or a highway, either by a natural barricade or by an artificial barricade. To be properly barricaded, a straight line from the top of any sidewall of the magazine containing explosive materials to the eave line of any other magazine or building, or to a point 12 feet above the center of a railway or highway, will pass through the natural or artificial barricade.

Blasting agent. Any material or mixture, consisting of fuel and oxidizer, that is intended for blasting and not otherwise defined as an explosive; if the finished product, as mixed for use or shipment, cannot be detonated by means of a number 8 test blasting cap when unconfined. A number 8 test blasting cap is one containing 2 grams of a mixture of 80 percent mercury fulminate and 20 percent potassium chlorate, or a blasting cap of equivalent strength. An equivalent strength cap comprises 0.40–0.45 grams of PETN base charge pressed in an aluminum shell with bottom thickness not to exceed to 0.03 of an inch, to a specific gravity of not less than 1.4 g/cc., and primed with standard weights of primer depending on the manufacturer.

Bulk salutes. Salute components prior to final assembly into aerial shells, and finished salute shells held separately prior to being packed with other types of display fireworks.

Bullet-sensitive explosive materials. Explosive materials that can be exploded by 150-grain M2 ball ammunition having a nominal muzzle velocity of 2700 fps (824 mps) when fired from a .30 caliber rifle at a distance of 100 ft (30.5 m), measured perpendicular. The test material is at a temperature of 70 to 75 degrees F (21 to 24 degrees C) and is placed against a ½ inch (12.4 mm) steel backing plate.

Bureau. (a) Prior to January 24, 2003. The Bureau of Alcohol, Tobacco and Firearms, Department of the Treasury, Washington, DC.

(b) On and after January 24, 2003. The Bureau of Alcohol, Tobacco, Firearms and Explosives, Department of Justice, Washington, DC.

Business premises. When used with respect to a manufacturer, importer, or dealer, the property on which explosive materials are manufactured, imported, stored or distributed. The premises include the property where the records of a manufacturer, importer, or dealer are kept if different than the premises where explosive materials are manufactured, imported, stored or distributed. When used with respect to a user of explosive materials, the property on which the explosive materials are received or stored. The premises includes the property where the records of the users are kept if different than the premises where explosive materials are received or stored.

Chief, Firearms and Explosives Licensing Center. The ATF official responsible for the issuance and renewal of licenses and permits under this part.

Committed to a mental institution. A formal commitment of a person to a mental institution by a court, board, commission, or other lawful authority. The term includes a commitment to a mental institution involuntarily. The term includes commitment for mental defectiveness or mental illness. It also includes commitments for other reasons, such as for drug use. The term does not include a person in a mental institution for observation or a voluntary admission to a mental institution.

Common or contract carrier. Any individual or organization engaged in the business of transporting passengers or goods.

Consumer fireworks. Any small firework device designed to produce visible effects by combustion and which must comply with the construction, chemical composition, and labeling regulations of the U.S. Consumer Product Safety Commission, as set forth in title 16, Code of Federal Regulations, parts 1500 and 1507. Some small devices designed to produce audible effects are included, such as whistling devices, ground devices containing 50 mg or less of explosive materials, and aerial devices containing 130 mg or less of explosive materials. Consumer fireworks are classified as fireworks UN0336, and UN0337 by the U.S. Department of Transportation at 49 CFR 172.101. This term does not include fused setpieces containing components which together exceed 50 mg of salute powder.

Controlled substance. A drug or other substance, or immediate precursor, as defined in section 102 of the Controlled Substances Act, 21 U.S.C. 802. The term includes, but is not limited to, marijuana, depressants, stimulants, and narcotic drugs. The term does not include distilled spirits, wine, malt beverages, or tobacco, as those terms are defined or used in Subtitle E of the Internal Revenue Code of 1986, as amended.

Crime punishable by imprisonment for a term exceeding one year. Any offense for which the maximum penalty, whether or not imposed, is capital punishment or imprisonment in excess of one year. The term does not include (a) any Federal or State offenses pertaining to antitrust violations, unfair trade practices, restraints of trade, or (b) any State offense (other than one involving a firearm or explosive) classified by the laws of the State as a misdemeanor and punishable by a term of imprisonment of two years or less.

Customs officer. Any officer of the Customs Service or any commissioned, warrant, or petty officer of the Coast Guard, or any agent or other person authorized to perform the duties of an officer of the Customs Service.

Dealer. Any person engaged in the business of distributing explosive materials at wholesale or retail.

Detonator. Any device containing a detonating charge that is used for initiating detonation in an explosive. The term includes, but is not limited to, electric blasting caps of instantaneous and delay types, blasting caps for use with safety fuses, detonating-cord delay connectors, and nonelectric instantaneous and delay blasting caps.

Director. (a) Prior to January 24, 2003. The Director, Bureau of Alcohol, Tobacco and Firearms, Department of the Treasury, Washington, DC. (b) On and after January 24, 2003. The Director, Bureau of Alcohol, Tobacco, Firearms and Explosives, Department of Justice, Washington, DC.

Discharged under dishonorable conditions. Separation from the U.S. Armed Forces resulting from a dishonorable discharge or dismissal adjudged by general court-martial. The term does not include any separation from the Armed Forces resulting from any other discharge, e.g., a bad conduct discharge.

Display fireworks. Large fireworks designed primarily to produce visible or audible effects by combustion, deflagration, or detonation. This term includes, but is not limited to, salutes containing more than 2 grains (130 mg) of explosive materials, aerial shells containing more than 40 grams of pyrotechnic compositions, and other display pieces which exceed the limits of explosive materials for classification as "consumer fireworks." Display fireworks are classified as fireworks UN0333, UN0334 or UN0335 by the U.S. Department of Transportation at 49 CFR 172.101. This term also includes fused setpieces containing components which together exceed 50 mg of salute powder.

Distribute. To sell, issue, give, transfer, or otherwise dispose of. The term does not include a mere change of possession from a person to his agent or employee in connection with the agency or employment.

Executed under penalties of perjury. Signed with the required declaration under the penalties of perjury as provided on or with respect to the return, form, or other document or, where no form of declaration is required, with the declaration:

"I declare under the penalties of perjury that this —(insert type of document, such as, statement, application, request, certificate), including the documents submitted in support thereof, has been examined by me and, to the best of my knowledge and belief, is true, correct, and complete".

Explosive actuated device. Any tool or special mechanized device which is actuated by explosives, but not a propellent actuated device.

Explosive materials. Explosives, blasting agents, water gels and detonators. Explosive materials include, but are not limited to all items in the "List of Explosive Materials" provided for in §555.23.

Explosives. Any chemical compound, mixture, or device, the primary or common purpose of which is to function by explosion. The term includes, but is not limited to, dynamite and other high explosives, black powder, pellet powder, initiating explosives, detonators, safety fuses, squibs, detonating cord, igniter cord, and igniters.

Fireworks. Any composition or device designed to produce a visible or an audible effect by combustion, deflagration, or detonation, and which meets the definition of "consumer fireworks" or "display fireworks" as defined by this section.

Fireworks mixing building. Any building or area used for mixing and blending pyrotechnic compositions except wet sparkler mix.

Fireworks nonprocess building. Any office building or other building or area in a fireworks plant where no fireworks, pyrotechnic compositions or explosive materials are processed or stored.

Fireworks plant. All land and buildings thereon used for or in connection with the assembly or processing of fireworks, including warehouses used with or in connection with fireworks plant operations.

Fireworks plant warehouse. Any building or structure used exclusively for the storage of materials which are neither explosive materials nor pyrotechnic compositions used to manufacture or assemble fireworks.

Fireworks process building. Any mixing building; any building in which pyrotechnic compositions or explosive materials is pressed or otherwise prepared for finished and assembly; or any finishing or assembly building.

Fireworks shipping building. A building used for the packing of assorted display fireworks into shipping cartons for individual public displays and for the loading of packaged displays for shipment to purchasers.

Flash powder. An explosive material intended to produce an audible report and a flash of light when ignited which includes but is not limited to oxidizers such as potassium chlorate or potassium perchlorate, and fuels such as sulfur or aluminum powder.

Fugitive from justice. Any person who has fled from the jurisdiction of any court of record to avoid prosecution for any crime or to avoid giving testimony in any criminal proceeding. The term also includes any person who has been convicted of any crime and has fled to avoid imprisonment.

Hardwood. Oak, maple, ash, hickory, or other hard wood, free om loose knots, spaces, or similar defects.

Highway. Any public street, public alley, or public road, cluding a privately financed, constructed, or maintained road at is regularly and openly traveled by the general public.

Identification document. A document containing the name, sidence address, date of birth, and photograph of the holder and hich was made or issued by or under the authority of the United ates Government, a State, political subdivision of a State, a reign government, a political subdivision of a foreign govern- ent, an international governmental or an international quasi- vernmental organization which, when completed with informa- on concerning a particular individual, is of a type intended or mmonly accepted for the purpose of identification of individu- s.

Importer. Any person engaged in the business of importing or inging explosive materials into the United States for purposes sale or distribution.

Indictment. Includes an indictment or information in any court der which a crime punishable by imprisonment for a term ceeding one year may be prosecuted.

Inhabited building. Any building regularly occupied in whole in part as a habitation for human beings, or any church, hoolhouse, railroad station, store, or other structure where ople are accustomed to assemble, except any building occupied connection with the manufacture, transportation, storage, or e of explosive materials.

Interstate or foreign commerce. Commerce between any ace in a State and any place outside of that State, or within any ssession of the United States or the District of Columbia, and mmerce between places within the same State but through any ace outside of that State.

Licensed dealer. A dealer licensed under this part.

Licensed importer. An importer licensed under this part.

Licensed manufacturer. A manufacturer licensed under this rt to engage in the business of manufacturing explosive aterials for purposes of sale or distribution or for his own use.

Licensee. Any importer, manufacturer, or dealer licensed under is part.

Limited permit. A permit issued to a person authorizing him receive for his use explosive materials from a licensee or rmittee in his state of residence on no more than 6 occasions ring the 12-month period in which the permit is valid. A limited rmit does not authorize the receipt or transportation of explo- ve materials in interstate or foreign commerce.

Magazine. Any building or structure, other than an explosives anufacturing building, used for storage of explosive materials.

Manufacturer. Any person engaged in the business of manu- cturing explosive materials for purposes of sale or distribution for his own use.

Mass detonation (mass explosion). Explosive materials mass detonate (mass explode) when a unit or any part of a larger quantity of explosive material explodes and causes all or a substantial part of the remaining material to detonate or explode.

Mental institution. Includes mental health facilities, mental hospitals, sanitariums, psychiatric facilities, and other facilities that provide diagnoses by licensed professionals of mental retardation or mental illness, including a psychiatric ward in a general hospital.

Natural barricade. Natural features of the ground, such as hills, or timber of sufficient density that the surrounding exposures which require protection cannot be seen from the magazine when the trees are bare of leaves.

Number 8 test blasting cap. (See definition of "blasting agent.")

Permittee. Any user of explosives for a lawful purpose who has obtained either a user permit or a limited permit under this part.

Person. Any individual, corporation, company, association, firm, partnership, society, or joint stock company.

Plywood. Exterior, construction grade (laminated wood) plywood.

Propellant actuated device. Any tool or special mechanized device or gas generator system which is actuated by a propellant or which releases and directs work through a propellant charge.

Pyrotechnic compositions. A chemical mixture which, upon burning and without explosion, produces visible, brilliant displays, bright lights, or sounds.

Railway. Any steam, electric, or other railroad or railway which carries passengers for hire.

Region. A geographical region of the Bureau of Alcohol, Tobacco and Firearms.

Regional director (compliance). The principal regional official responsible for administering regulations in this part.

Renounced U.S. citizenship. (a) A person has renounced his U.S. citizenship if the person, having been a citizen of the United States, has renounced citizenship either—

(1) Before a diplomatic or consular officer of the United States in a foreign state pursuant to 8 U.S.C. 1481(a)(5); or

(2) Before an officer designated by the Attorney General when the United States is in a state of war pursuant to 8 U.S.C. 1481(a)(6).

(b) The term will not include any renunciation of citizenship that has been reversed as a result of administrative or judicial appeal.

Responsible person. An individual who has the power to direct the management and policies of the applicant pertaining to explosive materials. Generally, the term includes partners, sole proprietors, site managers, corporate officers and directors, and majority shareholders.

Salute. An aerial shell, classified as a display firework, that contains a charge of flash powder and is designed to produce a flash of light and a loud report as the pyrotechnic effect.

Screen barricade. Any barrier that will contain the embers and debris from a fire or deflagration in a process building, thus preventing propagation of fire to other buildings or areas. Such barriers shall be constructed of metal roofing, ¼ to ½ inch (6 to 13 mm) mesh screen, or equivalent material. The barrier extends from floor level to a height such that a straight line from the top of any side wall of the donor building to the eave line of any exposed building intercepts the screen at a point not less than 5 feet (1.5 m) from the top of the screen. The top 5 feet (1.5 m) of the screen is inclined towards the donor building at an angle of 30 to 45 degrees.

Softwood. Fir, pine, or other soft wood, free from loose knots, spaces, or similar defects.

State. A State of the United States. The term includes the District of Columbia, the Commonwealth of Puerto Rico, and the possessions of the United States.

State of residence. The State in which an individual regularly resides or maintains his home. Temporary stay in a State does not make the State of temporary stay the State of residence.

Theatrical flash powder. Flash powder commercially manufactured in premeasured kits not exceeding 1 ounce and mixed immediately prior to use and intended for use in theatrical shows, stage plays, band concerts, magic acts, thrill shows, and clown acts in circuses.

Unlawful user of or addicted to any controlled substance. A person who uses a controlled substance and has lost the power of self-control with reference to the use of a controlled substance; and any person who is a current user of a controlled substance in a manner other than as prescribed by a licensed physician. Such use is not limited to the use of drugs on a particular day, or within a matter of days or weeks before possession of the explosive materials, but rather that the unlawful use has occurred recently enough to indicate that the individual is actively engaged in such conduct. A person may be an unlawful current user of a control substance even though the substance is not being used at the precise time the person seeks to acquire explosive materials or receives or possesses explosive materials. An inference of curre use may be drawn from evidence of a recent use or possession controlled substance or a pattern of use or possession that reasonably covers the present time, e.g., a conviction for use or possession of a controlled substance within the past year; multi arrests for such offenses within the past 5 years if the most rece arrest occurred within the past year; or persons found through a drug test to use a controlled substance unlawfully, provided tha the test was administered within the past year. For a current or former member of the Armed Forces, an inference of current us may be drawn from recent disciplinary or other administrative action based on confirmed drug use, e.g., court-martial conviction, nonjudicial punishment, or an administrative discharge ba on drug use or drug rehabilitation failure.

U.S.C. The United States Code.

User-limited permit. A user permit valid only for a single purchase transaction, a new permit being required for a subsequent purchase transaction.

User permit. A permit issued to a person authorizing him (a acquire for his own use explosive materials from a licensee in a State other than the State in which he resides or from a foreign country, and (b) to transport explosive materials in interstate or foreign commerce.

Water gels. Explosives or blasting agents that contain a substantial proportion of water.

(18 U.S.C. 847 (84 Stat. 959); 18 U.S.C. 926 (82 Stat. 1226)
[T.D. ATF–87, 46 FR 40384, Aug. 7, 1981, as amended by T.D. ATF–92, 46 FR 46916, Sept. 23, 1981; T.D. ATF–200, 50 FR 10497, Mar. 15, 1985; T.D. ATF– 54 FR 53053, Dec. 27, 1989; T.D. ATF–293, 55 FR 3720, Feb. 5, 1990; T.D. AT 314, 56 FR 49140, Sept. 27, 1991; T.D. ATF–382, 61 FR 38084, July 23, 1996; T.D. ATF–400, 63 FR 45001, Aug. 24, 1998; T.D. ATF–446, 66 FR 16602, Mar 2001; ATF No. 1, 68 FR 13780, Mar. 20, 2003]

Subpart C—Administrative and Miscellaneous Provisions

555.21 Forms prescribed.

) The Director is authorized to prescribe all forms required by is part. All of the information called for in each form shall be rnished as indicated by the headings on the form and the structions on or pertaining to the form. In addition, information lled for in each form shall be furnished as required by this part.

) Requests for forms should be mailed to the ATF Distribution enter, 7943 Angus Court, Springfield, Virginia 22153.
D. ATF-92, 46 FR 46916, Sept. 23, 1981, as amended by T.D. ATF-249, 52 FR 61, Feb. 27, 1987; T.D. 372, 61 FR 20724, May 8, 1996]

555.22 Alternate methods or procedures; nergency variations from requirements.

) Alternate methods or procedures. The permittee or licensee, specific approval by the Director as provided by this para- aph, may use an alternate method or procedure in lieu of a ethod or procedure specifically prescribed in this part. The irector may approve an alternate method or procedure, subject stated conditions, when he finds that:

(1) Good cause is shown for the use of the alternate method procedure;

(2) The alternate method or procedure is within the purpose ; and consistent with the effect intended by, the specifically escribed method or procedure and that the alternate method or ocedure is substantially equivalent to that specifically escribed method or procedure; and

(3) The alternate method or procedure will not be contrary any provision of law and will not result in an increase in cost to e Government or hinder the effective administration of this part. here the permittee or licensee desires to employ an alternate ethod or procedure, he shall submit a written application to the gional director (compliance), for transmittal to the Director. The plication shall specifically describe the proposed alternate ethod or procedure and shall set forth the reasons for it. ternate methods or procedures may not be employed until the plication is approved by the Director. The permittee or licensee all, during the period of authorization of an alternate method or ocedure, comply with the terms of the approved application. uthorization of any alternate method or procedure may be thdrawn whenever, in the judgment of the Director, the fective administration of this part is hindered by the continua- n of the authorization. As used in this paragraph, alternate ethods or procedures include alternate construction or equip- ent.

) Emergency variations from requirements. The Director may prove construction, equipment, and methods of operation other an as specified in this part, where he finds that an emergency ists and the proposed variations from the specified require- ents are necessary and the proposed variations:

(1) Will afford security and protection that are substantially equivalent to those prescribed in this part;

(2) Will not hinder the effective administration of this part; and

(3) Will not be contrary to any provisions of law.

Variations from requirements granted under this paragraph are conditioned on compliance with the procedures, conditions, and limitations set forth in the approval of the application. Failure to comply in good faith with the procedures, conditions, and limitations shall automatically terminate the authority for the variations and the licensee or permittee shall fully comply with the prescribed requirements of regulations from which the variations were authorized. Authority for any variation may be withdrawn whenever, in the judgment of the Director, the effective administration of this part is hindered by the continuation of the variation. Where the licensee or permittee desires to employ an emergency variation, he shall submit a written application to the regional director (compliance) for transmittal to the Director. The application shall describe the proposed variation and set forth the reasons for it. Variations may not be employed until the application is approved, except when the emergency requires immediate action to correct a situation that is threatening to life or property. Corrective action may then be taken concurrent with the filing of the application and notification of the Director via telephone.

(c) Retention of approved variations. The licensee or permittee shall retain, as part of his records available for examination by ATF officers, any application approved by the Director under this section.

§ 555.23 List of explosive materials.

The Director shall compile a list of explosive materials, which shall be published and revised at least annually in the Federal Register. The "List of Explosive Materials" (ATF Publication 5400 8) is available at no cost upon request from the ATF Distribution Center, 7943 Angus Court, Springfield, Virginia 22153.
[T.D. ATF-290, 54 FR 53054, Dec. 27, 1989, as amended by T.D. ATF-446, 66 FR 16602, Mar. 27, 2001]

§ 555.24 Right of entry and examination.

(a) Any ATF officer may enter during business hours the premises, including places of storage, of any licensee or holder of a user permit for the purpose of inspecting or examining any records or documents required to be kept under this part, and any facilities in which explosive materials are kept or stored.

(b) Any ATF officer may inspect the places of storage for explosive materials of an applicant for a limited permit or, in the case of a holder of a limited permit, at the time of renewal of such permit.

(c) The provisions of paragraph (b) of this section do not apply to an applicant for the renewal of a limited permit if an ATF officer has, within the preceding 3 years, verified by inspection that the applicant's place of storage for explosive materials meets the requirements of subpart K of this part.
[ATF No. 1, 68 FR 13781, Mar. 20, 2003]

§ 555.25 Disclosure of information.

Upon receipt of written request from any State or any political subdivision of a State, the regional director (compliance) may make available to the State or political subdivision any information which the regional director (compliance) may obtain under the Act with respect to the identification of persons within the State or political subdivision, who have purchased or received explosive materials, together with a description of the explosive materials.

§ 555.26 Prohibited shipment, transportation, receipt, possession, or distribution of explosive materials.

(a) General. No person, other than a licensee or permittee knowingly may transport, ship, cause to be transported, or receive any explosive materials: Provided, That the provisions of this paragraph (a) do not apply to the lawful purchase by a nonlicensee or nonpermittee of commercially manufactured black powder in quantities not to exceed 50 pounds, if the black powder is intended to be used solely for sporting, recreational, or cultural purposes in antique firearms as defined in 18 U.S.C. 921(a)(16), or in antique devices as exempted from the term "destructive device" in 18 U.S.C. 921(a)(4).

(b) Holders of a limited permit. No person who is a holder of a limited permit may—

(1) Transport, ship, cause to be transported, or receive in interstate or foreign commerce any explosive materials;

(2) Receive explosive materials from a licensee or permittee, whose premises are located outside the State of residence of the limited permit holder; or

(3) Receive explosive materials on more than 6 separate occasions, during the period of the permit, from one or more licensees or permittees whose premises are located within the State of residence of the limited permit holder. (See §555.105(b) for the definition of "6 separate occasions.")

(c) Possession by prohibited persons. No person may ship or transport any explosive material in or affecting interstate or foreign commerce or receive or possess any explosive materials which have been shipped or transported in or affecting interstate or foreign commerce who:

(1) Is under indictment or information for, or who has been convicted in any court of, a crime punishable by imprisonment for a term exceeding one year;

(2) Is a fugitive from justice;

(3) Is an unlawful user of or addicted to any controlled substance (as defined in section 102 of the Controlled Substanc Act (21 U.S.C. 802) and §555.11);

(4) Has been adjudicated as a mental defective or has bee committed to a mental institution;

(5) Is an alien, other than an alien who—

(i) Is lawfully admitted for permanent residence (as that term is defined in section 101(a)(20) of the Immigration ar Nationality Act (8 U.S.C. 1101)); or

(ii) Is in lawful nonimmigrant status, is a refugee admitted under section 207 of the Immigration and Nationality Act (8 U.S.C. 1157), or is in asylum status under section 208 of the Immigration and Nationality Act (8 U.S.C. 1158), and—

(A) Is a foreign law enforcement officer of a friendly foreign government, as determined by the Attorney General in consultation with the Secretary of State, entering the United States on official law enforcement business, and the shipping, transporting, possession, or receipt of explosive materials is in furtherance of this official law enforcement business;

(B) Is a person having the power to direct or ca the direction of the management and policies of a corporation, partnership, or association licensed pursuant to section 843(a) c the Act, and the shipping, transporting, possession, or receipt o explosive materials is in furtherance of such power;

(C) Is a member of a North Atlantic Treaty Organization (NATO) or other friendly foreign military force, a determined by the Attorney General in consultation with the Secretary of Defense, (whether or not admitted in a nonimmigr status) who is present in the United States under military orders for training or other military purpose authorized by the United States, and the shipping, transporting, possession, or receipt of explosive materials is in furtherance of the military purpose; or

(D) Is lawfully present in the United States in cooperation with the Director of Central Intelligence, and the shipment, transportation, receipt, or possession of the explosive materials is in furtherance of such cooperation;

(6) Has been discharged from the armed forces under dishonorable conditions; or

(7) Having been a citizen of the United States, has renounced citizenship.

(d) Distribution to prohibited persons. No person may knowin distribute explosive materials to any individual who:

(1) Is under twenty-one years of age;

(2) Is under indictment or information for, or who has be convicted in any court of, a crime punishable by imprisonment a term exceeding one year;

(3) Is a fugitive from justice;

(4) Is an unlawful user of or addicted to any controlled substance (as defined in section 102 of the Controlled Substan Act (21 U.S.C. 802) and §555.11);

(5) Has been adjudicated as a mental defective or has been committed to a mental institution;

(6) Is an alien, other than an alien who—

(i) Is lawfully admitted for permanent residence (as that term is defined in section 101(a)(20) of the Immigration and Nationality Act (8 U.S.C. 1101)); or

(ii) Is in lawful nonimmigrant status, is a refugee admitted under section 207 of the Immigration and Nationality Act (8 U.S.C. 1157), or is in asylum status under section 208 of the Immigration and Nationality Act (8 U.S.C. 1158), and—

(A) Is a foreign law enforcement officer of a friendly foreign government, as determined by the Attorney General in consultation with the Secretary of State, entering the United States on official law enforcement business, and the shipping, transporting, possession, or receipt of explosive materials is in furtherance of this official law enforcement business;

(B) Is a person having the power to direct or cause the direction of the management and policies of a corporation, partnership, or association licensed pursuant to section 843(a) of the Act, and the shipping, transporting, possession, or receipt of explosive materials is in furtherance of such power;

(C) Is a member of a North Atlantic Treaty Organization (NATO) or other friendly foreign military force, as determined by the Attorney General in consultation with the Secretary of Defense, (whether or not admitted in a nonimmigrant status) who is present in the United States under military orders for training or other military purpose authorized by the United States, and the shipping, transporting, possession, or receipt of explosive materials is in furtherance of the military purpose; or

(D) Is lawfully present in the United States in cooperation with the Director of Central Intelligence, and the shipment, transportation, receipt, or possession of the explosive materials is in furtherance of such cooperation;

(7) Has been discharged from the armed forces under dishonorable conditions; or

(8) Having been a citizen of the United States, has renounced citizenship.

(e) See §555.180 for regulations concerning the prohibited manufacture, importation, exportation, shipment, transportation, receipt, transfer, or possession of plastic explosives that do not contain a detection agent.

[ATF No. 1, 68 FR 13781, Mar. 20, 2003]

§ 555.27 Out-of-State disposition of explosive materials.

(a) No nonlicensee or nonpermittee may distribute any explosive materials to any other nonlicensee or nonpermittee who the distributor knows or who has reasonable cause to believe does not reside in the State in which the distributor resides.

(b) The provisions of this section do not apply on and after May 24, 2003.

[ATF No. 1, 68 FR 13782, Mar. 20, 2003]

§ 555.28 Stolen explosive materials.

No person shall receive, conceal, transport, ship, store, barter, sell, or dispose of any stolen explosive materials knowing or having reasonable cause to believe that the explosive materials were stolen.

§ 555.29 Unlawful storage.

No person shall store any explosive materials in a manner not in conformity with this part.

§ 555.30 Reporting theft or loss of explosive materials.

(a) Any licensee or permittee who has knowledge of the theft or loss of any explosive materials from his stock shall, within 24 hours of discovery, report the theft or loss by telephoning 1–800–800–3855 (nationwide toll free number) and on ATF F 5400.5 (formerly Form 4712) in accordance with the instructions on the form. Theft or loss of any explosive materials shall also be reported to appropriate local authorities.

(b) Any other person, except a carrier of explosive materials, who has knowledge of the theft or loss of any explosive materials from his stock shall, within 24 hours of discovery, report the theft or loss by telephoning 1–800–800–3855 (nationwide toll free number) and in writing to the nearest ATF office. Theft or loss shall be reported to appropriate local authorities.

(c) Reports of theft or loss of explosive materials under paragraphs (a) and (b) of this section must include the following information, if known:

(1) The manufacturer or brand name.

(2) The manufacturer's marks of identification (date and shift code).

(3) Quantity (applicable quantity units, such as pounds of explosives, number of detonators, etc.).

(4) Description (dynamite, blasting agents, detonators, etc.) and United Nations (UN) identification number, hazard division number, and classification letter, e.g., 1.1D, as classified by the U.S. Department of Transportation at 49 CFR 172.101 and 173.52.

(5) Size (length and diameter).

(d) A carrier of explosive materials who has knowledge of the theft or loss of any explosive materials shall, within 24 hours of discovery, report the theft or loss by telephoning 1–800–800–3855 (nationwide toll free number). Theft or loss shall also be reported to appropriate local authorities. Reports of theft or loss of explosive materials by carriers shall include the following information, if known:

(1) The manufacturer or brand name.

(2) Quantity (applicable quantity units, such as pounds of explosives, number of detonators, etc.).

(3) Description (United Nations (UN) identification number, hazard division number, and classification letter, e.g., 1.1D) as classified by the U.S. Department of Transportation at 49 CFR 172.101 and 173.52.

[T.D. ATF–87, 46 FR 40384, Aug. 7, 1981, as amended by T.D. ATF–400, 63 FR 45002, Aug. 24, 1998]

§ 555.31 Inspection of site accidents or fires; right of entry.

Any ATF officer may inspect the site of any accident or fire in which there is reason to believe that explosive materials were involved. Any ATF officer may enter into or upon any property where explosive materials have been used, are suspected of having been used, or have been found in an otherwise unauthorized location.

§ 555.32 Special explosive devices.

The Director may exempt certain explosive actuated devices, explosive actuated tools, or similar devices from the requirements of this part. A person who desires to obtain an exemption under this section for any special explosive device, which as designed does not constitute a public safety or security hazard, shall submit a written request to the Director. Each request shall be executed under the penalties of perjury and contain a complete and accurate description of the device, the name and address of the manufacturer or importer, the purpose of and use for which it is intended, and any photographs, diagrams, or drawings as may be necessary to enable the Director to make a determination. The Director may require that a sample of the device be submitted for examination and evaluation. If it is not possible to submit the device, the person requesting the exemption shall advise the Director and designate the place where the device will be available for examination and evaluation.

§ 555.33 Background checks and clearances (effective May 24, 2003).

(a) Background checks.

(1) If the Director receives from a licensee or permittee the names and appropriate identifying information of responsible persons and employees who will be authorized by the employer to possess explosive materials in the course of employment with the employer, the Director will conduct a background check in accordance with this section.

(2) The Director will determine whether the responsible person or employee is one of the persons described in any paragraph of section 842(i) of the Act (see §555.26). In making such determination, the Director may take into account a letter or document issued under paragraph (a)(3) of this section.

(3)(i) If the Director determines that the responsible person or the employee is not one of the persons described in any paragraph of section 842(i) of the Act (see §555.26), the Director will notify the employer in writing or electronically of the determination and issue, to the responsible person or employee, as the case may be, a letter of clearance which confirms the determination.

(ii) If the Director determines that the responsible person or employee is one of the persons described in any paragraph of section 842(i) of the Act (see §555.26), ATF will notify the employer in writing or electronically of the determination and issue to the responsible person or the employee, as the case may be, a document that confirms the determination; explains the grounds for the determination; provides information on how the disability may be relieved; and explains how the determination may be appealed. The employer will retain the notification as part of his permanent records in accordance with §555.121. The employer will take immediate steps to remove the responsible person from his position directing the management or policies of the business or operations as they relate to explosive materials or, as the case may be, to remove the employee from a position requiring the possession of explosive materials. Also, if the employer has listed the employee as a person authorized to accept delivery of explosive materials, as specified in §555.103 or §555.105, the employer must remove the employee from such list and immediately, and in no event later than the second business day after such change, notify distributors of such change.

(b) Appeals and correction of erroneous system information—

(1) In general. A responsible person or employee may challenge the adverse determination set out in the letter of denial, in writing and within 45 days of issuance of the determination, by directing his or her challenge to the basis for the adverse determination, or to the accuracy of the record upon which the adverse determination is based, to the Director. The appeal request must include appropriate documentation or record(s) establishing the legal and/or factual basis for the challenge. Any record or document of a court or other government entity or official furnished in support of an appeal must be certified by the court or other government entity or official as a true copy. In the case of an employee, or responsible person who did not submit fingerprints, such appeal must be accompanied by two properly completed FBI Forms FD–258 (fingerprint card). The Director will advise the individual in writing of his decision and the reasons for the decision.

(2) Employees. The letter of denial, among other things, will advise an employee who elects to challenge an adverse determination to submit the fingerprint cards as described above. The employee also will be advised of the agency name and address that originated the record containing the information causing the adverse determination ("originating agency"). At that time, and where appropriate, an employee is encouraged to apply to the originating agency to challenge the accuracy of the record(s) upon which the denial is based. The originating agency may respond to the individual's application by addressing the individual's specific reasons for the challenge, and by indicating hether additional information or documents are required. If the cord is corrected as a result of the application to the originating gency, the individual may so notify ATF which will, in turn, rify the record correction with the originating agency and take necessary steps to contact the agency responsible for the cord system and correct the record. The employee may provide ATF additional and appropriate documentation or record(s) tablishing the legal and/or factual basis for the challenge to TF's decision to uphold the initial denial. If ATF does not ceive such additional documentation or record(s) within 45 days the date of the decision upholding the initial denial, ATF will ose the appeal.

(3) Responsible persons. The letter of denial, among other ings, will advise a responsible person of the agency name and dress which originated the record containing the information using the adverse determination ("originating agency"). A sponsible person who elects to challenge the adverse determina-n, where appropriate, is encouraged to apply to the originating ency to challenge the accuracy of the record(s) upon which the nial is based. The originating agency may respond to the dividual's application by addressing the individual's specific asons for the challenge, and by indicating whether additional formation or documents are required. If the record is corrected a result of the application to the originating agency, the dividual may so notify ATF which will, in turn, verify the record correction with the originating agency and take all necessary steps to contact the agency responsible for the record system and correct the record. A responsible person may provide additional documentation or records as specified for employees in paragraph (b)(2) of this section.

(Approved by the Office of Management and Budget under control number 1140–0081)

[ATF No. 1, 68 FR 13783, Mar. 20, 2003]

§ 555.34 Replacement of stolen or lost ATF Form 5400.30 (Intrastate Purchase of Explosives Coupon (IPEC)).

When any Form 5400.30 is stolen, lost, or destroyed, the person losing possession will, upon discovery of the theft, loss, or destruction, immediately, but in all cases before 24 hours have elapsed since discovery, report the matter to the Director by telephoning 1–888–ATF-BOMB (nationwide toll free number). The report will explain in detail the circumstances of the theft, loss, or destruction and will include all known facts that may serve to identify the document. Upon receipt of the report, the Director will make such investigation as appears appropriate and may issue a duplicate document upon such conditions as the circumstances warrant.

(Approved by the Office of Management and Budget under control number 1140–0077)

[ATF No. 1, 68 FR 13783, Mar. 20, 2003]

Subpart D—Licenses and Permits

555.41 General.

Licenses and permits issued prior to May 24, 2003.

(1) Each person intending to engage in business as an porter or manufacturer of, or a dealer in, explosive materials, cluding black powder, must, before commencing business, tain the license required by this subpart for the business to be erated. Each person who intends to acquire for use explosive aterials from a licensee in a State other than the State in which resides, or from a foreign country, or who intends to transport plosive materials in interstate or foreign commerce, must tain a permit under this subpart; except that it is not necessary obtain a permit if the user intends to lawfully purchase:

(i) Explosive materials from a licensee in a State ntiguous to the user's State of residence and the user's State of idence has enacted legislation, currently in force, specifically horizing a resident of that State to purchase explosive materi- in a contiguous State; or

(ii) Commercially manufactured black powder in antities not to exceed 50 pounds, intended to be used solely for rting, recreational, or cultural purposes in antique firearms or antique devices.

(2) Each person intending to engage in business as an plosive materials importer, manufacturer, or dealer must file an plication, with the required fee (see §555.42), with ATF in accordance with the instructions on the form (see §555.45). A license will, subject to law, entitle the licensee to transport, ship, and receive explosive materials in interstate or foreign commerce, and to engage in the business specified by the license, at the location described on the license. A separate license must be obtained for each business premises at which the applicant is to manufacture, import, or distribute explosive materials except under the following circumstances:

(i) A separate license will not be required for storage facilities operated by the licensee as an integral part of one business premises or to cover a location used by the licensee solely for maintaining the records required by this part.

(ii) A separate license will not be required of a licensed manufacturer with respect to his on-site manufacturing.

(iii) It will not be necessary for a licensed importer or a licensed manufacturer (for purposes of sale or distribution) to also obtain a dealer's license in order to engage in business on his licensed premises as a dealer in explosive materials.

(iv) A separate license will not be required of licensed manufacturers with respect to their on-site manufacture of theatrical flash powder.

(3) Except as provided in paragraph (a)(1) of this section, each person intending to acquire explosive materials from a licensee in a State other than a State in which he resides, or from a foreign country, or who intends to transport explosive materials in interstate or foreign commerce, must file an application, with the required fee (see §555.43), with ATF in accordance with the instructions on the form (see §555.45). A permit will, subject to law, entitle the permittee to acquire, transport, ship, and receive in interstate or foreign commerce explosive materials. Only one permit is required under this part.

(b) Licenses and permits issued on and after May 24, 2003—

(1) In general.

(i) Each person intending to engage in business as an importer or manufacturer of, or a dealer in, explosive materials, including black powder, must, before commencing business, obtain the license required by this subpart for the business to be operated.

(ii) Each person who intends to acquire for use explosive materials within the State in which he resides on no more than 6 separate occasions during the 12-month period in which the permit is valid must obtain a limited permit under this subpart. (See §555.105(b) for definition of "6 separate occasions.")

(iii) Each person who intends to acquire for use explosive materials from a licensee or permittee in a State other than the State in which he resides, or from a foreign country, or who intends to transport explosive materials in interstate or foreign commerce, or who intends to acquire for use explosive materials within the State in which he resides on more than 6 separate occasions during a 12-month period, must obtain a user permit under this subpart.

(iv) It is not necessary to obtain a permit if the user intends only to lawfully purchase commercially manufactured black powder in quantities not to exceed 50 pounds, intended to be used solely for sporting, recreational, or cultural purposes in antique firearms or in antique devices.

(2) Importers, manufacturers, and dealers. Each person intending to engage in business as an explosive materials importer, manufacturer, or dealer must file an application, with the required fee (see §555.42), with ATF in accordance with the instructions on the form (see §555.45). A license will, subject to law, entitle the licensee to transport, ship, and receive explosive materials in interstate or foreign commerce, and to engage in the business specified by the license, at the location described on the license. A separate license must be obtained for each business premises at which the applicant is to manufacture, import, or distribute explosive materials except under the following circumstances:

(i) A separate license will not be required for storage facilities operated by the licensee as an integral part of one business premises or to cover a location used by the licensee solely for maintaining the records required by this part.

(ii) A separate license will not be required of a licensed manufacturer with respect to his on-site manufacturing.

(iii) It will not be necessary for a licensed importer a licensed manufacturer (for purposes of sale or distribution) tc also obtain a dealer's license in order to engage in business on licensed premises as a dealer in explosive materials. No license will be required to obtain a user permit to lawfully transport, sh or receive explosive materials in interstate or foreign commerce

(iv) A separate license will not be required of licens manufacturers with respect to their on-site manufacture of theatrical flash powder.

(3) Users of explosive materials.

(i) A limited permit will, subject to law, entitle the holder of such permit to receive for his use explosive materials from a licensee or permittee in his state of residence on no mor than 6 separate occasions during the 12-month period in which the permit is valid. A limited permit does not authorize the rec or transportation of explosive materials in interstate or foreign commerce. Holders of limited permits who need to receive explosive materials on more than 6 separate occasions during a 12-month period must obtain a user permit in accordance with this subpart.

(ii) Each person intending to acquire explosive materials from a licensee in a State other than a State in which resides, or from a foreign country, or who intends to transport explosive materials in interstate or foreign commerce, must fil an application for a user permit, with the required fee (see §555.43), with ATF in accordance with the instructions on the form (see §555.45). A user permit will, subject to law, entitle t permittee to transport, ship, and receive in interstate or foreign commerce explosive materials. Only one user permit per perso required under this part, irrespective of the number of location relating to explosive materials operated by the holder of the us permit.

(Approved by the Office of Management and Budget under control number 1140–0083)

[ATF No. 1, 68 FR 13783, Mar. 20, 2003, as amended by ATF 5F, 70 FR 3063 May 27, 2005]

§ 555.42 License fees.

(a) Each applicant shall pay a fee for obtaining a three year license, a separate fee being required for each business premis as follows:

(1) Manufacturer—$200.

(2) Importer—$200.

(3) Dealer—$200.

(b) Each applicant for a renewal of a license shall pay a fee f three year license as follows:

(1) Manufacturer—$100.

(2) Importer—$100.

(3) Dealer—$100.

[T.D. ATF–400, 63 FR 45002, Aug. 24, 1998]

555.43 Permit fees.

Each applicant must pay a fee for obtaining a permit as llows:

(1) User—$100 for a three-year period.

(2) User-limited (nonrenewable)—$75.

(3) Limited—$25 for a one-year period.

) (1) Each applicant for renewal of a user permit must pay a : of $50 for a three-year period.

(2) Each applicant for renewal of a limited permit must pay ee of $12 for a one-year period.

F No. 1, 68 FR 13785, Mar. 20, 2003]

555.44 License or permit fee not refundable.

No refund of any part of the amount paid as a license or permit : will be made where the operations of the licensee or permittee :, for any reason, discontinued during the period of an issued ense or permit. However, the license or permit fee submitted th an application for a license or permit will be refunded if that plication is denied, withdrawn, or abandoned, or if a license is ncelled subsequent to having been issued through administra- e error.

555.45 Original license or permit.

Licenses issued prior to May 24, 2003. Any person who ends to engage in business as an explosive materials importer, nufacturer, or dealer, or who has not timely submitted applica- n for renewal of a previous license issued under this part, shall e with ATF an application for License, Explosives, ATF F 00.13 with ATF in accordance with the instructions on the m. The application must be executed under the penalties of rjury and the penalties imposed by 18 U.S.C. 844(a). The plication is to be accompanied by the appropriate fee in the m of a money order or check made payable to the Bureau of cohol, Tobacco and Firearms. ATF F 5400.13 may be obtained m any ATF office. The Chief, Firearms and Explosives censing Center, will not approve an application postmarked on after March 20, 2003, unless it is submitted with a Responsible rson Questionnaire, ATF Form 5400.28. Form 5400.28 must be mpleted in accordance with the instructions on the form.

Permits issued prior to May 24, 2003. Any person, except as ovided in §555.41(a), who intends to acquire explosive materi- s from a licensee in a state other than the State in which that rson resides, or from a foreign country, or who intends to nsport explosive materials in interstate or foreign commerce, or no has not timely submitted application for renewal of a evious permit issued under this part, shall file an application for rmit, Explosives, ATF F 5400.16 or Permit, User Limited ecial Fireworks, ATF F 5400.21 with ATF in accordance with : instructions on the form. The application must be executed der the penalties of perjury and the penalties imposed by U.S.C. 844(a). The application is to be accompanied by the appropriate fee in the form of a money order or check made payable to the Bureau of Alcohol, Tobacco and Firearms. ATF F 5400.16 and ATF F 5400.21 may be obtained from any ATF office. The Chief, Firearms and Explosives Licensing Center, will not approve an application postmarked on or after March 20, 2003, unless it is submitted with a Responsible Person Questionnaire, ATF Form 5400.28. Form 5400.28 must be completed in accordance with the instructions on the form.

(c) Licenses and permits issued on and after May 24, 2003—

(1) License. Any person who intends to engage in the business as an importer of, manufacturer of, or dealer in explosive materials, or who has not timely submitted an application for renewal of a previous license issued under this part, must file an application for License, Explosives, ATF F 5400.13, with ATF in accordance with the instructions on the form. ATF Form 5400.13 may be obtained by contacting any ATF office. The application must:

(i) Be executed under the penalties of perjury and the penalties imposed by 18 U.S.C. 844(a);

(ii) Include appropriate identifying information concerning each responsible person;

(iii) Include a photograph and fingerprints for each responsible person;

(iv) Include the names of and appropriate identifying information regarding all employees who will be authorized by the applicant to possess explosive materials by submitting ATF F 5400.28 for each employee; and

(v) Include the appropriate fee in the form of money order or check made payable to the Bureau of Alcohol, Tobacco, Firearms and Explosives.

(2) User permit and limited permit. Except as provided in 555.41(b)(1)(iv), any person who intends to acquire explosive materials in the State in which that person resides or acquire explosive materials from a licensee or holder of a user permit in a State other than the State in which that person resides, or from a foreign country, or who intends to transport explosive materials in interstate or foreign commerce, or who has not timely submitted an application for renewal of a previous permit issued under this part, must file an application for Permit, Explosives, ATF F 5400.16 or Permit, User Limited Display Fireworks, ATF F 5400.21 with ATF in accordance with the instructions on the form. ATF Form 5400.16 and ATF Form 5400.21 may be obtained by contacting any ATF office. The application must:

(i) Be executed under the penalties of perjury and the penalties imposed by 18 U.S.C. 844(a);

(ii) Include a photograph, fingerprints, and appropriate identifying information for each responsible person;

(iii) Include the names of and appropriate identifying information regarding all employees who will be authorized by the applicant to possess explosive materials by submitting ATF F 5400.28 for each employee; and

(iv) Include the appropriate fee in the form of money order or check made payable to the Bureau of Alcohol, Tobacco, Firearms and Explosives.

(3) The Chief, Firearms and Explosives Licensing Center, will conduct background checks on responsible persons and employees authorized by the applicant to possess explosive materials in accordance with §555.33. If it is determined that any responsible person or employee is described in any paragraph of section 842(i) of the Act, the applicant must submit an amended application indicating removal or reassignment of that person before the license or permit will be issued.

(Approved by the Office of Management and Budget under control number 1140–0083)

(18 U.S.C. 847 (84 Stat. 959); 18 U.S.C. 926 (82 Stat. 1226))
[T.D. ATF–200, 50 FR 10497, Mar. 15, 1985, as amended by T.D. ATF–400, 63 FR 45002, Aug. 24, 1998; ATF No. 1, 68 FR 13785, Mar. 20, 2003]

§ 555.46 Renewal of license or permit.

(a) If a licensee or permittee intends to continue the business or operation described on a license or permit issued under this part after the expiration date of the license or permit, he shall, unless otherwise notified in writing by the Chief, Firearms and Explosives Licensing Center, execute and file prior to the expiration of his license or permit an application for license renewal, ATF F 5400.14 (Part III), or an application for permit renewal, ATF F 5400.15 (Part III), accompanied by the required fee, with ATF in accordance with the instructions on the form. In the event the licensee or permittee does not timely file a renewal application, he shall file an original application as required by §555.45, and obtain the required license or permit in order to continue business or operations.

(b) A user-limited permit is not renewable and is valid for a single purchase transaction. Applications for all user-limited permits must be filed on ATF F 5400.16 or ATF F 5400.21, as required by §555.45.

(18 U.S.C. 847 (84 Stat. 959); 18 U.S.C. 926 (82 Stat. 1226))
[T.D. ATF–87, 46 FR 40384, Aug. 7, 1981, as amended by T.D. ATF–200, 50 FR 10497, Mar. 15, 1985; T.D. ATF–290, 54 FR 53054, Dec. 27, 1989; T.D. ATF–400, 63 FR 45002, Aug. 24, 1998]

§ 555.47 Insufficient fee.

If an application is filed with an insufficient fee, the application and fee submitted will be returned to the applicant.

(18 U.S.C. 847 (84 Stat. 959); 18 U.S.C. 926 (82 Stat. 1226))
[T.D. ATF–200, 50 FR 10498, Mar. 15, 1985]

§ 555.48 Abandoned application.

Upon receipt of an incomplete or improperly executed application, the applicant will be notified of the deficiency in the application. If the application is not corrected and returned within 30 days following the date of notification, the application will be considered as having been abandoned and the license or permit fee returned.

§ 555.49 Issuance of license or permit.

(a) Issuance of license or permit prior to May 24, 2003. (1) Th Chief, Firearms and Explosives Licensing Center, will issue a license or permit if—

(i) A properly executed application for the license permit is received; and

(ii) Through further inquiry or investigation, or otherwise, it is found that the applicant is entitled to the license permit.

(2) The Chief, Firearms and Explosives Licensing Cente will approve a properly executed application for a license or permit, if:

(i) The applicant is 21 years of age or over;

(ii) The applicant (including, in the case of a corpo tion, partnership, or association, any individual possessing, directly or indirectly, the power to direct or cause the direction the management and policies of the corporation, partnership, o association) is not a person to whom distribution of explosive materials is prohibited under the Act;

(iii) The applicant has not willfully violated any provisions of the Act or this part;

(iv) The applicant has not knowingly withheld information or has not made any false or fictitious statement intended or likely to deceive, in connection with his applicatio

(v) The applicant has in a State, premises from whi he conducts business or operations subject to license or permit under the Act or from which he intends to conduct business or operations;

(vi) The applicant has storage for the class (as described in §555.202) of explosive materials described on the application, unless he establishes to the satisfaction of the Chie Firearms and Explosives Licensing Center, that the business or operations to be conducted will not require the storage of explosive materials;

(vii) The applicant has certified in writing that he is familiar with and understands all published State laws and loca ordinances relating to explosive materials for the location in which he intends to do business; and

(viii) The applicant for a license has submitted the certificate required by section 21 of the Federal Water Pollutio Control Act, as amended (33 U.S.C. 1341).

(3) The Chief, Firearms and Explosives Licensing Cente will approve or the regional director (compliance) will deny an application for a license or permit within the 45-day period beg ning on the date a properly executed application was received. However, when an applicant for license or permit renewal is a person who is, under the provisions of §555.83 or §555.142, conducting business or operations under a previously issued license or permit, action regarding the application will be held abeyance pending the completion of the proceedings against th

applicant's existing license or permit, or renewal application, or final action by the Director on an application for relief submitted under §555.142, as the case may be.

(4) The license or permit and one copy will be forwarded to the applicant, except that in the case of a user-limited permit, the original only will be issued.

(5) Each license or permit will bear a serial number and this number may be assigned to the licensee or permittee to whom issued for as long as he maintains continuity of renewal in the same region.

(b) Issuance of license or permit on and after May 24, 2003.

(1) The Chief, Firearms and Explosives Licensing Center, will issue a license or permit if:

(i) A properly executed application for the license or permit is received; and

(ii) Through further inquiry or investigation, or otherwise, it is found that the applicant is entitled to the license or permit.

(2) The Chief, Firearms and Explosives Licensing Center, will approve a properly executed application for a license or permit, if:

(i) The applicant (or, if the applicant is a corporation, partnership, or association, each responsible person with respect to the applicant) is not a person described in any paragraph of section 842(i) of the Act;

(ii) The applicant has not willfully violated any provisions of the Act or this part;

(iii) The applicant has not knowingly withheld information or has not made any false or fictitious statement intended or likely to deceive, in connection with his application;

(iv) The applicant has in a State, premises from which he conducts business or operations subject to license or permit under the Act or from which he intends to conduct business or operations;

(v) The applicant has storage for the class (as described in §555.202) of explosive materials described on the application;

(vi) The applicant has certified in writing that he is familiar with and understands all published State laws and local ordinances relating to explosive materials for the location in which he intends to do business;

(vii) The applicant for a license has submitted the certificate required by section 21 of the Federal Water Pollution Control Act, as amended (33 U.S.C. 1341);

(viii) None of the employees of the applicant who will be authorized by the applicant to possess explosive materials is a person described in any paragraph of section 842(i) of the Act; and

(ix) In the case of an applicant for a limited permit, the applicant has certified in writing that the applicant will not receive explosive materials on more than 6 separate occasions during the 12-month period for which the limited permit is valid.

(3) The Chief, Firearms and Explosives Licensing Center, will approve or the regional director (compliance) will deny any application for a license or permit within the 90-day period beginning on the date a properly executed application was received. However, when an applicant for license or permit renewal is a person who is, under the provisions of §555.83 or §555.142, conducting business or operations under a previously issued license or permit, action regarding the application will be held in abeyance pending the completion of the proceedings against the applicant's existing license or permit, or renewal application, or final action by the Director on an application for relief submitted under §555.142, as the case may be.

(4) The license or permit and one copy will be forwarded to the applicant, except that in the case of a user-limited permit, the original only will be issued.

(5) Each license or permit will bear a serial number and this number may be assigned to the licensee or permittee to whom issued for as long as he maintains continuity of renewal in the same region.

(Approved by the Office of Management and Budget under control number 1140–0082)

[ATF No. 1, 68 FR 13785, Mar. 20, 2003]

§ 555.50 Correction of error on license or permit.

(a) Upon receipt of a license or permit issued under this part, each licensee or permittee shall examine the license or permit to insure that the information on it is accurate. If the license or permit is incorrect, the licensee or permittee shall return the license or permit to the Chief, Firearms and Explosives Licensing Center, with a statement showing the nature of the error. The Chief, Firearms and Explosives Licensing Center, shall correct the error, if the error was made in his office, and return the license or permit. However, if the error resulted from information contained in the licensee's or permittee's application for the license or permit, the Chief, Firearms and Explosives Licensing Center, shall require the licensee or permittee to file an amended application setting forth the correct information and a statement explaining the error contained in the application. Upon receipt of the amended application and a satisfactory explanation of the error, the Chief, Firearms and Explosives Licensing Center, shall make the correction on the license or permit and return it to the licensee or permittee.

(b) When the Chief, Firearms and Explosives Licensing Center, finds through any means other than notice from the licensee or permittee that an incorrect license or permit has been issued, (1) the Chief, Firearms and Explosives Licensing Center, may require the holder of the incorrect license or permit to return the license or permit for correction, and (2) if the error resulted from information contained in the licensee's or permittee's application for the license or permit, the Chief, Firearms and Explosives Licensing Center, shall require the licensee or permittee to file an amended application setting forth the correct information, and a statement satisfactorily explaining the error contained in the application. The Chief, Firearms and Explosives Licensing Center, then shall make the correction on the license or permit and return it to the licensee or permittee.

[T.D. ATF–87, 46 FR 40384, Aug. 7, 1981, as amended by T.D. ATF–290, 54 FR 53054, Dec. 27, 1989]

§ 555.51 Duration of license or permit.

(a) Prior to May 24, 2003. An original license or permit is issued for a period of three years. A renewal license or permit is issued for a period of three years. However, a user-limited permit is valid only for a single purchase transaction.

(b) On and after May 24, 2003.

(1) An original license or user permit is issued for a period of three years. A renewal license or user permit is also issued for a period of three years. However, a user-limited permit is valid only for a single purchase transaction.

(2) A limited permit is issued for a period of one year. A renewal limited permit is also issued for a period of one year.

[ATF No. 1, 68 FR 13786, Mar. 20, 2003]

§ 555.52 Limitations on license or permit.

(a) The license covers the business of explosive materials specified in the license at the licensee's business premises (see §555.41(b)).

(b) The permit is valid with respect to the type of operations of explosive materials specified in the permit.

[T.D. ATF–87, 46 FR 40384, Aug. 7, 1981, as amended by T.D. ATF–387, 62 FR 8376, Feb. 25, 1997; ATF 5F, 70 FR 30633, May 27, 2005]

§ 555.53 License and permit not transferable.

Licenses and permits issued under this part are not transferable to another person. In the event of the lease, sale, or other transfer of the business or operations covered by the license or permit, the successor must obtain the license or permit required by this part before commencing business or operations. However, for rules on right of succession, see §555.59.

§ 555.54 Change of address.

(a) During the term of a license or permit, a licensee or permittee may move his business or operations to a new address at which he intends to regularly carry on his business or operations, without procuring a new license or permit. However, in every case, the licensee or permittee shall—

(1) Give notification of the new location of the business or operations to the Chief, Firearms and Explosives Licensing Center at least 10 days before the move; and

(2) Submit the license or permit to the Chief, Firearms and Explosives Licensing Center. The Chief, Firearms and Explosives Licensing Center will issue an amended license or permit, which will contain the new address (and new license or permit number, if any).

(b) Licensees and permittees whose mailing address will change must notify the Chief, Firearms and Explosives Licensing Center, at least 10 days before the change.

(Paragraph (b) approved by the Office of Management and Budget under control number 1140–0080)

[T.D. ATF–87, 46 FR 40384, Aug. 7, 1981, as amended by T.D. ATF–290, 54 FR 53054, Dec. 27, 1989; ATF No. 1, 68 FR 13786, Mar. 20, 2003]

§ 555.56 Change in trade name.

A licensee or permittee continuing to conduct business or operations at the location shown on his license or permit is not required to obtain a new license or permit by reason of a mere change in trade name under which he conducts his business or operations. However, the licensee or permittee shall furnish his license or permit and any copies furnished with the license or permit for endorsement of the change to the Chief, Firearms and Explosives Licensing Center, within 30 days from the date the licensee or permittee begins his business or operations under the new trade name.

[T.D. ATF–87, 46 FR 40384, Aug. 7, 1981, as amended by T.D. ATF–290, 54 FR 53054, Dec. 27, 1989]

§ 555.57 Change of control, change in responsible persons, and change of employees.

(a) In the case of a corporation or association holding a license or permit under this part, if actual or legal control of the corporation or association changes, directly or indirectly, whether by reason of change in stock ownership or control (in the corporation holding a license or permit or in any other corporation), by operation of law, or in any other manner, the licensee or permittee shall, within 30 days of the change, give written notification executed under the penalties of perjury, to the Chief, Firearms and Explosives Licensing Center. Upon expiration of the license or permit, the corporation or association shall file an ATF F 5400.13 or an ATF F 5400.16 as required by §555.45, and pay the fee prescribed in §555.42(b) or §555.43(b).

(b) For all licenses or permits issued on and after May 24, 2003, each person holding the license or permit must report to the Chief, Firearms and Explosives Licensing Center, any change in responsible persons or employees authorized to possess explosive materials. Such report must be submitted within 30 days of the change and must include appropriate identifying information for each responsible person. Reports relating to newly hired employees authorized to possess explosive materials must be submitted on ATF F 5400.28 for each employee.

(c) Upon receipt of a report, the Chief, Firearms and Explosives Licensing Center, will conduct a background check, if appropriate, in accordance with §555.33.

(d) The reports required by paragraph (b) of this section must be retained as part of a licensee's or permittee's permanent records for the period specified in §555.121.

(Approved by the Office of Management and Budget under ntrol number 1140–0074)

D. ATF–87, 46 FR 40384, Aug. 7, 1981, as amended by T.D. ATF–290, 54 FR)54, Dec. 27, 1989; ATF No. 1, 68 FR 13786, Mar. 20, 2003]

555.58 Continuing partnerships.

Where, under the laws of the particular State, the partnership is t terminated on death or insolvency of a partner, but continues til the winding up of the partnership affairs is completed, and surviving partner has the exclusive right to the control and ssession of the partnership assets for the purpose of liquidation d settlement, the surviving partner may continue to conduct business or operations under the license or permit of the rtnership. If the surviving partner acquires the business or erations on completion of settlement of the partnership, he all obtain a license or permit in his own name from the date of quisition, as provided in §555.45. The rule set forth in this ction will also apply where there is more than one surviving rtner.

555.59 Right of succession by certain persons.

Certain persons other than the licensee or permittee may cure the right to carry on the same explosive materials business operations at the same business premises for the remainder of term of license or permit. These persons are:

(1) The surviving spouse or child, or executor, administra-, or other legal representative of a deceased licensee or rmittee; and

(2) A receiver or trustee in bankruptcy, or an assignee for nefit of creditors.

In order to secure the right of succession, the person or rsons continuing the business or operations shall submit the ense or permit and all copies furnished with the license or rmit for endorsement of the succession to the Chief, Firearms l Explosives Licensing Center, within 30 days from the date on ich the successor begins to carry on the business or operations.

D. ATF–87, 46 FR 40384, Aug. 7, 1981, as amended by T.D. ATF–290, 54 FR 54, Dec. 27, 1989]

555.60 Certain continuances of business or erations.

A licensee or permittee who furnishes his license or permit to Chief, Firearms and Explosives Licensing Center, for correc-n, amendment, or endorsement, as provided in this subpart, y continue his business or operations while awaiting its return.

). ATF–87, 46 FR 40384, Aug. 7, 1981, as amended by T.D. ATF–290, 54 FR 54, Dec. 27, 1989]

555.61 Discontinuance of business or erations.

Where an explosive materials business or operations is either continued or succeeded by a new owner, the owner of the siness or operations discontinued or succeeded shall, within 30 ys, furnish notification of the discontinuance or succession and submit his license or permit and any copies furnished with the license or permit to the Chief, Firearms and Explosives Licensing Center. (See also §555.128.)

[T.D. ATF–87, 46 FR 40384, Aug. 7, 1981, as amended by T.D. ATF–290, 54 FR 53054, Dec. 27, 1989]

§ 555.62 State or other law.

A license or permit issued under this part confers no right or privilege to conduct business or operations, including storage, contrary to State or other law. The holder of a license or permit issued under this part is not, by reason of the rights and privileges granted by that license or permit, immune from punishment for conducting an explosive materials business or operations in violation of the provisions of any State or other law. Similarly, compliance with the provisions of any State or other law affords no immunity under Federal law or regulations.

§ 555.63 Explosives magazine changes.

(a) General.

(1) The requirements of this section are applicable to magazines used for other than temporary (under 24 hours) storage of explosives.

(2) A magazine is considered suitable for the storage of explosives if the construction requirements of this part are met during the time explosives are stored in the magazine.

(3) A magazine is considered suitable for the storage of explosives if positioned in accordance with the applicable table of distances as specified in this part during the time explosives are stored in the magazine.

(4) For the purposes of this section, notification of the regional director (compliance) may be by telephone or in writing. However, if notification of the regional director (compliance) is in writing it must be at least three business days in advance of making changes in construction to an existing magazine or constructing a new magazine, and at least five business days in advance of using any reconstructed magazine or added magazine for the storage of explosives.

(b) Exception. Mobile or portable type 5 magazines are exempt from the requirements of paragraphs (c) and (d) of this section, but must otherwise be in compliance with paragraphs (a) (2) and (3) of this section during the time explosives are stored in such magazines.

(c) Changes in magazine construction. A licensee or permittee who intends to make changes in construction of an existing magazine shall notify the regional director (compliance) describing the proposed changes prior to making any changes. Unless otherwise advised by the regional director (compliance), changes in construction may commence after explosives are removed from the magazine. Explosives may not be stored in a reconstructed magazine before the regional director (compliance) has been notified in accordance with paragraph (a)(4) of this section that the changes have been completed.

(d) Magazines acquired or constructed after permit or license is issued. A licensee or permittee who intends to construct or acquire additional magazines shall notify the regional director (compliance) in accordance with paragraph (a)(4) of this section describing the additional magazines and the class and quantity of explosives to be stored in the magazine. Unless otherwise advised by the regional director (compliance), additional magazines may be constructed, or acquired magazin may be used for the storage of explosives. Explosives must not stored in a magazine under construction. The regional director (compliance) must be notified that construction has been completed.

[T.D. ATF–87, 46 FR 40384, Aug. 7, 1981, as amended by T.D. ATF–400, 63 F 45002, Aug. 24, 1998]

Subpart E—License and Permit Proceedings

§ 555.71 Opportunity for compliance.

Except in cases of willfulness or those in which the public interest requires otherwise, and the regional director (compliance) so alleges in the notice of denial of an application or revocation of a license or permit, no license or permit will be revoked or renewal application denied without first calling to the attention of the licensee or permittee the reasons for the contemplated action and affording him an opportunity to demonstrate or achieve compliance with all lawful requirements and to submit facts, arguments, or proposals of adjustment. The notice of contemplated action, ATF F 5400.12, will afford the licensee or permittee 15 days from the date of receipt of the notice to respond. If no response is received within the 15 days, or if after consideration of relevant matters presented by the licensee or permittee, the regional director (compliance) finds that the licensee or permittee is not likely to abide by the law and regulations, he will proceed as provided in §555.74.

[T.D. ATF–87, 46 FR 40384, Aug. 7 1981, as amended by T.D. ATF–446, 66 FR 16602, Mar. 27, 2001]

§ 555.72 Denial of initial application.

Whenever the regional director (compliance) has reason to believe that an applicant for an original license or permit is not eligible to receive a license or permit under the provisions of §555.49, he shall issue a notice of denial on ATF F 5400.11. The notice will set forth the matters of fact and law relied upon in determining that the application should be denied, and will afford the applicant 15 days from the date of receipt of the notice in which to request a hearing to review the denial. If no request for a hearing is filed within that time, a copy of the application, marked *"Disapproved"*, will be returned to the applicant.

§ 555.73 Hearing after initial application is denied.

If the applicant for an original license or permit desires a hearing, he shall file a request with the regional director (compliance) within 15 days after receipt of the notice of denial. The request should include a statement of the reasons for a hearing. On receipt of the request, the regional director (compliance) shall refer the matter to an administrative law judge who shall set a time and place (see §555.77) for a hearing and shall serve notice of the hearing upon the applicant and the regional director (compliance) at least 10 days in advance of the hearing date. The hearing will be conducted in accordance with the hearing procedures prescribed in part 71 of this chapter (see §555.82). Within a reasonable time after the conclusion of the hearing, a as expeditiously as possible, the administrative law judge shall render his recommended decision. He shall certify to the complete record of the proceedings before him and shall immediat forward the complete certified record, together with four copie of his recommended decision, to the regional director (compliance) for decision.

§ 555.74 Denial of renewal application or revocation of license or permit.

If following the opportunity for compliance under §555.71, without opportunity for compliance under §555.71, as circumstances warrant, the regional director (compliance) finds that t licensee or permittee is not likely to comply with the law or regulations or is otherwise not eligible to continue operations authorized under his license or permit, the regional director (compliance) shall issue a notice of denial of the renewal application or revocation of the license or permit, ATF F 5400. or ATF F 5400.10, as appropriate. In either case, the notice wi set forth the matters of fact constituting the violations specifie dates, places, and the sections of law and regulations violated. The notice will, in the case of revocation of a license or permi specify the date on which the action is effective, which date w be on or after the date the notice is served on the licensee or permittee. The notice will also advise the licensee or permittee that he may, within 15 days after receipt of the notice, request hearing and, if applicable, a stay of the effective date of the revocation of his license or permit.

§ 555.75 Hearing after denial of renewal application or revocation of license or permit.

If a licensee or permittee whose renewal application has bee denied or whose license or permit has been revoked desires a hearing, he shall file a request for a hearing with the regional director (compliance). In the case of the revocation of a licens permit, he may include a request for a stay of the effective dat the revocation. On receipt of the request the regional director (compliance) shall advise the licensee or permittee whether th stay of the effective date of the revocation is granted. If the sta of the effective date of the revocation is granted, the regional director (compliance) shall refer the matter to an administrativ law judge who shall set a time and place (see §555.77) for a hearing and shall serve notice of the hearing upon the licensee ermittee and the regional director (compliance) at least 10 days advance of the hearing date. If the stay of the effective date of e revocation is denied, the licensee or permittee may request an nmediate hearing. In this event, the regional director (compli-nce) shall immediately refer the matter to an administrative law dge who shall set a date and place for a hearing, which date nall be no later than 10 days from the date the licensee or ermittee requested an immediate hearing. The hearing will be eld in accordance with the applicable provisions of part 71 of is chapter. Within a reasonable time after the conclusion of the aring, and as expeditiously as possible, the administrative law dge shall render his decision. He shall certify to the complete cord of the proceeding before him and shall immediately rward the complete certified record, together with two copies of s decision, to the regional director (compliance), serve one copy his decision on the licensee or permittee or his counsel, and ansmit a copy to the attorney for the Government.

555.76 Action by regional director (compliance).

) Initial application proceedings. If, upon receipt of the cord and the recommended decision of the administrative law dge, the regional director (compliance) decides that the license permit should be issued, the regional director (compliance) nall cause the application to be approved, briefly stating, for the cord, his reasons. If he contemplates that the denial should and, he shall serve a copy of the administrative law judge's commended decision on the applicant, informing the applicant his contemplated action and affording the applicant not more an 10 days in which to submit proposed findings and conclu-ons or exceptions to the recommended decision with supporting asons. If the regional director (compliance), after consideration the record of the hearing and of any proposed findings, nclusions, or exceptions filed with him by the applicant, proves the findings, conclusions and recommended decision of e administrative law judge, the regional director (compliance) all cause the license or permit to be issued or disapproved the plication accordingly. If he disapproves the findings, conclu-ons, and recommendation of the administrative law judge, in hole or in part, he shall by order make such findings and nclusions as in his opinion are warranted by the law and the cts in the record. Any decision of the regional director (compli-ce) ordering the disapproval of an initial application for a ense or permit shall state the findings and conclusions upon hich it is based, including his ruling upon each proposed nding, conclusion, and exception to the administrative law dge's recommended decision, together with a statement of his ndings and conclusions, and reasons or basis for his findings d conclusions, upon all material issues of fact, law or discretion esented on the record. A signed duplicate original of the cision will be served upon the applicant and the original copy ntaining certificate of service will be placed in the official record of the proceedings. If the decision of the regional director (compliance) is in favor of the applicant, he shall issue the license or permit, to be effective on issuance.

(b) Renewal application and revocation proceedings. Upon receipt of the complete certified records of the hearing, the regional director (compliance) shall enter an order confirming the revocation of the license or permit, or disapproving the application, in accordance with the administrative law judge's findings and decision, unless he disagrees with the findings and decision. A signed duplicate original of the order, ATF F 5400.9, will be served upon the licensee or permittee and the original copy containing certificate of service will be placed in the official record of the proceedings. If the regional director (compliance) disagrees with the findings and decision of the administrative law judge, he shall file a petition with the Director for review of the findings and decision, as provided in §555.79. In either case, if the renewal application denial is sustained, a copy of the application marked "Disapproved" will be returned to the applicant. If the renewal application denial is reversed, a license or permit will be issued to become effective on expiration of the license or permit being renewed, or on the date of issuance, whichever is later. If the proceedings involve the revocation of a license or permit which expired before a decision is in favor of the licensee or permittee, the regional director (compliance) shall:

(1) If renewal application was timely filed and a stay of the effective date of the revocation was granted, cause to be issued a license or permit effective on the date of issuance;

(2) If renewal application was not timely filed but a stay of the effective date of the revocation had been granted, request that a renewal application be filed and, following that, cause to be issued a license or permit to be effective on issuance; or

(3) If a stay of the effective date of the revocation had not been granted, request that an application be filed as provided in §555.45, and process it in the same manner as for an application for an original license or permit.

(c) Curtailment of stay of revocation effective date. If, after approval of a request for a stay of the effective date of an order revoking a license or permit but before actions are completed under this subpart, the regional director (compliance) finds that it is contrary to the public interest for the licensee or permittee to continue the operations or activities covered by his license or permit, the regional director (compliance) may issue a notice of withdrawal of the approval, effective on the date of issuance. Notice of withdrawal will be served upon the licensee or permittee in the manner provided in §555.81.

[T.D. ATF–87, 46 FR 40384, Aug. 7, 1981, as amended by T.D. ATF–290, 54 FR 53054, Dec. 27, 1989]

§ 555.77 Designated place of hearing.

The designated place of hearing set as provided in §555.73 or §555.75, will be at the location convenient to the aggrieved party.

§ 555.78 Representation at a hearing.

An applicant, licensee, or permittee may be represented by an attorney, certified public accountant, or other person recognized to practice before the Bureau of Alcohol, Tobacco and Firearms as provided in 31 CFR Part 8, if he has otherwise complied with the applicable requirements of 26 CFR 601.521 through 601.527. The regional director (compliance) shall be represented in proceedings under §§555.73 and 555.75 by an attorney in the office of the chief counsel or regional counsel who is authorized to execute and file motions, briefs, and other papers in the proceedings, on behalf of the regional director (compliance), in his own name as *"Attorney for the Government"*.

[T.D. ATF-87, 46 FR 40384, Aug. 7, 1981, as amended by T.D. ATF-92, 46 FR 46916, Sept. 23, 1981]

§ 555.79 Appeal on petition to the Director.

An appeal to the Director is not required prior to filing an appeal with the U.S. Court of Appeals for judicial review. An appeal may be taken by the applicant, licensee, or permittee to the Director from a decision resulting from a hearing under §555.73 or §555.75. An appeal may also be taken by a regional director (compliance) from a decision resulting from a hearing under §555.75 as provided in §555.76(b). The appeal shall be taken by filing a petition for review on appeal with the Director within 15 days of the service of an administrative law judge's decision or an order. The petition will set forth facts tending to show (a) action of an arbitrary nature, (b) action without reasonable warrant in fact, or (c) action contrary to law and regulations. A copy of the petition will be filed with the regional director (compliance) or served on the applicant, licensee, or permittee, as the case may be. In the event of appeal, the regional director (compliance) shall immediately forward the complete original record, by certified mail, to the Director for his consideration, review, and disposition as provided in subpart I of part 71 of this chapter. When, on appeal, the Director affirms the initial decision of the regional director (compliance) or the administrative law judge, as the case may be, the initial decision will be final.

§ 555.80 Court review.

An applicant, licensee, or permittee may, within 60 days after receipt of the decision of the administrative law judge or the final order of the regional director (compliance) or the Director, file a petition for a judicial review of the decision, with the U.S. Court of Appeals for the district in which he resides or has his principal place of business. The Director, upon notification that a petition has been filed, shall have prepared a complete transcript of the record of the proceedings. The regional director (compliance) or the Director, as the case may be, shall certify to the correctness of the transcript of the record, forward one copy to the attorney fo the Government in the review of the case, and file the original record of the proceedings with the original certificate in the U.S Court of Appeals.

§ 555.81 Service on applicant, licensee, or permittee.

All notices and other formal documents required to be serve on an applicant, licensee, or permittee under this subpart will b served by certified mail or by personal delivery. Where service by personal delivery, the signed duplicate original copy of the formal document will be delivered to the applicant, licensee, or permittee, or, in the case of a corporation, partnership, or assoc tion, by delivering it to an officer, manager, or general agent, o its attorney of record.

§ 555.82 Provisions of part 200 made applicable.

The provisions of subpart G of part 200 of this chapter, as we as those provisions of part 71 relative to failure to appear, withdrawal of an application or surrender of a permit, the conduct o hearings before an administrative law judge, and record of testimony, are hereby made applicable to application, license, a permit proceedings under this subpart to the extent that they ar not contrary to or incompatible with this subpart.

§ 555.83 Operations by licensees or permittees after notice of denial or revocation.

In any case where a notice of revocation has been issued and request for a stay of the effective date of the revocation has not been granted, the licensee or permittee shall not engage in the activities covered by the license or permit pending the outcome proceedings under this subpart. In any case where notice of revocation has been issued but a stay of the effective date of the revocation has been granted, the licensee or permittee may continue to engage in the activities covered by his license or permit unless, or until, formally notified to the contrary: Provid That in the event the license or permit would have expired befo proceedings under this subpart are completed, timely renewal application must have been filed to continue the license or perm beyond its expiration date. In any case where a notice of denial a renewal application has been issued, the licensee or permittee may continue to engage in the activities covered by the existing license or permit after the date of expiration of the license or permit until proceedings under this subpart are completed.

Subpart F—Conduct of Business or Operations

§ 555.101 Posting of license or user permit.

A license or user permit issued under this part, or a copy of a license or user permit, will be posted and available for inspection on the business premises at each place where explosive materials are manufactured, imported, or distributed.

[T.D. ATF–87, 46 FR 40384, Aug. 7, 1981. Redesignated by T.D. ATF–487, 68 FR 3748, Jan. 24, 2003. ATF No. 1, 68 FR 13786, Mar. 20, 2003]

§ 555.102 Authorized operations by permittees.

(a) In general. A permit issued under this part does not authorize the permittee to engage in the business of manufacturing, importing, or dealing in explosive materials. Accordingly, if a permittee's operations bring him within the definition of manufacturer, importer, or dealer under this part, he shall qualify for the appropriate license.

(b) Distributions of surplus stocks—

(1) Distributions of surplus stocks prior to May 24, 2003. Permittees are not authorized to engage in the business of sale or distribution of explosive materials. However, permittees may dispose of surplus stocks of explosive materials to other licensees or permittees in accordance with §555.103, and to nonlicensees or to nonpermittees in accordance with §555.105(a)(4).

(2) Distributions of surplus stocks on and after May 24, 2003. Permittees are not authorized to engage in the business of sale or distribution of explosive materials. However, permittees may dispose of surplus stocks of explosive materials to other licensees or permittees in accordance with §555.103 and §555.105.

[T.D. ATF–400, 63 FR 45002, Aug. 24, 1998, as amended by ATF No. 1, 68 FR 13787, Mar. 20, 2003]

§ 555.103 Transactions among licensees/permittees and transactions among licensees and holders of user permits.

(a) Transactions among licensees/permittees prior to May 24, 2003—

(1) General.

(i) A licensed importer, licensed manufacturer or licensed dealer selling or otherwise distributing explosive materials (or a permittee disposing of surplus stock to a licensee or another permittee) who has the certified information required by this section may sell or distribute explosive materials to a licensee or permittee for not more than 45 days following the expiration date of the distributee's license or permit, unless the distributor knows or has reason to believe that the distributee's authority to continue business or operations under this part has been terminated.

(ii) A licensed importer, licensed manufacturer or licensed dealer selling or otherwise distributing explosive materials (or a permittee disposing of surplus stock to another licensee or permittee) must verify the license or permit status of the distributee prior to the release of explosive materials ordered, as required by this section.

(iii) Licensees or permittees desiring to return explosive materials to a licensed manufacturer may do so without obtaining a certified copy of the manufacturer's license.

(iv) Where possession of explosive materials is transferred at the distributor's premises, the distributor must in all instances verify the identity of the person accepting possession on behalf of the distributee before relinquishing possession. Before the delivery at the distributor's premises of explosive materials to an employee of a licensee or permittee, or to an employee of a common or contract carrier transporting explosive materials to a licensee or permittee, the distributor delivering explosive materials must obtain an executed ATF F 5400.8, Explosives Delivery Record, from the employee before releasing the explosive materials. The ATF F 5400.8 must contain all of the information required on the form and required by this part.

Example 1. An ATF F 5400.8 is required when:

a. An employee of the purchaser takes possession at the distributor's premises.

b. An employee of a common or contract carrier hired by the purchaser takes possession at the distributor's premises.

Example 2. An ATF F 5400.8 is not required when:

a. An employee of the distributor takes possession of the explosives for the purpose of transport to the purchaser.

b. An employee of a common or contract carrier hired by the distributor takes possession of the explosives for the purpose of transport to the purchaser.

(2) License/permit verification of individuals.

(i) The distributee must furnish a certified copy (or, in the case of a user-limited, the original) of the license or permit. The certified copy need be furnished only once during the current term of the license or permit. Also, a licensee need not furnish certified copies of licenses to other licensed locations operated by such licensee.

(ii) The distributor may obtain any additional verification as the distributor deems necessary.

(3) License/permit verification of business organizations.

(i) A business organization may (in lieu of furnishing a certified copy of a license) furnish the distributor a certified list which contains the name, address, license number and date of license expiration of each licensed location. The certified list need be furnished only once during the current term of the license or permit. Also, a business organization need not furnish a certified list to other licensed locations operated by such business organization.

(ii) A business organization must, prior to ordering explosive materials, furnish the licensee or permittee a current certified list of the representatives or agents authorized to order explosive materials on behalf of the business organization showing the name, address, and date and place of birth of each representative or agent. A licensee or permittee may not distribute explosive materials to a business organization on the order of a person who does not appear on the certified list of representatives or agents and, if the person does appear on the certified list, the licensee or permittee must verify the identity of such person.

(4) Licensee/permittee certified statement.

(i) A licensee or permittee ordering explosive materials from another licensee or permittee must furnish a current, certified statement of the intended use of the explosive materials, e.g., resale, mining, quarrying, agriculture, construction, sport rocketry, road building, oil well drilling, seismographic research, to the distributor.

(ii) For individuals, the certified statement of intended use must specify the name, address, date and place of birth, and social security number of the distributee.

(iii) For business organizations, the certified statement of intended use must specify the taxpayer identification number, the identity and the principal and local places of **business.**

(iv) The licensee or permittee purchasing explosive materials must revise the furnished copy of the certified statement only when the information is no longer current.

(5) User-limited permit transactions. A user-limited permit issued under the provisions of this part is valid for only a single purchase transaction and is not renewable (see §555.51). Accordingly, at the time a user-limited permittee orders explosive materials, the licensed distributor must write on the front of the user-limited permit the transaction date, his signature, and the distributor's license number prior to returning the permit to the user-limited permittee.

(b) Transactions among licensees/permittees on and after May 24, 2003—

(1) General.

(i) A licensed importer, licensed manufacturer or licensed dealer selling or otherwise distributing explosive materials (or a holder of a user permit disposing of surplus stock to a licensee; a holder of a user permit; or a holder of a limited permit who is within the same State as the distributor) who has the certified information required by this section may sell or distribute explosive materials to a licensee or permittee for not more than 45 days following the expiration date of the distributee's license or permit, unless the distributor knows or has reason to believe that the distributee's authority to continue business or operations under this part has been terminated.

(ii) A licensed importer, licensed manufacturer or licensed dealer selling or otherwise distributing explosive materials (or a holder of a user permit disposing of surplus stock to another licensee or permittee) must verify the license or permit status of the distributee prior to the release of explosive materials ordered, as required by this section.

(iii) Licensees or permittees desiring to return explosive materials to a licensed manufacturer may do so without obtaining a certified copy of the manufacturer's license.

(2) Verification of license/user permit.

(i) Prior to or with the first order of explosive materials, the distributee must provide the distributor certified copy (or, in the case of a user-limited, the original) of the distributee's license or user permit. However, licensees or holders of user permits that are business organizations may (in lieu of a certified copy of a license or user permit) provide the distributor with a certified list that contains the name, address, license or user permit number, and date of the license or user permit expiration of each location.

(ii) The distributee must also provide the distributor with a current list of the names of persons authorized to accept delivery of explosive materials on behalf of the distributee. The distributee ordering explosive materials must keep the list current and provide updated lists to licensees and holders of user permits on a timely basis. A distributor may not transfer possession of explosive materials to any person whose name does not appear on the current list of names of persons authorized to accept delivery of explosive materials on behalf of the distributee. Except as provided in paragraph (b)(3) of this section, in all instances the distributor must verify the identity of the distributee, or the employee of the distributee accepting possession of explosive materials on behalf of the distributee, by examining an identification document (as defined in §555.11) before relinquishing possession.

(iii) A licensee or holder of a user permit ordering explosive materials from another licensee or permittee must provide to the distributor a current, certified statement of the intended use of the explosive materials, e.g., resale, mining, quarrying, agriculture, construction, sport rocketry, road building oil well drilling, seismographic research, etc.

(A) For individuals, the certified statement of intended use must specify the name, address, date and place of birth, and social security number of the distributee.

(B) For business organizations, the certified statement of intended use must specify the taxpayer identification number, the identity and the principal and local places of business.

(C) The licensee or holder of a user permit purchasing explosive materials must revise the furnished copy of the certified statement only when the information is no longer current.

(3) Delivery of explosive materials by common or contract carrier. When a common or contract carrier will transport explosive materials from a distributor to a distributee who is a licensee or holder of a user permit, the distributor must take the following actions before relinquishing possession of the explosive materials:

(i) Verify the identity of the person accepting ›ssession for the common or contract carrier by examining such rson's valid, unexpired driver's license issued by any State, ınada, or Mexico; and

(ii) Record the name of the common or contract rrier (i.e., the name of the driver's employer) and the full name the driver. This information must be maintained in the distribu-r's permanent records in accordance with §555.121.

(4) User-limited permit transactions. A user-limited rmit issued under the provisions of this part is valid for only a ıgle purchase transaction and is not renewable (see §555.51). ccordingly, at the time a user-limited permittee orders explosive aterials, the licensed distributor must write on the front of the er-limited permit the transaction date, his signature, and the stributor's license number prior to returning the permit to the er-limited permittee.

(Approved by the Office of Management and Budget under ntrol number 1140–0079)

TF No. 1, 68 FR 13787, Mar. 20, 2003, as amended by ATF No. 2, 68 FR 512, Sept. 11, 2003]

555.104 Certified copy of license or permit.

Except as provided in §555.49(a), each person issued a license permit under this part shall be furnished together with his ense or permit a copy for his certification. If a person desires additional copy of his license or permit for certification and use under §555.103, he shall:

Make a reproduction of the copy of his license or permit and ecute the certification on it;

) Make a reproduction of his license or permit, enter on the production the statement: "I certify that this is a true copy of a sert the word license or permit) issued to me to engage in the ecified business or operations", and sign his name next to the tement; or

Submit a request, in writing, for certified copies of his ense or permit to the Chief, Firearms and Explosives Licensing nter. The request will show the name, trade name (if any), and dress of the licensee or permittee and the number of copies of license or permit desired. There is a fee of $1 for each copy of icense or permit issued by the Chief, Firearms and Explosives censing Center under this paragraph. Fee payment must company each request for additional copies of a license or rmit. The fee must be paid by (1) cash, or (2) money order or eck made payable to the Bureau of Alcohol, Tobacco and earms.

D. ATF–87, 46 FR 40384, Aug. 7, 1981, as amended by T.D. ATF–290, 54 FR 54, Dec. 27, 1989]

555.105 Distributions to nonlicensees, npermittees, and limited permittees.

Distributions to nonlicensees and nonpermittees prior to ay 24, 2003.

(1) This section will apply in any case where distribution of explosive materials to the distributee is not otherwise prohibited by the Act or this part.

(2) Except as provided in paragraph (a)(3) of this section, a licensed importer, licensed manufacturer, or licensed dealer may distribute explosive materials to a nonlicensee or nonpermittee if the nonlicensee or nonpermittee is a resident of the same State in which the licensee's business premises are located, and the nonlicensee or nonpermittee furnishes to the licensee the explosives transaction record, ATF F 5400.4, required by §555.126. Disposition of ATF F 5400.4 will be made in accordance with §555.126.

(3) A licensed importer, licensed manufacturer, or licensed dealer may sell or distribute explosive materials to a resident of a State contiguous to the State in which the licensee's place of business is located if the purchaser's State of residence has enacted legislation, currently in force, specifically authorizing a resident of that State to purchase explosive materials in a contiguous State and the purchaser and the licensee have, prior to the distribution of the explosive materials, complied with all the requirements of paragraphs (a)(2), (a)(5), and (a)(6) of this section applicable to intrastate transactions occurring on the licensee's business premises.

(4) A permittee may dispose of surplus stocks of explosive materials to a nonlicensee or nonpermittee if the nonlicensee or nonpermittee is a resident of the same State in which the permittee's business premises or operations are located, or is a resident of a State contiguous to the State in which the permittee's place of business or operations are located, and if the requirements of paragraphs (a)(2), (a)(3), (a)(5), and (a)(6) of this section are fully met.

(5) A licensed importer, licensed manufacturer, or licensed dealer selling or otherwise distributing explosive materials to a business entity must verify the identity of the representative or agent of the business entity who is authorized to order explosive materials on behalf of the business entity. Each business entity ordering explosive materials must furnish the distributing licensee prior to or with the first order of explosive materials a current certified list of the names of representatives or agents authorized to order explosive materials on behalf of the business entity. The business entity ordering explosive materials is responsible for keeping the certified list current. A licensee may not distribute explosive materials to a business entity on the order of a person whose name does not appear on the certified list.

(6) Where the possession of explosive materials is transferred at the distributor's premises, the distributor must in all instances verify the identity of the person accepting possession on behalf of the distributee before relinquishing possession. Before the delivery at the distributor's premises of explosive materials to an employee of a nonlicensee or nonpermittee, or to an employee of a common or contract carrier transporting explosive materials to a nonlicensee or nonpermittee, the distributor delivering explosive materials must obtain an executed ATF F 5400.8 from the employee before releasing the explosive materials. The ATF F 5400.8 must contain all of the information required on the form and by this part. (See examples in **§555.103(a)).**

(7) A licensee or permittee disposing of surplus stock may sell or distribute commercially manufactured black powder in quantities of 50 pounds or less to a nonlicensee or nonpermittee if the black powder is intended to be used solely for sporting, recreational, or cultural purposes in antique firearms as defined in 18 U.S.C. 921(a)(16), or in antique devices as exempted from the term "destructive device" in 18 U.S.C. 921(a)(4).

(b) Distributions to holders of limited permits on and after May 24, 2003.

(1) This section will apply in any case where distribution of explosive materials to the distributee is not otherwise prohibited by the Act or this part.

(2) A licensed importer, licensed manufacturer or a licensed dealer may distribute explosive materials to a holder of a limited permit if such permittee is a resident of the same State in which the licensee's business premises are located, the holder of the limited permit presents in person or by mail ATF Form 5400.4, Limited Permittee Transaction Report (LPTR), and the licensee completes Form 5400.4 in accordance with §555.126(b). In no event will a licensee distribute explosive materials to a holder of a limited permit unless the holder presents a Form 5400.4 with an original unaltered and unexpired Intrastate Purchase of Explosives Coupon (IPEC), ATF Form 5400.30, affixed. The coupon must bear the name, address, permit number, and the coupon number of the limited permittee seeking distribution of the explosives.

(3) A holder of a limited permit is authorized to receive explosive materials from a licensee or permittee whose premises are located in the same State of residence in which the premises of the holder of the limited permit are located on no more than 6 separate occasions during the one-year period of the permit. For purposes of this section, the term "6 separate occasions" means six deliveries of explosive materials. Each delivery must—

(i) Relate to a single purchase transaction made on one ATF F 5400.4;

(ii) Be referenced on one commercial invoice or purchase order; and

(iii) Be delivered to the holder of the limited permit in one shipment delivered at the same time.

(4) A holder of a user permit may dispose of surplus stocks of explosive materials to a licensee or holder of a user permit, or a holder of a limited permit who is a resident of the same State in which the premises of the holder of the user permit are located. A holder of a limited permit may dispose of surplus stocks of explosive materials to another holder of a limited permit who is a resident of the same State in which the premises of the distributor are located, if the transaction complies with the requirements of paragraph (b)(2) of this section and §555.126(b). A holder of a limited permit may also dispose of surplus stocks of explosive materials to a licensee or holder of a user permit if the disposit occurs in the State of residence of the holder of the limited permit. (See §555.103.)

(5) Each holder of a limited permit ordering explosive materials must furnish the distributing licensee prior to or with the first order of the explosive materials a current list of the names of employees authorized to accept delivery of explosive materials on behalf of the limited permittee. The distributee ordering explosive materials must keep the list current and provide updated lists to licensees and holders of user permits o timely basis. A licensed importer, licensed manufacturer, licens dealer, or permittee, selling or otherwise distributing explosive materials to a holder of a limited permit must, prior to deliveri the explosive materials, obtain from the limited permittee a current list of persons who are authorized to accept deliveries c explosive materials on behalf of the limited permittee. A licens or permittee may not deliver explosive materials to a person whose name does not appear on the list.

(6)(i) Delivery at the distributor's premises. Where possession of explosive materials is transferred directly to the distributee at the distributor's premises, the distributor must obtain an executed Form 5400.4 in accordance with §555.126(and must in all instances verify the identity of the person accep ing possession on behalf of the distributee by examining an identification document (as defined in §555.11) before relinquishing possession.

(ii) Delivery by distributor. Where possession of explosive materials is transferred by the distributor to the distri tee away from the distributor's premises, the distributor must obtain an executed Form 5400.4 in accordance with §555.126(and must in all instances verify the identity of the person accep ing possession on behalf of the distributee by examining an identification document (as defined in §555.11) before relinquishing possession.

(iii) Delivery by common or contract carrier hired b the distributor. Where a common or contract carrier hired by th distributor will transport explosive materials from the distribut to a holder of a limited permit:

(A) The limited permittee must, prior to delive of the explosive materials, complete the appropriate section on Form 5400.4, affix to the Form 5400.4 one of the six IPECs he has been issued, and provide the form to the distributor in pers or by mail.

(B) The distributor must, before relinquishing possession of the explosive materials to the common or contra carrier:

(1) Verify the identity of the person accept possession for the common or contract carrier by examining su person's valid, unexpired driver's license issued by any State, Canada, or Mexico; and

(2) Record the name of the common or ntract carrier (i.e., the name of the driver's employer) and the ll name of the driver. This information must be maintained in e distributor's permanent records in accordance with §555.121.

(C) At the time of delivery of the explosive aterials, the common or contract carrier, as agent for the stributor, must verify the identity of the person accepting livery on behalf of the distributee, note the type and number of e identification document (as defined in §555.11) and provide is information to the distributor. The distributor must enter this formation in the appropriate section on Form 5400.4.

(iv) Delivery by common or contract carrier hired by e distributee. Where a common or contract carrier hired by the stributee will transport explosive materials from the distributor a holder of a limited permit:

(A) The limited permittee must, prior to delivery the explosive materials, complete the appropriate section on rm 5400.4, affix to the Form 5400.4 one of the six IPECs he s been issued, and provide the form to the distributor in person by mail.

(B) Before the delivery at the distributor's emises to the common or contract carrier who will transport plosive materials to the holder of a limited permit, the distribu- r must:

(1) Verify the identity of the person accepting ssession for the common or contract carrier by examining such rson's valid, unexpired driver's license issued by any State, nada, or Mexico; and

(2) Record the name of the common or ntract carrier (i.e., the name of the driver's employer) and the ll name of the driver. This information must be maintained in e distributor's permanent records in accordance with §555.121.

(7) A licensee or permittee disposing of surplus stock may ll or distribute commercially manufactured black powder in antities of 50 pounds or less to a holder of a limited permit, nlicensee, or nonpermittee if the black powder is intended to be ed solely for sporting, recreational, or cultural purposes in tique firearms as defined in 18 U.S.C. 921(a)(16), or in antique vices as exempted from the term "destructive device" in 18 S.C. 921(a)(4).

(Approved by the Office of Management and Budget under ntrol number 1140–0075)

TF No. 1, 68 FR 13788, Mar. 20, 2003, as amended by ATF No. 2, 68 FR 513, Sept. 11, 2003]

555.106 Certain prohibited distributions.

A licensee or permittee may not distribute explosive materi- s to any person except—

(1) A licensee;

(2) A holder of a user permit; or

(3) A holder of a limited permit who is a resident of the ate where distribution is made and in which the premises of the ansferor are located.

(b) A licensee shall not distribute any explosive materials to any person:

(1) Who the licensee knows is less than 21 years of age;

(2) In any State where the purchase, possession, or use by a person of explosive materials would be in violation of any State law or any published ordinance applicable at the place of distribution;

(3) Who the licensee has reason to believe intends to transport the explosive materials into a State where the purchase, possession, or use of explosive materials is prohibited or which does not permit its residents to transport or ship explosive materials into the State or to receive explosive materials in the State; or

(4) Who the licensee has reasonable cause to believe intends to use the explosive materials for other than a lawful purpose.

(c) A licensee shall not distribute any explosive materials to any person knowing or having reason to believe that the person:

(1) Is, except as provided under §555.142 (d) and (e), under indictment or information for, or was convicted in any court of, a crime punishable by imprisonment for a term exceeding 1 year;

(2) Is a fugitive from justice;

(3) Is an unlawful user of marijuana, or any depressant or stimulant drug, or narcotic drug (as these terms are defined in the Controlled Substances Act, 21 U.S.C. 802);

(4) Was adjudicated as a mental defective or was committed to a mental institution;

(5) Is an alien, other than an alien who—

(i) Is lawfully admitted for permanent residence (as that term is defined in section 101(a)(20) of the Immigration and Nationality Act (8 U.S.C. 1101));

(ii) Is in lawful nonimmigrant status, is a refugee admitted under section 207 of the Immigration and Nationality Act (8 U.S.C. 1157), or is in asylum status under section 208 of the Immigration and Nationality Act (8 U.S.C. 1158), and—

(A) Is a foreign law enforcement officer of a friendly foreign government, as determined by the Attorney General in consultation with the Secretary of State, entering the United States on official law enforcement business, and the shipping, transporting, possession, or receipt of explosive materials is in furtherance of this official law enforcement business;

(B) Is a person having the power to direct or cause the direction of the management and policies of a corporation, partnership, or association licensed pursuant to section 843(a), and the shipping, transporting, possession, or receipt of explosive materials is in furtherance of such power;

(C) Is a member of a North Atlantic Treaty Organization (NATO) or other friendly foreign military force, as determined by the Attorney General in consultation with the Secretary of Defense, (whether or not admitted in a nonimmigrant status) who is present in the United States under military orders for training or other military purpose authorized by the United States, and the shipping, transporting, possession, or receipt of explosive materials is in furtherance of the military purpose; or

(D) Is lawfully present in the United States in cooperation with the Director of Central Intelligence, and the shipment, transportation, receipt, or possession of the explosive materials is in furtherance of such cooperation;

(6) Has been discharged from the armed forces under dishonorable conditions; or

(7) Having been a citizen of the United States, has renounced citizenship.

(d) The provisions of this section do not apply to the purchase of commercially manufactured black powder in quantities not to exceed 50 pounds, intended to be used solely for sporting, recreational, or cultural purposes in antique firearms or in antique devices, if the requirements of §555.105(a)(7) or (b)(7) are fully met.

[T.D. ATF–87, 46 FR 40384, Aug. 7, 1981. Redesignated by T.D. ATF–487, 68 FR 3748, Jan. 24, 2003. ATF No. 1, 68 FR 13790, Mar. 20, 2003]

§ 555.107 Record of transactions.

Each licensee and permittee shall keep records of explosive materials as required by subpart G of this part.

§ 555.108 Importation.

(a) Explosive materials imported or brought into the United States by a licensed importer or holder of a user permit may be released from customs custody to the licensed importer or holder of a user permit upon proof of his status as a licensed importer or holder of a user permit. Proof of status must be made by the licensed importer or holder of a user permit furnishing to the customs officer a certified copy of his license or permit (see §555.103).

(b) A nonlicensee or nonpermittee may import or bring into the United States commercially manufactured black powder in quantities not to exceed 50 pounds. Upon submitting to the customs officer completed ATF F 5400.3, certifying that the black powder is intended to be used solely for sporting, recreational, or cultural purposes in antique firearms or in antique devices, black powder may be released from customs custody. The disposition of the executed ATF F 5400.3 will be in accordance with the instructions on the form.

(c) The provisions of this section are in addition to, and are not in lieu of, any applicable requirement under 27 CFR Part 447.

(d) For additional requirements relating to the importation of plastic explosives into the United States on or after April 24, 1997, see §555.183.

(e) For requirements relating to the marking of imported explosive materials, see §555.109.

[T.D. ATF–87, 46 FR 40384, Aug. 7, 1981, as amended by T.D. ATF–387, 62 FR 8376, Feb. 25, 1997; ATF No. 1, 68 FR 13790, Mar. 20, 2003; ATF 5F, 70 FR 30633, May 27, 2005]

§ 555.109 Identification of explosive materials.

(a) General. Explosive materials, whether manufactured in th United States or imported, must contain certain marks of identi cation.

(b) Required marks

(1) Licensed manufacturers. Licensed manufacturers wh manufacture explosive materials for sale or distribution must place the following marks of identification on explosive materi at the time of manufacture:

(i) The name of the manufacturer; and

(ii) The location, date, and shift of manufacture. Where a manufacturer operates his plant for only one shift duri the day, he does not need to show the shift of manufacture.

(2) Licensed importers.

(i) Licensed importers who import explosive mate als for sale or distribution must place the following marks of identification on the explosive materials they import:

(A) The name and address (city and state) of th importer; and

(B) The location (city and country) where the explosive materials were manufactured, date, and shift of manu facture. Where the foreign manufacturer operates his plant for only one shift during the day, he does not need to show the shif of manufacture.

(ii) Licensed importers must place the required ma on all explosive materials imported prior to distribution or shipment for use, and in no event later than 15 days after the da of release from Customs custody.

(c) General requirements.

(1) The required marks prescribed in this section must be permanent and legible.

(2) The required marks prescribed in this section must be the English language, using Roman letters and Arabic numerals

(3) Licensed manufacturers and licensed importers must place the required marks on each cartridge, bag, or other imme ate container of explosive materials that they manufacture or import, as well as on any outside container used for the packag of such explosive materials.

(4) Licensed manufacturers and licensed importers may any method, or combination of methods, to affix the required marks to the immediate container of explosive materials, or outside containers used for the packaging thereof, provided the identifying marks are legible, permanent, show all the required information, and are not rendered unreadable by extended peri of storage.

(5) If licensed manufacturers or licensed importers desir use a coding system and omit printed markings on the containe that show all the required information specified in paragraphs (b)(1) and (2) of this section, they must file with ATF a letterhe application displaying the coding that they plan to use and explaining the manner of its application. The Director must approve the application before the proposed coding can be usec

(d) Exceptions —

(1) Blasting caps. Licensed manufacturers or licensed importers are only required to place the identification marks prescribed in this section on the containers used for the packaging of blasting caps.

(2) Alternate means of identification. The Director may authorize other means of identifying explosive materials, including fireworks, upon receipt of a letter application from the licensed manufacturer or licensed importer showing that such other identification is reasonable and will not hinder the effective administration of this part.

(Paragraph (b)(2) approved by the Office of Management and Budget under control number 1140–0055)

[ATF 5F, 70 FR 30633, May 27, 2005]

§ 555.110 Furnishing of samples (Effective on and after January 24, 2003).

(a) In general. Licensed manufacturers and licensed importers and persons who manufacture or import explosive materials or ammonium nitrate must, when required by letter issued by the Director, furnish—

(1) Samples of such explosive materials or ammonium nitrate;

(2) Information on chemical composition of those products; and

(3) Any other information that the Director determines is relevant to the identification of the explosive materials or to identification of the ammonium nitrate.

(b) Reimbursement. The Director will reimburse the fair market value of samples furnished pursuant to paragraph (a) of this section, as well as reasonable costs of shipment.

(Approved by the Office of Management and Budget under control number 1140–0073)

[ATF No. 1, 68 FR 13790, Mar. 20, 2003]

Subpart G—Records and Reports

§ 555.121 General.

(a) **(1)** Licensees and permittees shall keep records pertaining to explosive materials in permanent form (i.e., commercial invoices, record books) and in the manner required in this subpart.

(2) Licensees and permittees shall keep records required by this part on the business premises for five years from the date a transaction occurs or until discontinuance of business or operations by the licensee or permittee. (See also §555.128 for discontinuance of business or operations.)

(b) ATF officers may enter the premises of any licensee or holder of a user permit for the purpose of examining or inspecting any record or document required by or obtained under this part (see §555.24). Section 843(f) of the Act requires licensees and holders of user permits to make all required records available for examination or inspection at all reasonable times. Section 843(f) of the Act also requires licensees and permittees (including holders of limited permits) to submit all reports and information relating to all required records and their contents, as the regulations in this part prescribe.

(c) Each licensee and permittee shall maintain all records of importation, production, shipment, receipt, sale, or other disposition, whether temporary or permanent, of explosive materials as the regulations in this part prescribe. Sections 842(f) and 842(g) of the Act make it unlawful for any licensee or permittee knowingly to make any false entry in, or fail to make entry in, any record required to be kept under the Act and the regulations in this part.

(Approved by the Office of Management and Budget under control number 1512–0373)

[T.D. ATF–87, 46 FR 40384, Aug. 7, 1981, as amended by T.D. ATF–172, 49 FR 14941, Apr. 16, 1984; ATF No. 1, 68 FR 13790, Mar. 20, 2003]

§ 555.122 Records maintained by licensed importers.

(a) Each licensed importer shall take true and accurate physical inventories which will include all explosive materials on hand required to be accounted for in the records kept under this part. The licensed importer shall take a special inventory

(1) At the time of commencing business, which is the effective date of the license issued upon original qualification under this part;

(2) At the time of changing the location of his business to another region;

(3) At the time of discontinuing business; and

(4) At any time the regional director (compliance) may in writing require. Each special inventory is to be prepared in duplicate, the original of which is submitted to the regional director (compliance), and the duplicate retained by the licensed importer. If a special inventory specified by paragraphs (a) (1) through (4) of this section has not been taken during the calendar year, at least one physical inventory will be taken. However, the record of the yearly inventory, other than a special inventory required by paragraphs (a) (1) through (4) of this section, will remain on file for inspection instead of being sent to the regional director (compliance). (See also §555.127.)

(b) Each licensed importer shall, not later than the close of the next business day following the date of importation or other acquisition of explosive materials, enter the following information in a separate record:

(1) Date of importation or other acquisition.

(2) Name or brand name of manufacturer and country of manufacture.

(3) Manufacturer's marks of identification.

(4) Quantity (applicable quantity units, such as pounds of explosives, number of detonators, number of display fireworks, etc.).

(5) Description (dynamite (dyn), blasting agents (ba), detonators (det), display fireworks (df), etc.) and size (length and diameter or diameter only of display fireworks).

(c) Each licensed importer shall, not later than the close of the next business day following the date of distribution of any explosive materials to another licensee or a permittee, enter in a separate record the following information:

(1) Date of disposition.

(2) Name or brand name of manufacturer and country of manufacture.

(3) Manufacturer's marks of identification.

(4) Quantity (applicable quantity units, such as pounds of explosives, number of detonators, number of display fireworks, etc.).

(5) Description (dynamite (dyn), blasting agents (ba), detonators (det), display fireworks (df), etc.) and size (length and diameter or diameter only of display fireworks).

(6) License or permit number of licensee or permittee to whom the explosive materials are distributed.

(d) The regional director (compliance) may authorize alternate records to be maintained by a licensed importer to record his distribution of explosive materials when it is shown by the licensed importer that alternate records will accurately and readily disclose the information required by paragraph (c) of this section. A licensed importer who proposes to use alternate records shall submit a letter application to the regional director (compliance) and shall describe the proposed alternate records and the need for them. Alternate records are not to be employed by the licensed importer until approval is received from the regional director (compliance).

(e) Each licensed importer shall maintain separate records of the sales or other distribution made of explosive materials to nonlicensees or nonpermittees. These records are maintained as prescribed by §555.126.

(Approved by the Office of Management and Budget under control number 1512–0373)

[T.D. ATF–87, 46 FR 40384, Aug. 7, 1981, as amended by T.D. ATF–172, 49 FR 14941, Apr. 16, 1984; T.D. ATF–293, 55 FR 3721, Feb. 5, 1990; T.D. ATF–400, 63 FR 45003, Aug. 24, 1998]

§ 555.123 Records maintained by licensed manufacturers.

(a) Each licensed manufacturer shall take true and accurate physical inventories which will include all explosive materials on hand required to be accounted for in the records kept under this part. The licensed manufacturer shall take a special inventory

(1) At the time of commencing business, which is the effective date of the license issued upon original qualification under this part;

(2) At the time of changing the location of his premises to another region;

(3) At the time of discontinuing business; and

(4) At any other time the regional director (compliance) may in writing require. Each special inventory is to be prepared in duplicate, the original of which is submitted to the regional director (compliance), and the duplicate retained by the licensed manufacturer. If a special inventory required by paragraphs (a) (1) through (4) of this section has not been taken during the calendar year, at least one physical inventory will be taken. However, the record of the yearly inventory, other than a special inventory required by paragraphs (a) (1) through (4) of this section, will remain on file for inspection instead of being sent to the regional director (compliance). (See also §555.127.)

(b) Each licensed manufacturer shall not later than the close of the next business day following the date of manufacture or other acquisition of explosive materials, enter the following information in a separate record:

(1) Date of manufacture or other acquisition.

(2) Manufacturer's marks of identification.

(3) Quantity (applicable quantity units, such as pounds of explosives, number of detonators, number of display fireworks, etc.).

(4) Name, brand name or description (dynamite (dyn), blasting agents (ba), detonators (det), display fireworks (df), etc.) and size (length and diameter or diameter only of display fireworks).

(c) Each licensed manufacturer shall, not later than the close of the next business day following the date of distribution of any explosive materials to another licensee or a permittee, enter in a separate record the following information:

(1) Date of disposition.

(2) Name or brand name of manufacturer or name of importer, as applicable, if acquired other than by his own manufacture.

(3) Manufacturer's marks of identification.

(4) Quantity (applicable quantity units, such as pounds of explosives, number of detonators, number of display fireworks, etc.).

(5) Description (dynamite (dyn), blasting agents (ba), tonators (det), display fireworks (df), etc.) and size (length and ameter or diameter only of display fireworks).

(6) License or permit number of licensee or permittee to iom the explosive materials are distributed.

) Each licensed manufacturer who manufactures explosive aterials for his own use shall, not later than the close of the next siness day following the date of use, enter in a separate record e following information:

(1) Date of use.

(2) Quantity (applicable quantity units, such as pounds of plosives, number of detonators, number of special fireworks, :.).

(3) Description (dynamite (dyn), blasting agents (ba), tonators (det), display fireworks (df), etc.) and size (length and ameter or diameter only of display fireworks).

Exception: A licensed manufacturer is exempt from the cordkeeping requirements of this subsection if the explosive aterials are manufactured for his own use and used within a 24 ur period at the same site.

The regional director (compliance) may authorize alternate cords to be maintained by a licensed manufacturer to record his stribution or use of explosive materials when it is shown by the ensed manufacturer that alternate records will accurately and adily disclose the information required by paragraph (c) of this ction. A licensed manufacturer who proposes to use alternate cords shall submit a letter application to the regional director ompliance) and shall describe the proposed alternate records d the need for them. Alternate records are not to be employed the licensed manufacturer until approval is received from the gional director (compliance).

Each licensed manufacturer shall maintain separate records the sales or other distribution made of explosive materials to nlicensees or nonpermittees. These records are maintained as scribed by §555.126.

(Approved by the Office of Management and Budget under ntrol number 1512–0373)

D. ATF–87, 46 FR 40384, Aug. 7, 1981, as amended by T.D. ATF–172, 49 FR 941, Apr. 16, 1984; T.D. ATF–293, 55 FR 3721, Feb. 5, 1990; T.D. ATF–400, 63 45003, Aug. 24, 1998]

555.124 Records maintained by licensed dealers.

Each licensed dealer shall take true and accurate physical entories which will include all explosive materials on hand quired to be accounted for in the records kept under this part. e licensed dealer shall take a special inventory

(1) At the time of commencing business, which is the ective date of the license issued upon original qualification der this part;

(2) At the time of changing the location of his premises to other region;

(3) At the time of discontinuing business; and

(4) At any other time the regional director (compliance) may in writing require. Each special inventory is to be prepared in duplicate, the original of which is submitted to the regional director (compliance), and the duplicate retained by the licensed dealer. If a special inventory required by paragraphs (a)(1) through (4) of this section has not been taken during the calendar year, at least one physical inventory will be taken. However, the record of the yearly inventory, other than a special inventory required by paragraphs (a)(1) through (4) of this section, will remain on file for inspection instead of being sent to the regional director (compliance). (See also §555.127.)

(b) Each licensed dealer shall, not later than the close of the next business day following the date of purchase or other acquisition of explosive materials (except as provided in paragraph (d) of this section), enter the following information in a separate record:

(1) Date of acquisition.

(2) Name or brand name of manufacturer and name of importer (if any).

(3) Manufacturer's marks of identification.

(4) Quantity (applicable quantity units, such as pounds of explosives, number of detonators, number of display fireworks, etc.).

(5) Description (dynamite (dyn), blasting agents (ba), detonators (det), display fireworks (df), etc.) and size (length and diameter or diameter only of display fireworks).

(6) Name, address, and license or permit number of the person from whom the explosive materials are received.

(c) Each licensed dealer shall, not later than the close of the next business day following the date of use (if the explosives are used by the dealer) or the date of distribution of any explosive materials to another licensee or a permittee (except as provided in paragraph (d) of this section), enter in a separate record the following information:

(1) Date of disposition.

(2) Name or brand name of manufacturer and name of importer (if any).

(3) Manufacturer's marks of identification.

(4) Quantity (applicable quantity units, such as pounds of explosives, number of detonators, number of display fireworks, etc.)

(5) Description (dynamite (dyn), blasting agents (ba), detonators (det), display fireworks (df), etc.) and size (length and diameter or diameter only of display fireworks).

(6) License or permit number of licensee or permittee to whom the explosive materials are distributed.

(d) When a commercial record is kept by a licensed dealer showing the purchase or other acquisition information required for the permanent record prescribed by paragraph (b) of this section, or showing the distribution information required for the permanent record prescribed by paragraph (c) of this section, the licensed dealer acquiring or distributing the explosive materials may, for a period not exceeding seven days following the date of acquisition of distribution of the explosive materials, delay making the required entry into the permanent record of acquisition or distribution. However, until the required entry of acquisition or disposition is made in the permanent record, the commercial record must be (1) kept by the licensed dealer separate from other commercial documents kept by the licensee, and (2) readily available for inspection on the licensed premises.

(e) The regional director (compliance) may authorize alternate records to be maintained by a licensed dealer to record his acquisition or disposition of explosive materials, when it is shown by the licensed dealer that alternate records will accurately and readily disclose the required information. A licensed dealer who proposes to use alternate records shall submit a letter application to the regional director (compliance) and shall describe the proposed alternate records and the need for them. Alternate records are not to be employed by the licensed dealer until approval is received from the regional director (compliance).

(f) Each licensed dealer shall maintain separate records of the sales or other distribution made of explosive materials to nonlicensees or nonpermittees. These records are maintained as prescribed by §555.126.

(Approved by the Office of Management and Budget under control number 1512–0373)

[T.D. ATF–87, 46 FR 40384, Aug. 7, 1981, as amended by T.D. ATF–172, 49 FR 14941, Apr. 16, 1984; T.D. ATF–293, 55 FR 3721, Feb. 5, 1990; T.D. ATF–400, 63 FR 45003, Aug. 24, 1998]

§ 555.125 Records maintained by permittees.

(a) Records maintained by permittees prior to May 24, 2003.

(1) Each permittee must take true and accurate physical inventories that will include all explosive materials on hand required to be accounted for in the records kept under this part. The permittee must take a special inventory—

(i) At the time of commencing business, which is the effective date of the permit issued upon original qualification under this part;

(ii) At the time of changing the location of his premises to another region;

(iii) At the time of discontinuing business; and

(iv) At any other time the regional director (compliance) may in writing require. Each special inventory is to be prepared in duplicate, the original of which is submitted to the regional director (compliance) and the duplicate retained by the permittee. If a special inventory required by paragraphs (a)(1)(i) through (iv) of this section has not been taken during the calendar year, a permittee is required to take at least one physical inventory. However, the record of the yearly inventory, other than a special inventory required by paragraphs (a)(1)(i) through (iv) of this section, will remain on file for inspection instead of being sent to the regional director (compliance). (See also §555.127.)

(2) Each permittee must, not later than the close of the ne business day following the date of acquisition of explosive materials, enter the following information in a separate record:

(i) Date of acquisition;

(ii) Name or brand name of manufacturer;

(iii) Manufacturer's marks of identification;

(iv) Quantity (applicable quantity units, such as pounds of explosives, number of detonators, number of display fireworks, etc.);

(v) Description (dynamite (dyn), blasting agents (b detonators (det), display fireworks (df), etc., and size (length ar diameter or diameter only of display fireworks)); and

(vi) Name, address, and license number of the perso from whom the explosive materials are received.

(3) Each permittee must, not later than the close of the ne business day following the date of disposition of surplus explos materials to another permittee or a licensee, enter in a separate record the information prescribed in §555.124(c).

(4) Each permittee must maintain separate records of disposition of surplus stocks of explosive materials to nonlicen ees or nonpermittees as prescribed in §555.126.

(5) The regional director (compliance) may authorize alternate records to be maintained by a permittee to record his acquisition of explosive materials, when it is shown by the permittee that alternate records will accurately and readily disclose the required information. A permittee who proposes to use alternate records must submit a letter application to the regional director (compliance) and must describe the proposed alternate records and the need for them. Alternate records are n to be employed by the permittee until approval is received from the regional director (compliance).

(b) Records maintained by permittees on and after May 24, 2003.

(1) Each holder of a user permit must take true and accura physical inventories that will include all explosive materials on hand required to be accounted for in the records kept under this part. The permittee must take a special inventory—

(i) At the time of commencing business, which is effective date of the permit issued upon original qualification under this part;

(ii) At the time of changing the location of his premises;

(iii) At the time of discontinuing business; and

(iv) At any other time the regional director (compliance) may in writing require. Each special inventory is to be prepared in duplicate, the original of which is submitted to the regional director (compliance) and the duplicate retained by th permittee. If a special inventory required by paragraphs (b)(1)(through (iv) of this section has not been taken during the calen year, a permittee is required to take at least one physical invent ıe record of the yearly inventory, other than a special inventory quired by paragraphs (b)(1)(i) through (iv) of this section, will main on file for inspection instead of being sent to the regional rector (compliance). (See also §555.127.)

(2) Each holder of a limited permit must take true and curate physical inventories, at least annually, that will include explosive materials on hand required to be accounted for in the cords kept under this part.

(3) Each holder of a user permit or a limited permit must, t later than the close of the next business day following the date acquisition of explosive materials, enter the following information in a separate record:

(i) Date of acquisition;

(ii) Name or brand name of manufacturer;

(iii) Manufacturer's marks of identification;

(iv) Quantity (applicable quantity units, such as unds of explosives, number of detonators, number of display eworks, etc.);

(v) Description (dynamite (dyn), blasting agents (ba), tonators (det), display fireworks (df), etc., and size (length and ameter or diameter only of display fireworks)); and

(vi) Name, address, and license number of the persons om whom the explosive materials are received.

(4) Each holder of a user permit or a limited permit must, t later than the close of the next business day following the date disposition of surplus explosive materials to another permittee a licensee, enter in a separate record the information prescribed §555.124(c).

(5) When a record book is used as a permittee's permanent cord the permittee may delay entry of the required information a period not to exceed seven days if the commercial record ntains all of the required information prescribed by paragraphs)(3) and (b)(4) of this section. However, the commercial record ay be used instead of a record book as a permanent record ovided that the record contains all of the required information escribed by paragraphs (b)(3) and (b)(4) of this section.

(6) Each holder of a user permit or a limited permit must aintain separate records of disposition of surplus stocks of plosive materials to holders of a limited permit as prescribed in 55.126.

(7) The regional director (compliance) may authorize ernate records to be maintained by a holder of a user permit or imited permit to record his acquisition of explosive materials, en it is shown by the permittee that alternate records will curately and readily disclose the required information. A rmittee who proposes to use alternate records must submit a ter application to the regional director (compliance) and must scribe the proposed alternate records and the need for them. ternate records are not to be employed by the permittee until proval is received from the regional director (compliance).

(Approved by the Office of Management and Budget under ntrol number 1140–0030)

[F No. 1, 68 FR 13790, Mar. 20, 2003]

§ 555.126 Explosives transaction record for distribution of explosive materials prior to May 24, 2003 and Limited Permittee Transaction Report for distribution of explosive materials on and after May 24, 2003.

(a) Explosives transaction record for distribution of explosive materials prior to May 24, 2003.

(1) A licensee or permittee shall not temporarily or permanently distribute explosive materials to any person, other than another licensee or permittee, unless he records the transaction on an explosives transaction record, ATF F 5400.4.

(2) Before the distribution of explosive materials to a nonlicensee or nonpermittee who is a resident of the State in which the licensee or permittee maintains his business premises, or to a nonlicensee or nonpermittee who is not a resident of the State in which the licensee or permittee maintains his business premises and is acquiring explosive materials under §555.105(a)(3), the licensee or permittee distributing the explosive materials shall obtain an executed ATF F 5400.4 from the distributee which contains all of the information required on the form and by the regulations in this part.

(3) Completed ATF F 5400.4 is to be retained by the licensee or permittee as part of his permanent records in accordance with paragraph (a)(4) of this section.

(4) Each ATF F 5400.4 is retained in numerical (by transaction serial number) order commencing with "1" and continuing in regular sequence. When the numbering of any series reaches "1,000,000," the licensee or permittee may recommence the series. The recommenced series is to be given an alphabetical prefix or suffix. Where there is a change in proprietorship, or in the individual, firm, corporate name or trade name, the series in use at the time of the change may be continued.

(5) The requirements of this section are in addition to any other recordkeeping requirement contained in this part.

(6) A licensee or permittee may obtain, upon request, a supply of ATF F 5400.4 from the Director.

(b) Limited Permittee Transaction Report for distribution of explosive materials on and after May 24, 2003.

(1) A licensee or permittee may not distribute explosive materials to any person who is not a licensee or permittee. A licensee or permittee may not distribute explosive materials to a limited permittee unless the distributor records the transaction on ATF Form 5400.4, Limited Permittee Transaction Report.

(2) Before distributing explosive materials to a limited permittee, the licensee or permittee must obtain an executed Form 5400.4 from the limited permittee with an original unaltered and unexpired Intrastate Purchase of Explosives Coupon (IPEC) affixed. Except when delivery of explosive materials is made by a common or contract carrier who is an agent of the limited permittee, the licensee, permittee, or an agent of the licensee or permittee, must verify the identity of the of the holder of the limited permit by examining an identification document (as defined in §555.11) and noting on the Form 5400.4 the type of document presented. The licensee or permittee must complete the appropriate section on Form 5400.4 to indicate the type and quantity of explosive materials distributed, the license or permit number of the seller, and the date of the transaction. The licensee or permittee must sign and date the form and include any other information required by the instructions on the form and the regulations in this part.

(3) One copy of Form 5400.4 must be retained by the distributor as part of his permanent records in accordance with paragraph (b)(4) of this section and for the period specified in §555.121. The distributor must mail the other copy of Form 5400.4 to the Bureau of Alcohol, Tobacco, Firearms and Explosives in accordance with the instructions on the form.

(4) Each Form 5400.4 must be retained in chronological order by date of disposition, or in alphabetical order by name of limited permittee. A licensee may not, however, use both methods in a single recordkeeping system. Where there is a change in proprietorship by a limited permittee, the forms may continue to be filed together after such change.

(5) The requirements of this section are in addition to any other recordkeeping requirement contained in this part.

(Approved by the Office of Management and Budget under control number 1140–0078)

[T.D. ATF–87, 46 FR 40384, Aug. 7, 1981, as amended by T.D. ATF–93, 46 FR 50787, Oct. 15, 1981; T.D. ATF–172, 49 FR 14941, Apr. 16, 1984; T.D. ATF–446, 66 FR 16602, Mar. 27, 2001; ATF No. 1, 68 FR 13791, Mar. 20, 2003]

§ 555.127 Daily summary of magazine transactions.

In taking the inventory required by §§555.122, 555.123, 555.124, and 555.125, a licensee or permittee shall enter the inventory in a record of daily summary transactions to be kept at each magazine of an approved storage facility; however, these records may be kept at one central location on the business premises if separate records of daily transactions are kept for each magazine. Not later than the close of the next business day, each licensee and permittee shall record by manufacturer's name or brand name, the total quantity received in and removed from each magazine during the day, and the total remaining on hand at the end of the day. Quantity entries for display fireworks may be expressed as the number and size of individual display fireworks in a finished state or as the number of packaged display segments or packaged displays. Information as to the number and size of display fireworks contained in any one packaged display segme or packaged display shall be provided to any ATF officer on request. Any discrepancy which might indicate a theft or loss o explosive materials is to be reported in accordance with §555.3

[T.D. ATF–293, 55 FR 3722, Feb. 5, 1990, as amended by T.D. ATF–400, 63 FI 45003, Aug. 24, 1998]

§ 555.128 Discontinuance of business.

Where an explosive materials business or operations is disco tinued and succeeded by a new licensee or new permittee, the records prescribed by this subpart shall appropriately reflect su facts and shall be delivered to the successor. Where discontinu- ance of the business or operations is absolute, the records required by this subpart must be delivered within 30 days following the business or operations discontinuance to any ATI office located in the region in which the business was located, (to the ATF Out-of-Business Records Center, Spring Mills Offic Park, 882 T.J. Jackson Drive, Falling Waters, West Virginia 254 Where State law or local ordinance requires the delivery of records to other responsible authority, the Chief, Firearms and Explosives Licensing Center may arrange for the delivery of th records required by this subpart to such authority. (See also, §555.61.)

[T.D. ATF–290, 54 FR 53054, Dec. 27, 1989, as amended by T.D. ATF–446a, 6 FR 19089, Apr. 13, 2001; ATF No. 1, 68 FR 13792, Mar. 20, 2003]

§ 555.129 Exportation.

Exportation of explosive materials is to be in accordance wit the applicable provisions of section 38 of the Arms Export Control Act (22 U.S.C. 2778) and implementing regulations. However, a licensed importer, licensed manufacturer, or license dealer exporting explosive materials shall maintain records showing the manufacture or acquisition of explosive materials required by this part and records showing the quantity, the manufacturer's name or brand name of explosive materials, the name and address of the foreign consignee of the explosive materials, and the date the explosive materials were exported. S §555.180 for regulations concerning the exportation of plastic explosives.

[T.D. ATF–87, 46 FR 40384, Aug. 7, 1981, as amended by T.D. ATF–387, 62 F 8377, Feb. 25, 1997]

§ 555.130 [Reserved]

Subpart H—Exemptions

§ 555.141 Exemptions.

(a) General. Except for the provisions of §§555.180 and 555.181, this part does not apply to:

(1) Any aspect of the transportation of explosive materials via railroad, water, highway, or air which is regulated by the U.S. Department of Transportation and its agencies, and which pertains to safety. For example, regulations issued by the Department of Transportation addressing the security risk of aliens transporting explosives by commercial motor or railroad carrier from Canada preclude the enforcement of 18 U.S.C. 842(i)(5) against persons shipping, transporting, receiving, or possessing explosives incident to and in connection with the commercial transportation of explosives by truck or rail from Canada into the United States. Questions concerning this exception should be directed to ATF's Public Safety Branch in Washington, DC.

(2) The use of explosive materials in medicines and medicinal agents in the forms prescribed by the official United States Pharmacopeia or the National Formulary. "The United States Pharmacopeia and The National Formulary," USP and NF Compendia, are available from the United States Pharmacopeial Convention, Inc., 12601 Twinbrook Parkway, Rockville, Maryland 20852.

(3) The transportation, shipment, receipt, or importation of explosive materials for delivery to any agency of the United States or to any State or its political subdivision.

(4) Small arms ammunition and components of small arms ammunition.

(5) The manufacture under the regulation of the military department of the United States of explosive materials for, or their distribution to or storage or possession by, the military or naval services or other agencies of the United States.

(6) Arsenals, navy yards, depots, or other establishments owned by, or operated by or on behalf of, the United States.

(7) The importation, distribution, and storage of fireworks classified as UN0336, UN0337, UN0431, or UN0432 explosives by the U.S. Department of Transportation at 49 CFR 172.101 and generally known as "consumer fireworks" or "articles pyrotechnic."

(8) Gasoline, fertilizers, propellant actuated devices, or propellant actuated industrial tools manufactured, imported, or distributed for their intended purposes.

(9) Industrial and laboratory chemicals which are intended for use as reagents and which are packaged and shipped pursuant to U.S. Department of Transportation regulations, 49 CFR Parts 100 to 177, which do not require explosives hazard warning labels.

(10) Model rocket motors that meet all of the following criteria—

(i) Consist of ammonium perchlorate composite propellant, black powder, or other similar low explosives;

(ii) Contain no more than 62.5 grams of total propellant weight; and

(iii) Are designed as single-use motors or as reload kits capable of reloading no more than 62.5 grams of propellant into a reusable motor casing.

(b) Black powder. Except for the provisions applicable to persons required to be licensed under subpart D, this part does not apply with respect to commercially manufactured black powder in quantities not to exceed 50 pounds, percussion caps, safety and pyrotechnic fuses, quills, quick and slow matches, and friction primers, if the black powder is intended to be used solely for sporting, recreational, or cultural purposes in antique firearms, as defined in 18 U.S.C. 921(a)(16) or antique devices, as exempted from the term "destructive devices" in 18 U.S.C. 921(a)(4).

[T.D. ATF–87, 46 FR 40384, Aug. 7, 1981 as amended by T.D. ATF–87, 46 FR 46916, Sept. 23, 1981; T.D. ATF–293, 55 FR 3722, Feb. 5, 1990; T.D. ATF–387, 62 FR 8377, Feb. 25, 1997; T.D. ATF–400, 63 FR 45003, Aug. 24, 1998; ATF No. 1, 68 FR 13792, Mar. 20, 2003; ATF 6F, 71 FR 46101, Aug. 11, 2006]

§ 555.142 Relief from disabilities (effective January 24, 2003).

(a) Any person prohibited from shipping or transporting any explosive in or affecting interstate or foreign commerce or from receiving or possessing any explosive which has been shipped or transported in or affecting interstate or foreign commerce may make application for relief from disabilities under section 845(b) of the Act.

(b) An application for relief from disabilities must be filed with the Director by submitting ATF Form 5400.29, Application for Restoration of Explosives Privileges, in accordance with the instructions on the form. The application must be supported by appropriate data, including the information specified in paragraph (f) of this section. Upon receipt of an incomplete or improperly executed application for relief, the applicant will be notified of the deficiency in the application. If the application is not corrected and returned within 30 days following the date of notification, the application will be considered abandoned.

(c) **(1)** The Director may grant relief to an applicant if it is established to the satisfaction of the Director that the circumstances regarding the disability and the applicant's record and reputation are such that the applicant will not be likely to act in a manner dangerous to public safety and that the granting of such relief is not contrary to the public interest.

(2) Except as provided in paragraph (c)(3) of this section, the Director will not grant relief if the applicant—

(i) Has not been discharged from parole or probation for a period of at least 2 years;

(ii) Is a fugitive from justice;

(iii) Is a prohibited alien;

(iv) Is an unlawful user of or addicted to any controlled substance;

(v) Has been adjudicated a mental defective or committed to a mental institution, unless the applicant was subsequently determined by a court, board, commission, or other lawful authority to have been restored to mental competency, to be no longer suffering from a mental disorder, and to have had all rights restored; or

(vi) Is prohibited by the law of the State where the applicant resides from receiving or possessing explosive materials.

(3)(i) The Director may grant relief to aliens who have been lawfully admitted to the United States or to persons who have not been discharged from parole or probation for a period of at least 2 years if he determines that the applicant has a compelling need to possess explosives, such as for purposes of employment.

(ii) The Director may grant relief to the persons identified in paragraph (c)(2) of this section in extraordinary circumstances where the granting of such relief is consistent with the public interest.

(d) A person who has been granted relief under this section is relieved of all disabilities imposed by the Act for the disabilities disclosed in the application. The granting of relief will not affect any disabilities incurred subsequent to the date the application was filed. Relief from disabilities granted to aliens will be effective only so long as the alien retains his or her lawful immigration status.

(e)(1) A licensee or permittee who is under indictment or information for, or convicted of, a crime punishable by imprisonment for a term exceeding one year during the term of a current license or permit, or while he has pending a license or permit renewal application, shall not be barred from licensed or permit operations for 30 days after the date of indictment or information or 30 days after the date upon which his conviction becomes final. Also, if he files his application for relief under this section within such 30 day period, he may further continue licensed or permit operations while his application is pending. A licensee or permittee who does not file an application within 30 days from the date of his indictment or information, or within 30 days from the date his conviction becomes final, shall not continue licensed or permit operations beyond 30 days from the date of his indictment or information or beyond 30 days from the date his conviction becomes final.

(2) In the event the term of a license or permit of a person expires during the 30 day period following the date of indictment of information of during the 30 day period after the date upon which his conviction becomes final or while his application for relief is pending, he shall file a timely application for renewal of his license or permit in order to continue licensed or permit operations. The license or permit application is to show that the applicant has been indicted or under information for, or convicted of, a crime punishable by imprisonment for a term exceeding one year.

(3) A licensee or permittee shall not continue licensed or permit operations beyond 30 days following the date the Director issues notification that the licensee's or permittee's application for removal of the disabilities resulting from an indictment, information or conviction has been denied.

(4) When a licensee or permittee may no longer continue licensed or permit operations under this section, any application for renewal of license of permit filed by the licensee or permittee while his application for removal of disabilities resulting from an indictment, information or conviction is pending, will be denied by the regional director (compliance).

(f)(1) Applications for relief from disabilities must include the following information:

(i) In the case of a corporation, or of any person having the power to direct or control the management of the corporation, information as to the absence of culpability in the offense for which the corporation, or any such person, was indicted, formally accused or convicted;

(ii) In the case of an applicant who is an individual, two properly completed FBI Forms FD–258 (fingerprint card), and a written statement from each of three references who are not related to the applicant by blood or marriage and have known the applicant for at least 3 years, recommending the granting of relief;

(iii) Written consent to examine and obtain copies of records and to receive statements and information regarding the applicant's background, including records, statements and other information concerning employment, medical history, military service, immigration status, and criminal record;

(iv) In the case of an applicant having been convicted of a crime punishable by imprisonment for a term exceeding one year, a copy of the indictment or information on which the applicant was convicted, the judgment of conviction or record of any plea of nolo contendere or plea of guilty or finding of guilt by the court;

(v) In the case of an applicant under indictment, a copy of the indictment or information;

(vi) In the case of an applicant who has been adjudicated a mental defective or committed to a mental institution, a copy of the order of a court, board, commission, or other lawful authority that made the adjudication or ordered the commitment, any petition that sought to have the applicant so adjudicated or committed, any medical records reflecting the reasons for commitment and diagnoses of the applicant, and any court order or finding of a court, board, commission, or other lawful authority showing the applicant's discharge from commitment, restoration of mental competency and the restoration of rights;

(vii) In the case of an applicant who has been dis-
arged from the Armed Forces under dishonorable conditions, a py of the applicant's Certificate of Release or Discharge from ctive Duty (Department of Defense Form 214), Charge Sheet Department of Defense Form 458), and final court martial order;

(viii) In the case of an applicant who, having been a izen of the United States, has renounced his or her citizenship, copy of the formal renunciation of nationality before a diplo- atic or consular officer of the United States in a foreign state or fore an officer designated by the Attorney General when the ited States was in a state of war (see 8 U.S.C. 1481(a)(5) and)); and

(ix) In the case of an applicant who is an alien, cumentation that the applicant is an alien who has been wfully admitted to the United States; certification from the plicant including the applicant's INS-issued alien number or mission number, country/countries of citizenship, and immigration status, and certifying that the applicant is legally authorized to work in the United States, or other purposes for which possession of explosives is required; certification from an appropriate law enforcement agency of the applicant's country of citizenship stating that the applicant does not have a criminal record; and, if applicable, certification from a Federal explosives licensee or permittee or other employer stating that the applicant is employed by the employer and must possess explosive materials for purposes of employment. These certifications must be submitted in English.

(2) Any record or document of a court or other government entity or official required by paragraph (f)(1) of this section must be certified by the court or other government entity or official as a true copy.

(Approved by the Office of Management and Budget under control number 1140–0076)

[T.D. ATF–87, 46 FR 40384, Aug. 7, 1981. Redesignated by T.D. ATF–487, 68 FR 3748, Jan. 24, 2003. ATF No. 1, 68 FR 13792, Mar. 20, 2003]

Subpart I—Unlawful Acts, Penalties, Seizures and Forfeitures

555.161 Engaging in business without a license.

Any person engaging in the business of importing, manufactur- g, or dealing in explosive materials without a license issued der the Act, shall be fined not more than $10,000 or impris- ed not more than 10 years, or both.

555.162 False statement or representation.

Any person who knowingly withholds information or makes y false or fictitious oral or written statement or furnishes or hibits any false, fictitious, or misrepresented identification, ended or likely to deceive for the purpose of obtaining plosive materials, or a license, permit, exemption, or relief m disability under the Act, shall be fined not more than 0,000 or imprisoned not more than 10 years, or both.

555.163 False entry in record.

Any licensed importer, licensed manufacturer, licensed dealer, permittee who knowingly makes any false entry in any record quired to be kept under subpart G of this part, shall be fined not re than $10,000 or imprisoned not more than 10 years, or both.

). ATF–87, 46 FR 40384, Aug. 7, 1981, as amended by T.D. ATF–400, 63 FR 03, Aug. 24, 1998]

555.164 Unlawful storage.

Any person who stores any explosive material in a manner not conformity with this part, shall be fined not more than $1,000 imprisoned not more than one year, or both.

§ 555.165 Failure to report theft or loss.

(a) Any person who has knowledge of the theft or loss of any explosive materials from his stock and fails to report the theft or loss within 24 hours of discovery in accordance with §555.30, shall be fined not more than $1,000 or imprisoned not more than one year, or both.

(b) On and after January 24, 2003, any licensee or permittee who fails to report a theft of explosive materials in accordance with §555.30 will be fined under title 18 U.S.C., imprisoned not more than 5 years, or both.

[T.D. ATF–87, 46 FR 40384, Aug. 7, 1981, as amended by ATF No. 1, 68 FR 13792, Mar. 20, 2003]

§ 555.166 Seizure or forfeiture.

Any explosive materials involved or used or intended to be used in any violation of the Act or of this part or in any violation of any criminal law of the United States are subject to seizure and forfeiture, and all provisions of title 26, U.S.C. relating to the seizure, forfeiture, and disposition of firearms, as defined in 26 U.S.C. 5845(a), will, so far as applicable, extend to seizures and forfeitures under the Act. (See §72.27 of this title for regulations on summary destruction of explosive materials which are impracticable or unsafe to remove to a place of storage.)

[T.D. ATF–87, 46 FR 40384, Aug. 7, 1981, as amended by T.D. ATF–363, 60 FR 17449, Apr. 6, 1995]

Subpart J—Marking of Plastic Explosives

§ 555.180 Prohibitions relating to unmarked plastic explosives.

(a) No person shall manufacture any plastic explosive that does not contain a detection agent.

(b) No person shall import or bring into the United States, or export from the United States, any plastic explosive that does not contain a detection agent. This paragraph does not apply to the importation or bringing into the United States, or the exportation from the United States, of any plastic explosive that was imported or brought into, or manufactured in the United States prior to April 24, 1996, by or on behalf of any agency of the United States performing military or police functions (including any military reserve component) or by or on behalf of the National Guard of any State, not later than 15 years after the date of entry into force of the Convention on the Marking of Plastic Explosives with respect to the United States, i.e., not later than June 21, 2013.

(c) No person shall ship, transport, transfer, receive, or possess any plastic explosive that does not contain a detection agent. This paragraph does not apply to:

(1) The shipment, transportation, transfer, receipt, or possession of any plastic explosive that was imported or brought into, or manufactured in the United States prior to April 24, 1996, by any person during the period beginning on that date and ending on April 24, 1999; or

(2) The shipment, transportation, transfer, receipt, or possession of any plastic explosive that was imported or brought into, or manufactured in the United States prior to April 24, 1996, by or on behalf of any agency of the United States performing a military or police function (including any military reserve component) or by or on behalf of the National Guard of any State, not later than 15 years after the date of entry into force of the Convention on the Marking of Plastic Explosives with respect to the United States, i.e., not later than June 21, 2013.

(d) When used in this subpart, terms are defined as follows:

(1) Convention on the Marking of Plastic Explosives means the Convention on the Marking of Plastic Explosives for the Purposes of Detection, Done at Montreal on 1 March 1991.

(2) "Date of entry into force" of the Convention on the Marking of Plastic Explosives means that date on which the Convention enters into force with respect to the U.S. in accordance with the provisions of Article XIII of the Convention on the Marking of Plastic Explosives. The Convention entered into force on June 21, 1998.

(3) Detection agent means any one of the substances specified in this paragraph when introduced into a plastic explosive or formulated in such explosive as a part of the manufacturing process in such a manner as to achieve homogeneous distribution in the finished explosive, including—

(i) Ethylene glycol dinitrate (EGDN), C[2]H[4](NO[3])[2], molecular weight 152, when the minimur concentration in the finished explosive is 0.2 percent by mass;

(ii) 2,3-Dimethyl-2,3-dinitrobutane (DMNB), C[6]H[12](NO[2])[2], molecular weight 176, when the minimu concentration in the finished explosive is 0.1 percent by mass;

(iii) Para-Mononitrotoluene (p-MNT), C[7]H[7]NO[2], molecular weight 137, when the minimum concentration in the finished explosive is 0.5 percent by mass;

(iv) Ortho-Mononitrotoluene (o-MNT), C[7]H[7]NO[2], molecular weight 137, when the minimum concentration in the finished explosive is 0.5 percent by mass; and

(v) Any other substance in the concentration specified by the Director, after consultation with the Secretary of St and Secretary of Defense, that has been added to the table in P 2 of the Technical Annex to the Convention on the Marking of Plastic Explosives.

(4) Plastic explosive means an explosive material in flexible or elastic sheet form formulated with one or more higl explosives which in their pure form has a vapor pressure less t 10<4>Pa at a temperature of 25 degrees C, is formulated with binder material, and is as a mixture malleable or flexible at normal room temperature. High explosives, as defined in §555.202(a), are explosive materials which can be caused to detonate by means of a blasting cap when unconfined.

[T.D. ATF–387, 62 FR 8376, Feb. 25, 1997, as amended by T.D. ATF–419, 64 55628, Oct. 14, 1999]

§ 555.181 Reporting of plastic explosives.

All persons, other than an agency of the United States (inclu ing any military reserve component) or the National Guard of State, possessing any plastic explosive on April 24, 1996, shall submit a report to the Director no later than August 22, 1996. report shall be in writing and mailed by certified mail (return receipt requested) to the Director at P.O. Box 50204, Washingt DC 20091–0204. The report shall include the quantity of plast explosives possessed on April 24, 1996; any marks of identific tion on such explosives; the name and address of the manufacturer or importer; the storage location of such explosives, including the city and State; and the name and address of the person possessing the plastic explosives.

(Approved by the Office of Management and Budget under control number 1512–0535)

[T.D. ATF–382, 61 FR 38085, July 23, 1996, as amended by T.D. ATF–387, 62 8377, Feb. 25, 1997]

§ 555.182 Exceptions.

It is an affirmative defense against any proceeding involving §§555.180 and 555.181 if the proponent proves by a preponde ance of the evidence that the plastic explosive—

Consisted of a small amount of plastic explosive intended for d utilized solely in lawful—

(1) Research, development, or testing of new or modified plosive materials;

(2) Training in explosives detection or development or ting of explosives detection equipment; or

(3) Forensic science purposes; or

Was plastic explosive that, by April 24, 1999, will be or is orporated in a military device within the territory of the United ates and remains an integral part of such military device, or is ended to be, or is incorporated in, and remains an integral part a military device that is intended to become, or has become, property of any agency of the United States performing litary or police functions (including any military reserve mponent) or the National Guard of any State, wherever such vice is located. For purposes of this paragraph, the term ilitary device" includes, but is not restricted to, shells, bombs, ojectiles, mines, missiles, rockets, shaped charges, grenades, rforators, and similar devices lawfully manufactured exclu-ely for military or police purposes.
D. ATF–387, 62 FR 8377, Feb. 25, 1997]

555.183 Importation of plastic explosives on or ter April 24, 1997.

Persons filing Form 6 applications for the importation of plastic plosives on or after April 24, 1997, shall attach to the applica-n the following written statement, prepared in triplicate, ecuted under the penalties of perjury:

"I declare under the penalties of perjury that the plastic plosive to be imported contains a detection agent as required by CFR 555.180(b)"; or

(b) "I declare under the penalties of perjury that the plastic explosive to be imported is a "small amount" to be used for research, training, or testing purposes and is exempt from the detection agent requirement pursuant to 27 CFR 555.182."
[T.D. ATF–387, 62 FR 8377, Feb. 25, 1997]

§ 555.184 Statements of process and samples.

(a) A complete and accurate statement of process with regard to any plastic explosive or to any detection agent that is to be introduced into a plastic explosive or formulated in such plastic explosive shall be submitted by a licensed manufacturer or licensed importer, upon request, to the Director.

(b) Samples of any plastic explosive or detection agent shall be submitted by a licensed manufacturer or licensed importer, upon request, to the Director.

(Paragraph (a) approved by the Office of Management and Budget under control number 1512–0539)
[T.D. ATF–387, 62 FR 8378, Feb. 25, 1997]

§ 555.185 Criminal sanctions.

Any person who violates the provisions of 18 U.S.C. 842(l)–(o) shall be fined under title 18, U.S.C., imprisoned for not more than 10 years, or both.
[T.D. ATF–387, 62 FR 8378, Feb. 25, 1997]

§ 555.186 Seizure or forfeiture.

Any plastic explosive that does not contain a detection agent in violation of 18 U.S.C. 842(l)–(n) is subject to seizure and forfeiture, and all provisions of 19 U.S.C. 1595a, relating to seizure, forfeiture, and disposition of merchandise introduced or attempted to be introduced into the U.S. contrary to law, shall extend to seizures and forfeitures under this subpart. See §72.27 of this chapter for regulations on summary destruction of plastic explosives that do not contain a detection agent.
[T.D. ATF–387, 62 FR 8378, Feb. 25, 1997]

Subpart K—Storage

555.201 General.

Section 842(j) of the Act and §555.29 of this part require t the storage of explosive materials by any person must be in ordance with the regulations in this part. Further, section 846 this Act authorizes regulations to prevent the recurrence of idental explosions in which explosive materials were involved. e storage standards prescribed by this subpart confer no right privileges to store explosive materials in a manner contrary to te or local law.

The Director may authorize alternate construction for plosives storage magazines when it is shown that the alternate gazine construction is substantially equivalent to the standards safety and security contained in this subpart. Any alternate plosive magazine construction approved by the Director prior to August 9, 1982, will continue as approved unless notified in writing by the Director. Any person intending to use alternate magazine construction shall submit a letter application to the regional director (compliance) for transmittal to the Director, specifically describing the proposed magazine. Explosive materials may not be stored in alternate magazines before the applicant has been notified that the application has been approved.

(c) A licensee or permittee who intends to make changes in his magazines, or who intends to construct or acquire additional magazines, shall comply with §555.63.

(d) The regulations set forth in §§555.221 through 555.224 pertain to the storage of display fireworks, pyrotechnic compositions, and explosive materials used in assembling fireworks and articles pyrotechnic.

(e) The provisions of §555.202(a) classifying flash powder and bulk salutes as high explosives are mandatory after March 7, 1990: Provided, that those persons who hold licenses or permits under this part on that date shall, with respect to the premises covered by such licenses or permits, comply with the high explosives storage requirements for flash powder and bulk salutes by March 7, 1991.

(f) Any person who stores explosive materials shall notify the authority having jurisdiction for fire safety in the locality in which the explosive materials are being stored of the type, magazine capacity, and location of each site where such explosive materials are stored. Such notification shall be made orally before the end of the day on which storage of the explosive materials commenced and in writing within 48 hours from the time such storage commenced.

(Paragraph (f) approved by the Office of Management and Budget under control number 1512–0536)

[T.D. ATF–87, 46 FR 40384, Aug. 7, 1981, as amended by T.D. ATF–293, 55 FR 3722, Feb. 5, 1990; T.D. ATF–400, 63 FR 45003, Aug. 24, 1998]

§ 555.202 Classes of explosive materials.

For purposes of this part, there are three classes of explosive materials. These classes, together with the description of explosive materials comprising each class, are as follows:

(a) High explosives. Explosive materials which can be caused to detonate by means of a blasting cap when unconfined, (for example, dynamite, flash powders, and bulk salutes). See also §555.201(e).

(b) Low explosives. Explosive materials which can be caused to deflagrate when confined (for example, black powder, safety fuses, igniters, igniter cords, fuse lighters, and "display fireworks" classified as UN0333, UN0334, or UN0335 by the U.S. Department of Transportation regulations at 49 CFR 172.101, except for bulk salutes).

(c) Blasting agents. (For example, ammonium nitrate-fuel oil and certain water-gels (see also §555.11).

[T.D. ATF–87, 46 FR 40384, Aug. 7, 1981, as amended by T.D. ATF–293, 55 FR 3722, Feb. 5, 1990; T.D. ATF–400, 63 FR 45003, Aug. 24, 1998]

§ 555.203 Types of magazines.

For purposes of this part, there are five types of magazines. These types, together with the classes of explosive materials, as defined in §555.202, which will be stored in them, are as follows:

(a) Type 1 magazines. Permanent magazines for the storage of high explosives, subject to the limitations prescribed by §§555.206 and 555.213. Other classes of explosive materials may also be stored in type 1 magazines.

(b) Type 2 magazines. Mobile and portable indoor and outdoor magazines for the storage of high explosives, subject to the limitations prescribed by §§555.206, 555.208(b), and 555.213. Other classes of explosive materials may also be stored in type 2 magazines.

(c) Type 3 magazines. Portable outdoor magazines for the temporary storage of high explosives while attended (for examp a "day-box"), subject to the limitations prescribed by §§555.20 and 555.213. Other classes of explosives materials may also be stored in type 3 magazines.

(d) Type 4 magazines. Magazines for the storage of low explosives, subject to the limitations prescribed by §§555.206(555.210(b), and 555.213. Blasting agents may be stored in type magazines, subject to the limitations prescribed by §§555.206(555.211(b), and 555.213. Detonators that will not mass detonat may also be stored in type 4 magazines, subject to the limitatio prescribed by §§555.206(a), 555.210(b), and 555.213.

(e) Type 5 magazines. Magazines for the storage of blasting agents, subject to the limitations prescribed by §§555.206(c), 555.211(b), and 555.213.

§ 555.204 Inspection of magazines.

Any person storing explosive materials shall inspect his magazines at least every seven days. This inspection need not b an inventory, but must be sufficient to determine whether there has been unauthorized entry or attempted entry into the magazines, or unauthorized removal of the contents of the magazine.

§ 555.205 Movement of explosive materials.

All explosive materials must be kept in locked magazines meeting the standards in this subpart unless they are:

(a) In the process of manufacture;

(b) Being physically handled in the operating process of a licensee or user;

(c) Being used; or

(d) Being transported to a place of storage or use by a license or permittee or by a person who has lawfully acquired explosiv materials under §555.106.

§ 555.206 Location of magazines.

(a) Outdoor magazines in which high explosives are stored m be located no closer to inhabited buildings, passenger railways, public highways, or other magazines in which high explosives stored, than the minimum distances specified in the table of distances for storage of explosive materials in §555.218.

(b) Outdoor magazines in which low explosives are stored mu be located no closer to inhibited buildings, passenger railways, public highways, or other magazines in which explosive materi are stored, than the minimum distances specified in the table o distances for storage of low explosives in §555.219, except tha the table of distances in §555.224 shall apply to the storage of display fireworks. The distances shown in §555.219 may not b reduced by the presence of barricades.

(c)(1) Outdoor magazines in which blasting agents in quantities of more than 50 pounds are stored must be located no closer to inhabited buildings, passenger railways, or public highways than the minimum distances specified in the table of distances for storage of explosive materials in §555.218.

(2) Ammonium nitrate and magazines in which blasting agents are stored must be located no closer to magazines in which high explosives or other blasting agents are stored than the minimum distances specified in the table of distances for the separation of ammonium nitrate and blasting agents in §555.220. However, the minimum distances for magazines in which explosives and blasting agents are stored from inhabited buildings, etc., may not be less than the distances specified in the table of distances for storage of explosives materials in §555.218.

[T.D. ATF–87, 46 FR 40384, Aug. 7, 1981, as amended by T.D. ATF–293, 55 FR 3722, Feb. 5, 1990; T.D. ATF–400, 63 FR 45003, Aug. 24, 1998]

§ 555.207 Construction of type 1 magazines.

A type 1 magazine is a permanent structure: a building, an igloo or "Army-type structure", a tunnel, or a dugout. It is to be bullet-resistant, fire-resistant, weather-resistant, theft-resistant, and ventilated.

(a) Buildings. All building type magazines are to be constructed of masonry, wood, metal, or a combination of these materials, and have no openings except for entrances and ventilation. The ground around building magazines must slope away for drainage or other adequate drainage provided.

(1) Masonry wall construction. Masonry wall construction is to consist of brick, concrete, tile, cement block, or cinder block and be not less than 6 inches in thickness. Hollow masonry units used in construction must have all hollow spaces filled with well-tamped, coarse, dry sand or weak concrete (at least a mixture of one part cement and eight parts of sand with enough water to dampen the mixture while tamping in place). Interior walls are to be constructed of, or covered with, a nonsparking material.

(2) Fabricated metal wall construction. Metal wall construction is to consist of sectional sheets of steel or aluminum not less than number 14-gauge, securely fastened to a metal framework. Metal wall construction is either lined inside with brick, solid cement blocks, hardwood not less than four inches thick, or will have at least a six inch sand fill between interior and exterior walls. Interior walls are to be constructed of, or covered with, a nonsparking material.

(3) Wood frame wall construction. The exterior of outer wood walls is to be covered with iron or aluminum not less than number 26-gauge. An inner wall of, or covered with nonsparking material will be constructed so as to provide a space of not less than six inches between the outer and inner walls. The space is to be filled with coarse, dry sand or weak concrete.

(4) Floors. Floors are to be constructed of, or covered with, a nonsparking material and shall be strong enough to bear the weight of the maximum quantity to be stored. Use of pallets covered with a nonsparking material is considered equivalent to a floor constructed of or covered with a nonsparking material.

(5) Foundations. Foundations are to be constructed of brick, concrete, cement block, stone, or wood posts. If piers or posts are used, in lieu of a continuous foundation, the space under the buildings is to be enclosed with metal.

(6) Roof. Except for buildings with fabricated metal roofs, the outer roof is to be covered with no less than number 26-guage iron or aluminum, fastened to at least ⅞ inch sheathing.

(7) Bullet-resistant ceilings or roofs. Where it is possible for a bullet to be fired directly through the roof and into the magazine at such an angle that the bullet would strike the explosives within, the magazine is to be protected by one of the following methods:

(i) A sand tray lined with a layer of building paper, plastic, or other nonporous material, and filled with not less than four inches of coarse, dry sand, and located at the tops of inner walls covering the entire ceiling area, except that portion necessary for ventilation.

(ii) A fabricated metal roof constructed of $\frac{3}{16}$-inch plate steel lined with four inches of hardwood. (For each additional ⅛ inch of plate steel, the hardwood lining may be decreased one inch.)

(8) Doors. All doors are to be constructed of not less than ¼ inch plate steel and lined with at least two inches of hardwood. Hinges and hasps are to be attached to the doors by welding, riveting or bolting (nuts on inside of door). They are to be installed in such a manner that the hinges and hasps cannot be removed when the doors are closed and locked.

(9) Locks. Each door is to be equipped with (i) two mortise locks; (ii) two padlock fastened in separate hasps and staples; (iii) a combination of a mortise lock and a padlock; (iv) a mortise lock that requires two keys to open; or (v) a three-point lock. Padlocks must have at least five tumblers and a casehardened shackle of at least ⅜ inch diameter. Padlocks must be protected with not less than ¼ inch steel hoods constructed so as to prevent sawing or lever action on the locks, hasps, and staples. These requirements do not apply to magazine doors that are adequately secured on the inside by means of a bolt, lock, or bar that cannot be actuated from the outside.

(10) Ventilation. Ventilation is to be provided to prevent dampness and heating of stored explosive materials. Ventilation openings must be screened to prevent the entrance of sparks. Ventilation openings in side walls and foundations must be offset or shielded for bullet-resistant purposes. Magazines having foundation and roof ventilators with the air circulating between the side walls and the floors and between the side walls and the ceiling must have a wooden lattice lining or equivalent to prevent the packages of explosive materials from being stacked against the side walls and blocking the air circulation.

(11) Exposed metal. No sparking material is to be exposed to contact with the stored explosive materials. All ferrous metal nails in the floor and side walls, which might be exposed to contact with explosive materials, must be blind nailed, countersunk, or covered with a nonsparking lattice work or other nonsparking material.

(b) Igloos, "Army-type structures", tunnels, and dugouts. Igloo, "Army-type structure", tunnel, and dugout magazines are to be constructed of reinforced concrete, masonry, metal, or a combination of these materials. They must have an earthmound covering of not less than 24 inches on the top, sides and rear unless the magazine meets the requirements of paragraph (a)(7) of this section. Interior walls and floors must be constructed of, or covered with, a nonsparking material. Magazines of this type are also to be constructed in conformity with the requirements of paragraph (a)(4) and paragraphs (a)(8) through (11) of this section.

§ 555.208 Construction of type 2 magazines.

A type 2 magazine is a box, trailer, semitrailer, or other mobile facility.

(a) Outdoor magazines—

(1) General. Outdoor magazines are to be bullet-resistant, fire-resistant, weather-resistant, theft-resistant, and ventilated. They are to be supported to prevent direct contact with the ground and, if less than one cubic yard in size, must be securely fastened to a fixed object. The ground around outdoor magazines must slope away for drainage or other adequate drainage provided. When unattended, vehicular magazines must have wheels removed or otherwise effectively immobilized by kingpin locking devices or other methods approved by the Director.

(2) Exterior construction. The exterior and doors are to be constructed of not less than ¼-inch steel and lined with at least two inches of hardwood. Magazines with top openings will have lids with water-resistant seals or which overlap the sides by at least one inch when in a closed position.

(3) Hinges and hasps. Hinges and hasps are to be attached to doors by welding, riveting, or bolting (nuts on inside of door). Hinges and hasps must be installed so that they cannot be removed when the doors are closed and locked.

(4) Locks. Each door is to be equipped with (i) two mortise locks; (ii) two padlocks fastened in separate hasps and staples; (iii) a combination of a mortise lock and a padlock; (iv) a mortise lock that requires two keys to open; or (v) a three-point lock. Padlocks must have at least five tumblers and a case-hardened shackle of at least ⅜-inch diameter. Padlocks must be protected with not less than ¼-inch steel hoods constructed so as to prevent sawing or lever action on the locks, hasps, and staples. These requirements do not apply to magazine doors that are adequately secured on the inside by means of a bolt, lock, or bar that cannot be actuated from the outside.

(b) Indoor magazines—

(1) General. Indoor magazines are to be fire-resistant and theft-resistant. They need not be bullet-resistant and weather-resistant if the buildings in which they are stored provide protection from the weather and from bullet penetration. No indoor magazine is to be located in a residence or dwelling. The indoor storage of high explosives must not exceed a quantity of 50 pounds. More than one indoor magazine may be located in the same building if the total quantity of explosive materials stored does not exceed 50 pounds. Detonators must be stored in a separate magazine (except as provided in §555.213) and the total quantity of detonators must not exceed 5,000.

(2) Exterior construction. Indoor magazines are to be constructed of wood or metal according to one of the following specifications:

(i) Wood indoor magazines are to have sides, bottom and doors constructed of at least two inches of hardwood and are to be well braced at the corners. They are to be covered with shee metal of not less than number 26-gauge (.0179 inches). Nails exposed to the interior of magazines must be countersunk.

(ii) Metal indoor magazines are to have sides, bottom and doors constructed of not less than number 12-gauge (.1046 inches) metal and be lined inside with a nonsparking material. Edges of metal covers must overlap sides at least one inch.

(3) Hinges and hasps. Hinges and hasps are to be attached to doors by welding, riveting, or bolting (nuts on inside of door). Hinges and hasps must be installed so that they cannot be removed when the doors are closed and locked.

(4) Locks. Each door is to be equipped with (i) two mortise locks; (ii) two padlocks fastened in separate hasps and staples; (iii) a combination of a mortise lock and a padlock; (iv) a mortise lock that requires two keys to open; or (v) a three-point lock. Padlocks must have at least five tumblers and a case-hardened shackle of at least ⅜-inch diameter. Padlocks must be protected with not less than ¼-inch steel hoods constructed so as to prevent sawing or lever action on the locks, hasps, and staples. Indoor magazines located in secure rooms that are locked as provided in this subparagraph may have each door locked with one steel padlock (which need not be protected by a steel hood) having at least five tumblers and a case-hardened shackle of at least ⅜-inch diameter, if the door hinges and lock hasp are securely fastened to the magazine.

These requirements do not apply to magazine doors that are adequately secured on the inside by means of a bolt, lock, or bar that cannot be actuated from the outside.

(c) Detonator boxes. Magazines for detonators in quantities of 100 or less are to have sides, bottoms and doors constructed of not less than number 12-gauge (.1046 inches) metal and lined with a nonsparking material. Hinges and hasps must be attached so they cannot be removed from the outside. One steel padlock hich need not be protected by a steel hood) having at least five mblers and a case-hardened shackle of at least ⅜-inch diameter sufficient for locking purposes.

555.209 Construction of type 3 magazines.

A type 3 magazine is a "day-box" or other portable magazine. must be fire-resistant, weather-resistant, and theft-resistant. A pe 3 magazine is to be constructed of not less than number 12-uge (.1046 inches) steel, lined with at least either ½-inch ,wood or ½-inch Masonite-type hardboard.

Doors must overlap sides by at least one inch. Hinges and hasps e to be attached by welding, riveting or bolting (nuts on inside). One steel padlock (which need not be protected by a steel hood) ving at least five tumblers and a case-hardened shackle of at ist ⅜-inch diameter is sufficient for locking purposes.

:plosive materials are not to be left unattended in type 3 igazines and must be removed to type 1 or 2 magazines for attended storage.

555.210 Construction of type 4 magazines.

A type 4 magazine is a building, igloo or "Army-type struc-e", tunnel, dugout, box, trailer, or a semitrailer or other mobile igazine.

Outdoor magazines—

(1) General. Outdoor magazines are to be fire-resistant, ather-resistant, and theft-resistant. The ground around outdoor igazines must slope away for drainage or other adequate ainage be provided. When unattended, vehicular magazines ist have wheels removed or otherwise be effectively immobi-ed by kingpin locking devices or other methods approved by Director.

(2) Construction. Outdoor magazines are to be constructed masonry, metal-covered wood, fabricated metal, or a combina-n of these materials. Foundations are to be constructed of brick, ncrete, cement block, stone, or metal or wood posts. If piers or sts are used, in lieu of a continuous foundation, the space under building is to be enclosed with fire-resistant material. The lls and floors are to be constructed of, or covered with, a nsparking material or lattice work. The doors must be metal or id wood covered with metal.

(3) Hinges and hasps. Hinges and hasps are to be attached doors by welding, riveting, or bolting (nuts on inside of door). nges and hasps must be installed so that they cannot be noved when the doors are closed and locked.

(4) Locks. Each door is to be equipped with (i) two mortise ks; (ii) two padlocks fastened in separate hasps and staples;) a combination of a mortise lock and a padlock; (iv) a mortise k that requires two keys to open; or (v) a three-point lock. dlocks must have at least five tumblers and case-hardened ickle of at least ⅜ inch diameter. Padlocks must be protected th not less than ¼ inch steel hoods constructed so as to prevent ving or lever action on the locks, hasps, and staples.

These requirements do not apply to magazine doors that are adequately secured on the inside by means of a bolt, lock, or bar that cannot be actuated from the outside.

(b) Indoor magazine—

(1) General. Indoor magazines are to be fire-resistant and theft-resistant. They need not be weather-resistant if the buildings in which they are stored provide protection from the weather. No indoor magazine is to be located in a residence or dwelling. The indoor storage of low explosives must not exceed a quantity of 50 pounds. More than one indoor magazine may be located in the same building if the total quantity of explosive materials stored does not exceed 50 pounds. Detonators that will not mass detonate must be stored in a separate magazine and the total number of electric detonators must not exceed 5,000.

(2) Construction. Indoor magazines are to be constructed of masonry, metal-covered wood, fabricated metal, or a combination of these materials. The walls and floors are to be constructed of, or covered with, a nonsparking material. The doors must be metal or solid wood covered with metal.

(3) Hinges and hasps. Hinges and hasps are to be attached to doors by welding, riveting, or bolting (nuts on inside of door). Hinges and hasps must be installed so that they cannot be removed when the doors are closed and locked.

(4) Locks. Each door is to be equipped with (i) two mortise locks; (ii) two padlocks fastened in separate hasps and staples; (iii) a combination of a mortise lock and padlock; (iv) a mortise lock that requires two keys to open; or (v) a three-point lock.

Padlocks must have at least five tumblers and a case-hardened shackle of at least ⅜ inch diameter. Padlocks must be protected with not less than ¼ inch steel hoods constructed so as to prevent sawing or lever action on the locks, hasps, and staples. Indoor magazines located in secure rooms that are locked as provided in this subparagraph may have each door locked with one steel padlock (which need not be protected by a steel hood) having at least five tumblers and a case-hardened shackle of at least ⅜ inch diameter, if the door hinges and lock hasp are securely fastened to the magazine.

These requirements do not apply to magazine doors that are adequately secured on the inside by means of a bolt, lock, or bar that cannot be actuated from the outside.

§ 555.211 Construction of type 5 magazines.

A type 5 magazine is a building, igloo or "Army-type structure", tunnel, dugout, bin, box, trailer, or a semitrailer or other mobile facility.

(a) Outdoor magazines—

(1) General. Outdoor magazines are to be weather-resistant and theft-resistant. The ground around magazines must slope away for drainage or other adequate drainage be provided. When unattended, vehicular magazines must have wheels removed or otherwise be effectively immobilized by kingpin locking devices or other methods approved by the Director.

(2) Construction. The doors are to be constructed of solid wood or metal.

(3) Hinges and hasps. Hinges and hasps are to be attached to doors by welding, riveting, or bolting (nuts on inside of door). Hinges and hasps must be installed so that they cannot be removed when the doors are closed and locked.

(4) Locks. Each door is to be equipped with (i) two mortise locks; (ii) two padlocks fastened in separate hasps and staples; (iii) a combination of a mortise lock and a padlock; (iv) a mortise lock that requires two keys to open; or (v) a three-point lock. Padlocks must have at least five tumblers and a case-hardened shackle of at least ⅜ inch diameter. Padlocks must be protected with not less than ¼ inch steel hoods constructed so as to prevent sawing or lever action on the locks, hasps, and staples. Trailers, semitrailers, and similar vehicular magazines may, for each door, be locked with one steel padlock (which need not be protected by a steel hood) having at least five tumblers and a case-hardened shackle of at least ⅜ inch diameter, if the door hinges and lock hasp are securely fastened to the magazine and to the door frame.

These requirements do not apply to magazine doors that are adequately secured on the inside by means of a bolt, lock, or bar that cannot be actuated from the outside.

(5) Placards. The placards required by Department of Transportation regulations at 49 CFR part 172, subpart F, for the transportation of blasting agents shall be displayed on all magazines.

(b) Indoor magazines—

(1) General. Indoor magazines are to be theft-resistant. They need not be weather-resistant if the buildings in which they are stored provide protection from the weather. No indoor magazine is to be located in a residence or dwelling. Indoor magazines containing quantities of blasting agents in excess of 50 pounds are subject to the requirements of §555.206 of this subpart.

(2) Construction. The doors are to be constructed of wood or metal.

(3) Hinges and hasps. Hinges and hasps are to be attached to doors by welding, riveting, or bolting (nuts on inside). Hinges and hasps must be installed so that they cannot be removed when the doors are closed and locked.

(4) Locks. Each door is to be equipped with (i) two mortise locks; (ii) two padlocks fastened in separate hasps and staples; (iii) a combination of a mortise lock and a padlock; (iv) a mortise lock that requires two keys to open; or (v) a three-point lock.

Padlocks must have at least five tumblers and a case-hardened shackle of at least ⅜ inch diameter. Padlocks must be protected with not less than ¼ inch steel hoods constructed so as to prevent sawing or lever action on the locks, hasps, and staples. Indoor magazines located in secure rooms that are locked as provided in this subparagraph may have each door locked with one steel padlock (which need not be protected by a steel hood) having at least five tumblers and a case-hardened shackle of at least ⅜ inch diameter, if the door hinges and lock hasps are securely fastened to the magazine and to the door frame.

These requirements do not apply to magazine doors that a adequately secured on the inside by means of a bolt, lock, or b that cannot be actuated from the outside.

[T.D. ATF–87, 46 FR 40384, Aug. 7, 1981, as amended by T.D. ATF–298, 55 F 21863, May 30, 1990]

§ 555.212 Smoking and open flames.

Smoking, matches, open flames, and spark producing device are not permitted:

(a) In any magazine;

(b) Within 50 feet of any outdoor magazine; or

(c) Within any room containing an indoor magazine.

§ 555.213 Quantity and storage restrictions.

(a) Explosive materials in excess of 300,000 pounds or deton tors in excess of 20 million are not to be stored in one magazir unless approved by the Director.

(b) Detonators are not to be stored in the same magazine wit other explosive materials, except under the following circumstances:

(1) In a type 4 magazine, detonators that will not mass detonate may be stored with electric squibs, safety fuse, ignite and igniter cord.

(2) In a type 1 or type 2 magazine, detonators may be stored with delay devices and any of the items listed in paragr (b)(1) of this section.

§ 555.214 Storage within types 1, 2, 3, and 4 magazines.

(a) Explosive materials within a magazine are not to be place directly against interior walls and must be stored so as not to interfere with ventilation. To prevent contact of stored explosi materials with walls, a nonsparking lattice work or other nons king material may be used.

(b) Containers of explosive materials are to be stored so that marks are visible. Stocks of explosive materials are to be store so they can be easily counted and checked upon inspection.

(c) Except with respect to fiberboard or other nonmetal cont ers, containers of explosive materials are not to be unpacked o repacked inside a magazine or within 50 feet of a magazine, a must not be unpacked or repacked close to other explosive materials. Containers of explosive materials must be closed w being stored.

(d) Tools used for opening or closing containers of explosive materials are to be of nonsparking materials, except that meta slitters may be used for opening fiberboard containers. A woo wedge and a fiber, rubber, or wooden mallet are to be used fo ening or closing wood containers of explosive materials. Metal ls other than nonsparking transfer conveyors are not to be red in any magazine containing high explosives.

555.215 Housekeeping.

Magazines are to be kept clean, dry, and free of grit, paper, pty packages and containers, and rubbish. Floors are to be ularly swept. Brooms and other utensils used in the cleaning d maintenance of magazines must have no spark-producing tal parts, and may be kept in magazines. Floors stained by kage from explosive materials are to be cleaned according to tructions of the explosives manufacturer. When any explosive terial has deteriorated it is to be destroyed in accordance with advice or instructions of the manufacturer. The area surround- magazines is to be kept clear of rubbish, brush, dry grass, or es (except live trees more than 10 feet tall), for not less than 25 t in all directions. Volatile materials are to be kept a distance of less than 50 feet from outdoor magazines. Living foliage ich is used to stabilize the earthen covering of a magazine need be removed.

§ 555.216 Repair of magazines.

Before repairing the interior of magazines, all explosive materials are to be removed and the interior cleaned. Before repairing the exterior of magazines, all explosive materials must be removed if there exists any possibility that repairs may produce sparks or flame. Explosive materials removed from magazines under repair must be

(a) placed in other magazines appropriate for the storage of those explosive materials under this subpart, or

(b) placed a safe distance from the magazines under repair where they are to be properly guarded and protected until the repairs have been completed.

§ 555.217 Lighting.

(a) Battery-activated safety lights or battery-activated safety lanterns may be used in explosives storage magazines.

(b) Electric lighting used in any explosives storage magazine must meet the standards prescribed by the "National Electrical Code," (National Fire Protection Association, NFPA 70–81), for the conditions present in the magazine at any time. All electrical switches are to be located outside of the magazine and also meet the standards prescribed by the National Electrical Code.

(c) Copies of invoices, work orders or similar documents which indicate the lighting complies with the National Electrical Code must be available for inspection by ATF officers.

§ 555.218 Table of distances for storage of explosive materials.

Quantity of Explosives		Distances in feet							
Pounds over	Pounds not over	Inhabited buildings		Public highways with traffic volume 3000 or fewer vehicles/day		Passenger railways-public highways with traffic volume more than 3,000 vehicles/day		Separation of magazines	
		Barricaded	Unbarricaded	Barricaded	Unbarricaded	Barricaded	Unbarricaded	Barricaded	Unbarricade
0	5	70	140	30	60	51	102	6	12
5	10	90	180	35	70	64	128	8	16
10	20	110	220	45	90	81	162	10	20
20	30	125	250	50	100	93	186	11	22
30	40	140	280	55	110	103	206	12	24
40	50	150	300	60	120	110	220	14	28
50	75	170	340	70	140	127	254	15	30
75	100	190	380	75	150	139	278	16	32
100	125	200	400	80	160	150	300	18	36
125	150	215	430	85	170	159	318	19	38
150	200	235	470	95	190	175	350	21	42
200	250	255	510	105	210	189	378	23	46
250	300	270	540	110	220	201	402	24	48
300	400	295	590	120	240	221	442	27	54
400	500	320	640	130	260	238	476	29	58
500	600	340	680	135	270	253	506	31	62
600	700	355	710	145	290	266	532	32	64
700	800	375	750	150	300	278	556	33	66
800	900	390	780	155	310	289	578	35	70
900	1,000	400	800	160	320	300	600	36	72
1,000	1,200	425	850	165	330	318	636	39	78
1,200	1,400	450	900	170	340	336	672	41	82
1,400	1,600	470	940	175	350	351	702	43	86
1,600	1,800	490	980	180	360	366	732	44	88
1,800	2,000	505	1,010	185	370	378	756	45	90
2,000	2,500	545	1,090	190	380	408	816	49	98
2,500	3,000	580	1,160	195	390	432	864	52	104
3,000	4,000	635	1,270	210	420	474	948	58	116
4,000	5,000	685	1,370	225	450	513	1,026	61	122
5,000	6,000	730	1,460	235	470	546	1,092	65	130
6,000	7,000	770	1,540	245	490	573	1,146	68	136
7,000	8,000	800	1,600	250	500	600	1,200	72	144
8,000	9,000	835	1,670	255	510	624	1,248	75	150
9,000	10,000	865	1,730	260	520	645	1,290	78	156
10,000	12,000	875	1,750	270	540	687	1,374	82	164
12,000	14,000	885	1,770	275	550	723	1,446	87	174
14,000	16,000	900	1,800	280	560	756	1,512	90	180
16,000	18,000	940	1,880	285	570	786	1,572	94	188
18,00	20,000	975	1,950	290	580	813	1,626	98	196
20,000	25,000	1,055	2,000	315	630	876	1,752	105	210
25,000	30,000	1,130	2,000	340	680	933	1,866	112	224
30,000	35,000	1,205	2,000	360	720	981	1,962	119	238
35,000	40,000	1,275	2,000	380	760	1,026	2,000	124	248
40,000	45,000	1,340	2,000	400	800	1,068	2,000	129	258
45,000	50,000	1,400	2,000	420	840	1,104	2,000	135	270
50,000	55,000	1,460	2,000	440	880	1,140	2,000	140	280
55,000	60,000	1,515	2,000	455	910	1,173	2,000	145	290
60,000	65,000	1,565	2,000	470	940	1,206	2,000	150	300
65,000	70,000	1,610	2,000	485	970	1,236	2,000	155	310
70,000	75,000	1,655	2,000	500	1,000	1,263	2,000	160	320
75,000	80,000	1,695	2,000	510	1,020	1,293	2,000	165	330
80,000	85,000	1,730	2,000	520	1,040	1,317	2,000	170	340
85,000	90,000	1,760	2,000	530	1,060	1,344	2,000	175	350
90,000	95,000	1,790	2,000	540	1,080	1,368	2,000	180	360
95,000	100,000	1,815	2,000	545	1,090	1,392	2,000	185	370
100,000	110,000	1,835	2,000	550	1,100	1,437	2,000	195	390
110,000	120,000	1,855	2,000	555	1,110	1,479	2,000	205	410
120,000	130,000	1,875	2,000	560	1,120	1,521	2,000	215	430
130,000	140,000	1,890	2,000	565	1,130	1,557	2,000	225	450
140,000	150,000	1,900	2,000	570	1,140	1,593	2,000	235	470
150,000	160,000	1,935	2,000	580	1,160	1,629	2,000	245	490
160,000	170,000	1,965	2,000	590	1,180	1,662	2,000	255	510
170,000	180,000	1,990	2,000	600	1,200	1,695	2,000	265	530
180,000	190,000	2,010	2,010	605	1,210	1,725	2,000	275	550
190,000	200,000	2,030	2,030	610	1,220	1,755	2,000	285	570
200,000	210,000	2,055	2,055	620	1,240	1,782	2,000	295	590
210,000	230,000	2,100	2,100	635	1,270	1,836	2,000	315	630
230,000	250,000	2,155	2,155	650	1,300	1,890	2,000	335	670
250,000	275,000	2,215	2,215	670	1,340	1,950	2,000	360	720
275,000	300,000	2,275	2,275	690	1,380	2,000	2,000	385	770

Table: American Table of Distances for Storage of Explosives (December 1910), as Revised and Approved by th Institute of Makers of Explosives—July, 1991.

Notes to the Table of Distances for Storage of Explosives

(1) Terms found in the table of distances for storage of explosive materials are defined in §555.11.

(2) When two or more storage magazines are located on the same property, each magazine must comply with the minimum distances specified from inhabited buildings, railways, and highways, and, in addition, they should be separated from each other by not less than the distances shown for "Separation of Magazines," except that the quantity of explosives contained in cap magazines shall govern in regard to the spacing of said cap magazines from magazines containing other explosives. If any two or more magazines are separated from each other by less than the specified "Separation of Magazines" distances, then such two or more magazines, as a group, must be considered as one magazine, and the total quantity of explosives stored in such group must be treated as if stored in a single magazine located on the site of any magazine of the group, and must comply with the minimum of distances specified from other magazines, inhabited buildings, railways, and highways.

(3) All types of blasting caps in strengths through No. 8 cap should be rated at 1½ lbs. (1.5 lbs.) of explosives per 1,000 caps. For strengths higher than No. 8 cap, consult the manufacturer.

(4) For quantity and distance purposes, detonating cord of 50 or 60 grains per foot should be calculated as equivalent to 9 lbs. of high explosives per 1,000 feet. Heavier or lighter core loads should be rated proportionately.

[T.D. ATF–87, 46 FR 40384, Aug. 7, 1981, as amended by T.D. ATF–400, 63 FR 45003, Aug. 24, 1998; T.D. ATF–446, 66 FR 16602, Mar. 27, 2001; T.D. ATF–446a, 66 FR 19089, Apr. 13, 2001]

§ 555.219 Table of distances for storage of low explosives.

Pounds		From Inhabited building distance (feet)	From public railroad and highway distance (feet)	From above ground magazine (feet)
Over	Not over			
0	1,000	75	75	50
1,000	5,000	115	115	75
5,000	10,000	150	150	100
10,000	20,000	190	190	125
20,000	30,000	215	215	145
30,000	40,000	235	235	155
40,000	50,000	250	250	165
50,000	60,000	260	260	175
60,000	70,000	270	270	185
70,000	80,000	280	280	190
80,000	90,000	295	295	195
90,000	100,000	300	300	200
100,000	200,000	375	375	250
200,000	300,000	450	450	300

Table: Department of Defense Ammunition and Explosives Standards, Table 5–4.1 Extract; 4145.27 M, March 1969

§ 555.220 Table of separation distances of ammonium nitrate and blasting agents from explosives or blasting agents.

Donor weight (pounds)		Minimum separation distance of acceptor from donor when barricaded (feet)		Minimum thickness of artificial barricades (inches)
Over	Not over	Ammonium nitrate	Blasting agent	
0	100	3	11	12
100	300	4	14	12
300	600	5	18	12
600	1,000	6	22	12
1,000	1,600	7	25	12
1,600	2,000	8	29	12
2,000	3,000	9	32	15
3,000	4,000	10	36	15
4,000	6,000	11	40	15
6,000	8,000	12	43	20
8,000	10,000	13	47	20
10,000	12,000	14	50	20
12,000	16,000	15	54	25
16,000	20,000	16	58	25
20,000	25,000	18	65	25
25,000	30,000	19	68	30
30,000	35,000	20	72	30
35,000	40,000	21	76	30
40,000	45,000	22	79	35
45,000	50,000	23	83	35
50,000	55,000	24	86	35
55,000	60,000	25	90	35
60,000	70,000	26	94	40
70,000	80,000	28	101	40
80,000	90,000	30	108	40
90,000	100,000	32	115	40
100,000	120,000	34	122	50
120,000	140,000	37	133	50
140,000	160,000	40	144	50
160,000	180,000	44	158	50
180,000	200,000	48	173	50
200,000	220,000	52	187	60
220,000	250,000	56	202	60
250,000	275,000	60	216	60
275,000	300,000	64	230	60

Table: National Fire Protection Association (NFPA) Official Standard No. 492, 1968

Notes of Table of Separation Distances of Ammonium Nitrate and Blasting Agents from Explosives or Blasting Agents

(1) This table specifies separation distances to prevent explosion of ammonium nitrate and ammonium nitrate-based blasting agents by propagation from nearby stores of high explosives or blasting agents referred to in the table as the "donor." Ammonium nitrate, by itself, is not considered to be a donor when applying this table. Ammonium nitrate, ammonium nitrate-fuel oil or combinations thereof are acceptors. If stores of ammonium nitrate are located within the sympathetic detonation distance of explosives or blasting agents, one-half the mass of the ammonium nitrate is to be included in the mass of the donor.

(2) When the ammonium nitrate and/or blasting agent is not barricaded, the distances shown in the table must be multiplied by six. These distances allow for the possibility of high velocity metal fragments from mixers, hoppers, truck bodies, sheet metal structures, metal containers, and the like which may enclose the "donor." Where explosives storage is in bullet-resistant magazines or where the storage is protected by a bullet-resistant wall, distances and barricade thicknesses in excess of those prescribed in the table in §555.218 are not required.

(3) These distances apply to ammonium nitrate that passes the insensitivity test prescribed in the definition of ammonium nitrate fertilizer issued by the Fertilizer Institute. 1 Ammonium nitrate failing to pass the test must be stored at separation distances in accordance with the table in §555.218.

[1] Definition and Test Procedures for Ammonium Nitrate Fertilizer, Fertilizer Institute 1015–18th St. N.W. Washington, DC 20036.

(4) These distances apply to blasting agents which pass the insensitivity test prescribed in regulations of the U.S. Department of Transportation (49 CFR part 173).

(5) Earth or sand dikes, or enclosures filled with the :scribed minimum thickness of earth or sand are acceptable ificial barricades. Natural barricades, such as hills or timber of ficient density that the surrounding exposures which require tection cannot be seen from the "donor" when the trees are e of leaves, are also acceptable.

(6) For determining the distances to be maintained from abited buildings, passenger railways, and public highways, use table in §555.218.

555.221 Requirements for display fireworks, rotechnic compositions, and explosive aterials used in assembling fireworks or articles rotechnic.

Display fireworks, pyrotechnic compositions, and explosive terials used to assemble fireworks and articles pyrotechnic ll be stored at all times as required by this Subpart unless they in the process of manufacture, assembly, packaging, or are ng transported.

No more than 500 pounds (227 kg) of pyrotechnic composins or explosive materials are permitted at one time in any eworks mixing building, any building or area in which the otechnic compositions or explosive materials are pressed or erwise prepared for finishing or assembly, or any finishing or embly building. All pyrotechnic compositions or explosive terials not in immediate use will be stored in covered, nonrous containers.

The maximum quantity of flash powder permitted in any eworks process building is 10 pounds (4.5 kg).

All dry explosive powders and mixtures, partially assembled play fireworks, and finished display fireworks shall be noved from fireworks process buildings at the conclusion of a 's operations and placed in approved magazines.

). ATF–293, 55 FR 3722, Feb. 5, 1990, as amended by T.D. ATF–400, 63 FR 04, Aug. 24, 1998]

555.222 Table of distances between fireworks ocess buildings and between fireworks process d fireworks nonprocess buildings.

Net weight of eworks1 (pounds)	Display fireworks2 (feet)	Consumer fireworks3 (feet)
0-100	57	37
101-200	69	37
201-300	77	37
301-400	85	37
401-500	91	37
Above 500	Not permitted4 5	Not permitted4 5

Net weight is the weight of all pyrotechnic compositions, and losive materials and fuse only.

The distances in this column apply only with natural or ficial barricades. If such barricades are not used, the distances st be doubled.

[3] While consumer fireworks or articles pyrotechnic in a finished state are not subject to regulation, explosive materials used to manufacture or assemble such fireworks or articles are subject to regulation. Thus, fireworks process buildings where consumer fireworks or articles pyrotechnic are being processed shall meet these requirements.

[4] A maximum of 500 pounds of in-process pyrotechnic compositions, either loose or in partially-assembled fireworks, is permitted in any fireworks process building. Finished display fireworks may not be stored in a fireworks process building.

[5] A maximum of 10 pounds of flash powder, either in loose form or in assembled units, is permitted in any fireworks process building. Quantities in excess of 10 pounds must be kept in an approved magazine.

[T.D. ATF–293, 55 FR 3723, Feb. 5, 1990, as amended by T.D. ATF–400, 63 FR 45004, Aug. 24, 1998]

§ 555.223 Table of distances between fireworks process buildings and other specified areas.

Net weight of fireworks1 (pounds)	Display fireworks1 (feet)	Consumer fireworks2 (feet)
0-100	200	25
101-200	200	50
201-300	200	50
301-400	200	50
401-500	200	50
Above 500	Not permitted	Not permitted

Distance from Passenger Railways, Public Highways, Fireworks Plant Buildings used to Store Consumer Fireworks and Articles Pyrotechnic, Magazines and Fireworks Shipping Buildings, and Inhabited Buildings. [3 4 5]

[1] Net weight is the weight of all pyrotechnic compositions, and explosive materials and fuse only.

[2] While consumer fireworks or articles pyrotechnic in a finished state are not subject to regulation, explosive materials used to manufacture or assemble such fireworks or articles are subject to regulation. Thus, fireworks process buildings where consumer fireworks or articles pyrotechnic are being processed shall meet these requirements.

[3] This table does not apply to the separation distances between fireworks process buildings (see § 555.222) and between magazines (see §§ 555.218 and 555.224).

[4] The distances in this table apply with or without artificial or natural barricades or screen barricades. However, the use of barricades is highly recommended.

[5] No work of any kind, except to place or move items other than explosive materials from storage, shall be conducted in any building designated as a warehouse. A fireworks plant warehouse is not subject to § 555.222 or this section, tables of distances.

[T.D. ATF–293, 55 FR 3723, Feb. 5, 1990, as amended by T.D. ATF–400, 63 FR 45004, Aug. 24, 1998]

§ 555.224 Table of distances for the storage of display fireworks (except bulk salutes).

Net weight of fireworks1 (pounds)	Distance between magazine and inhabited building, passenger railway, or public highway3 4 (feet)	Distance between magazines2 3 (feet)
0-1000	150	100
1,001-5,000	230	150
5,001-10,000	300	200
Above 10,000	Use Table § 55.218	

[1] Net weight is the weight of all pyrotechnic compositions, and explosive materials and fuse only.

[2] For the purposes of applying this table, the term "magazine" also includes fireworks shipping buildings for display fireworks.

[3] For fireworks storage magazines in use prior to (30 days from the date of publication of the final rule in the Federal Register), the distances in this table may be halved if properly barricaded between the magazine and potential receptor sites.

[4] This table does not apply to the storage of bulk salutes. Us table at § 555.218.

[T.D. ATF–293, 55 FR 3723, Feb. 5, 1990, as amended by T.D. ATF–400, 63 F 45004, Aug. 24, 1998]

Questions and Answers (Revised 1/07)
8 U.S.C. Chapter 40 and 27 CFR Part 555

ntroduction

he following list of Questions and Answers are tended to aid you in gaining a better understanding :

3 U.S.C. Chapter 40 – Importation, anufacture, Distribution and Storage of xplosive Materials and the implementing gulations issued within:

7 CFR Part 555 – Commerce in Explosives

nis listing is not all-inclusive. However it contains me of the most frequently asked questions that ATF ceives. These questions and answers are intended ly as a general overview. To determine how the law d regulations apply to your specific circumstances, u must refer directly to the applicable law and gulation or contact your local ATF Office of Industry Operations. Also, please be aware that both the law and regulations are subject to change. Please contact your local ATF office for the most up-to-date information. You can also find the latest Federal explosives regulations on the ATF website: www.atf.gov.

Unless otherwise stated, these Questions and Answers apply only to Federal law and regulations. States and local jurisdictions have, in many cases, enacted their own requirements relating to explosives. Check with appropriate State or local authorities for information on those requirements. Compliance with Federal law and regulations does not exempt any person from compliance with any State or local requirements.

A Table of Contents and a Subject Index has been included for your convenience. The Index is located at the end of the Questions and Answers section.

Table of Contents

General Questions

1. Who is affected by the Federal explosives law?

The law affects all persons who import, manufacture, deal in, purchase, use, store, or possess explosive materials. It also affects those who ship, transport or cause to be transported, or receive explosive materials. Also, see 18 U.S.C. 845 and 27 CFR 555.141 for exemptions.

2. What changes were brought about by the Safe Explosives Act?

Among other things, the Act mandated that all persons who wish to receive or transport explosive materials must first obtain a Federal explosives license or permit. In addition, the act imposed new restrictions on who may lawfully receive and possess explosive materials (See question 4). All Federal explosive licensees and permittees and their responsible persons and employees authorized to possess explosives are affected by the new requirements and background checks mandated by the Act.

3. Can I obtain general information from ATF on the Internet?

Yes. ATF maintains a website on the Internet at www.atf.gov.

4. Does the law make some classes of persons ineligible to receive a Federal license to import, manufacture, or deal in explosive materials or to receive a Federal explosives permit?

Yes. A license or permit will not be issued to any person who:

(a) Is under indictment for, or who has been convicted in any court, of a crime punishable by imprisonment for a term exceeding one year;

(b) Is a fugitive from justice;

(c) Is an unlawful user of or addicted to any controlled substance (as defined in section 102 of the Controlled Substances Act (21 U.S.C. 802));

(d) Has been adjudicated a mental defective or who has been committed to a mental institution;

(e) Is an alien (with certain exceptions);

(f) Has been discharged from the armed forces under dishonorable conditions; or,

(g) Having been a citizen of the United States, has renounced his citizenship. [18 U.S.C. 842(i), 843(b)(1); 27 CFR 555.49(b)(2)(i)]

5. Are there classes of persons to whom the distribution of explosive materials by licensees is prohibited?

Yes. No person shall knowingly distribute explosive materials to any individual listed in Question 4, or to an individual who is under 21 years of age. [18 U.S.C. 842(d); 27 CFR 555.26(d)]

6. What other distributions of explosive mat rials by licensees and permittees are prohibited

A licensee or permittee shall not knowingly distribute any explosive materials to any person who:

(a) Is not a licensee [18 U.S.C. 842(b), 27 CFR 555.105, 555.106];

(b) Is not a holder of a user permit [18 U.S.C. 842(b); 27 CFR 555.105, 555.106];

(c) Is not a holder of a limited permit who resides in the same State where distribution is made and in which premises of the transferor are located. [18 U.S.C. 842(b); 27 CFR 555.105, 555.106];

(d) The licensee has reason to believe intends to transport suc explosive materials into a State where the purchase, possessior or use of explosive materials is prohibited or which does not permit its residents to transport or ship explosive materials intc the State or to receive explosive materials in the State. [18 U.S 842(c); 27 CFR 555.106]

(e) Is in any State where the purchase, possession, or use by such person of such explosive materials would be in violation any State law or any published ordinance applicable at the plac of distribution. [18 U.S.C. 842(e); 27 CFR 555.106(b)(2)]

7. Does Federal law prohibit certain persons from receiving or possessing explosive material

Yes. The law prohibits the receipt or possession of explosive materials by any person listed in question 4. [18 U.S.C. 842(i); CFR 555.26, 555.49(b)]

8. May a licensed manufacturer, importer, or dealer distribute explosive materials to nonlicen ees and nonpermittees?

No. Every person who receives explosive materials must firs obtain a Federal explosives license or permit. Distribution of explosive materials to persons who do not hold a license or permit is unlawful. Also, see 18 U.S.C. 845 and 27 CFR 555.1 for exemptions. [18 U.S.C. 842(a), 842(b); 27 CFR 555.26(a), 555.106]

9. Does Federal law provide penalties for purchasers who give false information at the tim of purchasing explosive materials?

Yes. The penalty for knowingly providing false information misrepresented identification is a maximum 10 years' impriso ment and/or a fine not exceeding $250,000. [18 U.S.C. 842(a)(844(a)]

10. Is the theft of explosive materials, as well as the possession of stolen explosive materials, a Federal crime?

Yes. It is a Federal crime for any person to steal any explosive materials. It is also a Federal crime for any person to receive, possess, transport, ship, conceal, store, barter, sell, dispose of, or pledge or accept as security for a loan any stolen explosive materials. [18 U.S.C. 842(h), 844(k), (l)]

11. Are thefts and losses of explosive materials required to be reported to ATF?

Yes. Any licensee or permittee who has knowledge of the theft or loss of any explosive materials from his or her stock shall, within 24 hours of discovery, report the theft or loss by telephoning 800-461-8841 (Monday-Friday 8:00a.m.– 5:00 p.m. Eastern Time) or 888-283-2662 (after hours and weekends) (nationwide toll free number) and on ATF Form 5400.5, "Report of Theft or Loss – Explosive Materials", in accordance with the instructions on the form. The theft or loss shall also be reported to appropriate local authorities. The same requirements are imposed upon persons other than licensees and permittees, except that nonlicensees and nonpermittees, other than carriers, need not report a theft or loss on Form 5400.5, but must report the theft or loss by telephone, using the same numbers: 800-461-8841 (Monday-Friday 8:00a.m.–5:00 p.m. Eastern Time) or 888-283-2662 (after hours and weekends) and in writing to the nearest ATF office. The theft or loss shall also be reported to appropriate local authorities. Carriers of explosive materials must report a theft or loss by telephone but need not make the report on the ATF form or in writing. See 27 CFR 555.30 for the specific information required to be reported in connection with a theft or loss. [18 U.S.C. 842(k), 18 U.S.C. 844(p); 27 CFR 555.30]

12. My company holds a Federal explosives license and after conducting an inventory of our explosives on hand, we noticed one case of dynamite missing. After double-checking all Daily Summaries of Magazine Transactions, invoices, and delivery sheets, we still cannot reconcile the discrepancy. What are we required to do?

This should be considered a theft or loss of explosive materials. As stated in the answer to Question 11, you must report the theft or loss of explosive materials, within 24 hours of discovery, to ATF by telephone (toll free: 800-461-8841 (Monday-Friday 8:00 a.m.– 5:00 p.m. Eastern Time) or 888-283-2662 (after hours and weekends). ATF Form 5400.5, "Report of Theft or Loss – Explosive Materials", must then be completed and forwarded in accordance with the instructions on the form. [18 U.S.C. 842(k); 27 CFR 555.30]

13. May ATF conduct warrantless inspections of licensees' and permittees' records of explosives materials, stocks of such materials, and magazines?

Except for limited permit holders, any ATF officer may, without a warrant, enter during business hours the premises, including places of storage, of any licensee or permittee for the purpose of inspecting or examining any records or documents required to be kept by the law and regulations and any explosive materials kept or stored at the premises. For inspection purposes, "business hours" includes hours during which business is actually conducted, not just those hours stated on license applications. Any licensee or permittee who refuses to permit the inspection or examination is subject to having his or her license or permit revoked, as well as to denial of an application to renew the license or permit. For limited permit holders, an ATF officer may inspect the places of storage for explosive materials of either an applicant for a limited permit or at the time of renewal of such permit, but in no event shall such inspection occur more than once every three years. [18 U.S.C. 843(b)(4), 18 U.S.C. 843(f)]

14. Will ATF investigate accidents involving explosive materials?

ATF is authorized to inspect the site of any accident or fire where there is reason to believe that explosive materials were involved. Other Federal agencies, or State or local agencies, may also investigate such incidents, depending on the circumstances. [18 U.S.C. 846(a); 27 CFR 555.31]

15. Is black powder subject to regulation under Federal explosives laws?

Black powder is an explosive material for purposes of Federal explosives laws and regulations. However, the law exempts from regulation commercially manufactured black powder in quantities not exceeding 50 pounds (as well as percussion caps, safety and pyrotechnic fuses, quills, quick and slow matches, and friction primers) intended to be used solely for sporting, recreational, or cultural purposes in antique firearms as defined in 18 U.S.C. 921(a)(16) or in antique devices exempted from the term "destructive device" in 18 U.S.C. 921(a)(4). However, persons engaged in the business of importing, manufacturing, or dealing in black powder in any quantity must have a Federal explosives license. [18 U.S.C. 841(c), 841(d), 845(a)(5); 27 CFR 555.11: definitions of "explosives" and "explosive materials", 555.141(b)]

16. Is small arms ammunition subject to regulation under Federal explosives laws?

No. The law specifically exempts small arms ammunition and components thereof. (See also Question 81.) [18 U.S.C. 845(a)(4)]

17. Are binary explosives subject to regulation under Federal explosives laws?

Until the compounds are mixed, they are not classified as explosives and, therefore, are not subject to control. However, once mixed, binary explosives are "explosive materials" and are subject to all applicable Federal requirements. A person who mixes or combines compounds of binary explosives for the purpose of sale or distribution or for the person's own business use is a "manufacturer" of explosive materials and must be licensed as a manufacturer under the law. [18 U.S.C. 841(h); 27 CFR 555.11: definition of "manufacturer"]

18. Does ATF have any regulations governing the actual transportation of explosive materials?

Federal explosives laws and regulations generally prohibit any person from transporting explosive materials interstate or intrastate unless the person has a Federal explosives license or permit. Also, the transportation of stolen explosives materials is a Federal crime (see also Question 10). However, the law exempts from regulation under 18 U.S.C. Chapter 40 and 27 CFR Part 555 aspects of the transportation of explosive materials via railroad, water, highway, or air which are regulated by the United States Department of Transportation, and the Department of Homeland Security, and agencies thereof and which pertain to safety and security. [18 U.S.C. 842(a)(3), 842(h), 845(a)(1); 27 CFR 555.26, 555.28, 555.141(a)(1)]

19. Are common or contract carriers required to obtain a Federal explosives license or permit to transport explosive materials?

No. The actual transportation of explosive materials by carriers is subject to Department of Transportation or Department of Homeland Security regulations. [18 U.S.C. 845(a)(1); 27 CFR 555.141(a)(1)]

20. What is the "Explosives List"?

The Explosives List is a comprehensive (but not all-inclusive) listing of explosive materials which have been determined to be within the coverage of Chapter 40. The list is published annually by ATF (the most recent list can be found under the Explosives Laws and Regulations link on the ATF website). [18 U.S.C. 841(d); 27 CFR 555.23]

21. May a person under the age of 21 be lawfully employed by an explosives business and lawfully receive, possess, and use explosive materials on behalf of the business?

Yes. Federal explosives law prohibits any person from distributing explosive materials to persons under 21 years of age. However, it does not prohibit the delivery to or possession of explosive materials by persons under the age of 21 who are receiving or using the materials on behalf of their employers to whom the materials were lawfully sold. [18 U.S.C. 842(d), (i); 27 CFR 555.11 (definition of "distribute"), 555.26, 555.106(b)(1)]

22. ATF regulations require explosive materials to be stored at certain minimum distances from a "public highway". What is a "public highway" for purposes of the regulations?

The term "highway" is defined in 27 CFR 555.11 as "any public street, public alley, or public road, including a privately financed, constructed, or maintained road that is regularly and openly traveled by the general public." Privately financed, constructed, or maintained roads that are marked and barricaded in a manner that prevents access by the general public do not fall within the meaning of the term and would, therefore, be exempt from table of distance requirements. [27 CFR 555.11: definition of "highway"](See also ATF Ruling 2005-2)

23. Is an airport runway or taxiway considered a public highway for purposes of the Table of Distances for storage of explosive materials?

No. However, airport terminals are considered inhabited buildings for Table of Distance requirements.

24. How is shock tube regulated by ATF?

Shock tube contains highly explosive material. However, it may be stored as a low explosive when not attached to a detonator. [27 CFR 555.202(b), 555.213]

25. What is an EX number?

An EX number is a number, preceded by the prefix "EX-", which is issued and used by the Department of Transportation (DOT) to identify an explosive which has been tested and classified by DOT. See U.S. Department of Transportation regulations at 49 CFR 171.8 and 49 CFR 173.56.

26. What is a UN number?

A UN (United Nations) number is used by DOT as a method of identification and classification of products for shipping purposes. UN numbers are different from the hazard class or division designations used by DOT (for example, 1.1, 1.2, 1.3, 1.4, and 1.5). ATF regulations in 27 CFR Part 555 also use UN numbers to help identify certain explosives. [27 CFR 555.11]

27. Can Federal explosives disabilities resulting from a conviction of a crime punishable by imprisonment for a term exceeding one year be removed if the conviction is expunged or set aside or the convicted person has received a pardon for the offense or has had his or her civil rights restored?

A person convicted of, or under indictment for, a "crime punishable by imprisonment for a term exceeding one year" may not lawfully receive or possess explosive materials or be issued a Federal explosives license or permit. The term "crime punishable by imprisonment for a term exceeding one year" does not include offenses pertaining to antitrust violations, unfair trade practices, restraints of trade, or any State offense (other than one involving a firearm or explosive) classified as a misdemeanor and nishable by imprisonment for 2 years or less. There are only 3 ans by which Federal explosives disabilities resulting from a nviction of, or indictment for, a "crime punishable by imprison- nt for a term exceeding one year" can be removed:

- A decision of a court invalidating a conviction on the basis at the conviction was unconstitutional;
- In the case of a Federal conviction, a presidential pardon; and
- The granting of relief from Federal explosives disabilities by F pursuant to the filing of a relief application with the Director. ormation on how to apply for relief and contact information for Relief of Disabilities Section is available on the ATF website. 3 U.S.C. 841(l), 842(d),(i), 845(b); 27 CFR 555.11: definition "crime punishable by imprisonment for a term exceeding one ar", 555.26(c), 555.142]

icenses and Permits

28. Who needs a Federal explosives license or ermit?

All persons who wish to transport, ship, cause to be transported, receive explosive materials must first obtain a Federal explo- es license or permit. Certain exemptions apply. [18 U.S.C. 2(b); 18 U.S.C. 845; 27 CFR 555.26(a), 27 CFR 555.141]

29. Who is eligible for a Federal explosives ense or permit?

The Chief, Federal Explosives Licensing Center, will approve a operly completed application for a license or permit on ATF rm 5400.13/5400.16 if the applicant:

Is not a person prohibited from possessing or receiving olosive materials under 18 U.S.C. 842(i) and none of the olicant's "responsible persons" are prohibited under section 2(1); (see also Question 4 in General Q&A);

Has not willfully violated any provision of Chapter 40 or the ulations in 27 CFR Part 555;

Has not knowingly withheld information or has not made any se or fictitious statement intended or likely to deceive, in nnection with the application,

Has premises in a State from which he intends to conduct siness or operations;

Has storage for the class (as described in 27 CFR 555.202) of losive materials described on the application;

Has certified in writing that he is familiar with and under- nds all published State laws and local ordinances relating to olosive materials for the location in which he intends to do siness;

Has submitted the certificate required by section 21 of the leral Water Pollution Control Act, as amended (33 U.S.C. 41) [18 U.S.C. 843(b); 27 CFR 555.49(b)];

(h) None of the applicant's employees authorized to possess explosives are prohibited persons under 18 U.S.C. 842(i); and

(i) In the case of an applicant for a limited permit, the applicant has certified that the applicant will not receive explosive materi- als on more than 6 occasions during the 12-month period for which the limited permit is valid.

30. What activities are covered by licenses and permits?

Licenses allow persons to engage in the business of importing, manufacturing, or dealing in explosive materials. Any individual or business entity intending to engage in any of these activities must first obtain a license. A user permit allows the receipt and transportation of explosive materials. A limited permit allows the receipt of explosive materials from a licensee or permittee within the permittee's state of residence only, and on no more than six occasions in the 12-month period during which the limited permit is valid. A limited permit does not authorize the receipt or transportation of explosive materials in interstate or foreign commerce. [27 CFR 555.11: definitions of "importer", "manu- facturer", "dealer", "limited permit", and "user permit", 555.41]

A separate license is needed for each business premises where an explosives business or activity is conducted. Only a single user permit is needed by a permittee who uses explosives in more than one location. [27 CFR 555.41]

31. What is a Limited Permit?

A limited permit is for persons who wish to transport, ship, cause to be transported, or receive explosive materials in intra- state commerce only. This permit is designed for the infrequent receipt of explosive materials by intrastate users. The limited permit will allow a purchaser to receive explosive materials on no more than six separate occasions from in-state licensees or permittees during the 12-month period of the permit. The limited permit does not allow the holder to transport, ship, cause to be transported, or receive explosive materials in interstate commerce.

32. What is the duration of a license or permit?

(a) A user license or permit is valid for a period of 3 years.

(b) The user-limited permit is valid only for a single purchase transaction.

(c) Limited permits are valid for no more than six separate receipts of explosive materials during a 12-month period. [27 CFR 555.51]

33. What are the fees for licenses and permits?

Each license applicant must pay a fee of $200 for obtaining a 3-year license, a separate license and fee being required for each business premises. The fee for renewal of a license is $100 for a 3-year license. [27 CFR 555.42]

Each applicant for a user permit must pay a fee of $100 for a 3-year permit, and each applicant for a user-limited permit (nonrenewable) must pay a fee of $75. The fee for renewal of a user permit is $50 for a 3-year permit. [27 CFR 555.43]

Each applicant for a limited permit must pay a fee of $25 for a 1-year limited permit. The fee for renewal of a limited permit is $12 for a 1-year limited permit. [27 CFR 555.43]

34. Will the Government investigate an application for a license or permit?

ATF will investigate any applicant before issuing a license or permit. Additionally, ATF must inspect places of storage and conduct background checks on responsible persons and employee possessors authorized to possess explosives [18 U.S.C. 843(b); 18 U.S.C. 843(h); 27 CFR 555.33, 27 CFR 555.49(b)]

35. What may a licensed explosives dealer do?

A licensed dealer may engage in the business of distributing explosive materials at wholesale or retail [27 CFR 555.11: definition of "dealer"]

36. What may a licensed explosives importer do?

A licensed importer may engage in the business of importing or bringing explosive materials into the United States for purposes of sale or distribution. It is not necessary for a licensed importer to also obtain a dealer's license to engage in business on his or her licensed premises as a dealer in explosive materials (see also Question 52 and 53) [27 CFR 555.11: definition of "importer", 27 CFR 555.41(b)(2)]

37. When is a manufacturer's license required?

A manufacturer's license is required by persons engaged in the business of manufacturing explosive materials for sale, distribution, or for their own business use. For example, persons engaged in the business of providing a blasting service using explosives of their own manufacture would be required to have a manufacturer's license. Persons who manufacture explosives for their personal, non-business use are not required to have a manufacturer's license. However, no person may ship, transport, cause to be transported, or receive explosive materials unless such person holds a license or permit. [27 CFR 555.11: definition of "manufacturer", 555.41(b)] A separate manufacturer's license is not required by a licensed manufacturer for the purpose of on-site manufacture, for example, mixing binary explosives or making blasting agents at a quarry or other job site. It is not necessary for a licensed manufacturer to also obtain a dealer's license to engage in business on his or her licensed premises as a dealer in explosive materials (see also Question 52 and 53) [27 CFR 555.11: definition of "manufacturer", 555.41(b)(2)]

38. How do I apply for a Federal explosives license or permit?

You can request an application for a Federal explosives license or permit from the Federal Explosives Licensing Cente at 877-283-3352 or from the ATF Distribution Center at 703-455-7801. As part of the application process, you must co plete and submit an ATF Form 5400.13/5400.16, Application f Explosives License or Permit. You must also submit the names identifying information, fingerprints, and photographs of all responsible persons. In addition, you must submit the names a identifying information of all employees who are authorized to possess explosive materials in the course of their employment ATF Form 5400.28, Employee Possessor Questionnaire. [27 C 555.45(c)]

39. Who is a "responsible person"?

Federal explosives laws define a "responsible person" as an individual who has the power to direct the management and policies of the applicant pertaining to explosive materials. Responsible persons generally include sole proprietors and explosives facility site managers. In the case of a corporation, association, or similar organization, responsible persons gener include only those corporate directors/officers, and stockholde who have the power to direct management and policies as they pertain to explosive materials.

For example, a corporate vice president whose duties includ acquiring and approving contracts with explosives distributors would be considered a responsible person. Other corporate officials whose duties do not include the power to direct the management and policies of the applicant pertaining to explosi materials, for example, a vice president responsible solely for human resources, would not typically be considered a responsi person. Each applicant for a license or permit must assess the corporate and other management responsibilities for all key personnel and determine whether or not these duties place the individual in the position of being a responsible person. [18 U. 841(s), 27 CFR 555.11: definition of "responsible person"]

40. Who is a "possessor of explosives"?

A possessor of explosives is any employee of a license or pe holder or any employee of an applicant for a license or permit has or will have actual physical possession of explosive materi or who has or will have constructive possession of explosive materials. For example, persons who physically handle explosi materials would be considered to be actual possessors of explo materials. This would include employees who directly handle explosive materials as part of the production process; employe who handle explosive materials in order to ship, transport, or s them; and employees, such as blasters and their helpers who actually use explosive materials. A constructive possessor is a rson who has access to explosive materials, without physically ndling them. For example, a supervisor at a construction site o keeps keys for storage magazines in which explosives are red or who directs the use of explosive materials by other ployees has constructive possession of explosives.

41. Why is it necessary to provide new and ditional information on responsible persons d employee possessors of explosives?

The law requires this information for ATF to conduct back- ound checks on all responsible persons and employee possess- s to restrict the availability of explosives to authorized persons ly and to reduce the risk of prohibited persons acquiring plosive materials. [18 U.S.C. 843(h); 27 CFR 555.33, 5.45(c)]

42. When will I need to submit the identifying formation for my responsible persons and nployee possessors of explosives?

- All license and permit applicants and any renewal applicants ıst submit identifying information for responsible persons and ployee possessors (and fingerprints and photographs for sponsible persons) upon submission of an original or renewal plication.
- Any new responsible person added after a license or permit s been issued by ATF must be reported to ATF within 30 days. owever, the submission of fingerprints and photographs by the w responsible person is required only at the time of any bsequent renewal.
- For all licenses and permits (new and renewal), any new ployee possessors must be reported to ATF within 30 days of e on the Employee Possessor Questionnaire form (ATF F 00.28). [27 CFR 555.45(c), 27 CFR 555.57(b)]

43. How do I get my fingerprints taken?

Fingerprints must be submitted on Fingerprint Identification rds, FD-258 that have been issued by ATF. The fingerprint rds must contain the following ORI information: WVATF0900; F-NATL EXPL LIC, MARTINSBURG WV. These fingerprint ds may be obtained by contacting the Federal Explosives censing Center at 877-283-3352 or the ATF Distribution Center 703-455-7801. The fingerprint cards must be completed by ur local law enforcement authority.

44. Will ATF notify me whether or not my sponsible persons and employee possessors ve passed their background checks?

Yes. A "Notification of Clearance" will be issued directly to all ense or permit holders advising whether their responsible rsons and employee possessors have been cleared to possess plosive materials, or are or may be prohibited from possessing plosives. These notices must be retained as part of the license or permit holders permanent records. In addition, letters of clearance or denial will be issued directly to responsible persons and employee possessors. [27 CFR 555.33]

45. What notification will I receive if one of my responsible persons or employee possessors does not pass their ATF background check?

If an individual does not pass the background check, a letter will be sent to the licensee or permittee who submitted the individual's name indicating that the individual was denied. A letter will also be sent to that individual explaining the prohibition and outlining appeal and relief procedures, as may be applicable. Unless and until an appeal overturns the denial or relief from disabilities is granted, that individual may not lawfully possess explosives. [27 CFR 555.33]

46. Who will conduct the background checks on applicants, responsible persons, and possessors?

ATF will perform the background checks. If employers wish to require their own background checks as a condition of employment, they may do so. However, such a background check will not be accepted in place of the ATF background check. [27 CFR 555.33]

47. May I sell black powder without a license?

No. Anyone who engages in the business of selling black powder, regardless of quantity, must be licensed as an explosives dealer. [27 CFR 555.41(b)]

48. Is a manufacturer's license required to acquire and mix binary explosives?

If the individual purchasing the binary explosives is engaged in the business of manufacturing explosives, i.e., mixes and uses them in the operation of a commercial business (for example, operating a quarry, or providing the service of removing stumps or boulders from a farm field), then a manufacturer's license is required.

An individual farmer who merely wishes to mix the binary explosives to remove obstacles from his field and provides no other outside service would not need a manufacturer's license.

Please note, however: A Federal explosives license or permit would be required to obtain any explosive device, such as detonators, used to initiate the mixed binary explosives. In addition, transportation of any explosive material, including mixed binary explosives, without a Federal license or permit is prohibited. [27 CFR 555.11: Definition of "manufacturer"; 27 CFR 555.26, 555.41(b)]

49. What is theatrical flash powder and is there a license for its manufacture?

Theatrical flash powder is flash powder commercially manufactured in premeasured kits not exceeding 1 ounce in weight, and mixed immediately prior to use and intended for use in events such as theatrical shows, stage plays, band concerts, magic acts, thrill shows, and clown acts in circuses. A manufacturer's license allows on-site manufacturers to operate nationally on one license issued to their principal place of business. [27 CFR 555.11: definitions of "flash powder" and "theatrical flash powder", 555.41(b)]

50. Is a separate license required for each location where business is conducted?

Yes. A separate license is required for each location where business is conducted. However, a separate license is not required for:

(a) Facilities used only for the storage of explosive materials;

(b) Locations used solely for the storage of records relating to the business; and

(c) Licensed manufacturers' on-site manufacturing. [27 CFR 555.41(b)]

51. Must a person who engages in the business of both manufacturing and importing at the same location have both licenses?

Yes. The licenses for manufacturing and importing allow a person to engage in separate and distinct activities and a separate license is required for each activity. However, a manufacturer or an importer does not need a separate dealer's license to also distribute explosive materials from the licensed premises. [27 CFR 555.41(b)]

52. Does a licensed manufacturer, importer, or dealer need a permit to use explosive materials?

No. No licensee will be required to obtain a user permit to lawfully transport, ship, or receive explosive materials in interstate or foreign commerce. [27 CFR 555.41(b)(2)]

53. Does a Federal license or permit exempt the holder from State or local requirements?

No. A license or permit confers no right or privilege to conduct business or operations, including storage, contrary to State or other law. All legal requirements must be followed, whether Federal, State, or local. [18 U.S.C. 848; 27 CFR 555.62]

54. Who is authorized to import explosive materials?

Any licensed importer is authorized to engage in the business of importing explosive materials for sale, distribution, or their own use. Any licensed manufacturer, dealer, or holder of a user permit may import explosive materials for their own use only. Licensees and user permittees importing explosive materials must provide to the U.S. Customs and Border Protection (CBP) a copy of the license or permit. Note, however, that in the case of certain military explosives or propellant powder or other components of small arms ammunition, Federal firearms regulations require th importer to provide an approved ATF Form 6 to the CBP. [27 C 555.41(b)(2), 555.41(b)(3), 447.21, 555.108(a), 555.183, 478.113]

55. How may an employee of an explosives licensee or permittee qualify to accept delivery c explosive materials for the employer?

The employee must be on the current list of representatives o agents authorized to accept delivery of explosive materials on behalf of the employer and be an authorized employee possessc of explosives. [27 CFR 555.103(b), 555.105(b)]

56. When an explosives licensee or permittee sends one of their truck drivers to the distributo premises to pick up explosive materials that hav been purchased by the licensee or permittee, wi the driver be required to sign any forms?

No, however the driver is required to furnish the seller with a identification document as defined in 27 CFR Part 555.11. [27 CFR 555.103(b), 555.105(b)]

57. Will a licensee or permittee be notified in advance when the license or user permit needs t be renewed?

Generally, prior to expiration of the license or permit, a licen or permittee will be notified. The application form must be completed and filed with ATF before expiration of the current license or permit for the renewal to be considered timely. However, if a licensee or permittee does not receive a renewal notification, it is still that licensee's or permittee's responsibilit to ensure that an application is filed prior to expiration of the current license or permit. [27 CFR 555.46]

58. I have timely filed my application for renew al of my license (or user's permit) but I have not received my new license (or permit). May I continue in business even though the expiration date shown on my license or permit has passed? If sc how long?

Yes. You may continue to operate the business pursuant to yo current license or permit until the application for renewal is act upon. [5 U.S.C. 558]

59. Can a license or permit be revoked?

Yes. The Director, Industry Operations for the ATF Field Division in which a licensee or permittee is located may revoke license or permit if the holder has violated any provision of 18 U.S.C. Chapter 40 or its implementing regulations or has becon ineligible to receive explosive materials under 18 U.S.C. 842(i). [18 U.S.C. 843(d); 27 CFR 555.71, 555.74]

60. If a Federal explosives licensee or permittee is indicted for or convicted of a "crime punishable by imprisonment for a term exceeding one year", may he or she continue operations under the license or permit?

As stated in the answer to Question 4 in General Q&A, a person under indictment for, or convicted of, a crime punishable by imprisonment for a term exceeding one year is not eligible to be issued a license or permit. However, a licensee or permittee who is indicted for, or convicted of, such a crime during the term of his or her existing license or permit is not barred from licensed or permit operations for 30 days after the date of the indictment or the date the conviction becomes final. If the licensee or permittee files an application for relief from disabilities within such 30-day period, he or she may continue licensed or permit operations while the application is pending. If a relief application is not filed during that period, the licensee or permittee may not continue operations beyond such 30-day period. The right of a licensee to continue licensed or permitted operations beyond such 30-day period is also conditioned on the licensee or permittee timely filing a license or permit renewal application disclosing that the applicant has been indicted for, or convicted of, the crime. A licensee or permittee may not continue operations beyond 30 days following the date the Director issues notification that the relief application has been denied. [18 U.S.C. 845(b); 27 CFR 555.142]

61. May a licensed dealer make a sale to a holder of a limited permit in an adjoining State?

No. Sales may not be made to limited permittees who are out-of-State residents. [18 U.S.C. 842(a); 27 CFR 555.11: definition of "limited permit", 555.41(b)(3)]

Recordkeeping

62. Does a licensee or permittee have to keep records of the acquisition, distribution, and storage of explosive materials?

Yes. Licensees and permittees must keep records of acquisitions, dispositions, and storage of explosive materials. [18 U.S.C. 842(f), 847; 27 CFR 555.107, 555.122-.125, and 555.127, Subpart G]

63. How do licensees and permittees account for explosive quantities in their records?

If acquisitions are recorded by weight, then distribution must also be recorded by weight. If acquisitions are recorded by physical count (e.g., by units), then distribution must also be recorded by physical count. [27 CFR 555.122-.125]

64. Must a licensee or permittee maintain a daily summary of magazine transactions?

Yes. After the initial inventory required by regulations has been taken, the inventory shall be entered in a record of daily transactions. Not later than the close of the next business day, each licensee and permittee shall record by manufacturer's name or brand name the total quantity received in and removed from each magazine during the day and the total remaining on hand at the end of the day. [27 CFR 555.127]

65. Where must a licensee or permittee keep the daily summary of magazine transactions?

The records must either be kept at each magazine or at one central location on the business premises, provided a separate record of daily transactions is maintained for each magazine. [27 CFR 555.127]

66. How can I obtain additional copies of ATF Forms?

Forms are available on-line at www.atf.gov/forms/5000.htm#explosives. Requests for forms should be mailed to the ATF Distribution Center, 7943 Angus Court, Springfield, Virginia 22153. You may also have forms mailed to you by submitting an on-line request at www.atf.gov/dcof/index.htm, or by telephoning your request to 703-455-7801. [27 CFR 555.21(b)]

67. Does a purchaser of black powder have to sign any forms at the time of purchase?

If 50 pounds or less of commercially manufactured black powder is being purchased, and the powder is intended to be used solely for sporting, recreational, or cultural purposes in antique firearms as defined in 18 U.S.C. 921(a)(16) or in antique devices exempt from the term "destructive device" in 18 U.S.C. 921(a)(4), no form is required. However, if the black powder is being purchased for any other purpose (regardless of quantity), the purchaser or other transferee must possess a Federal explosives license or permit. [18 U.S.C. 845(a)(5); 18 U.S.C. 926(c); 27 CFR 555.141(b), 555.26(a)]

68. Is there a requirement for licensees and permittees to make an annual inventory of explosive materials on hand?

Yes. An inventory is required to be taken at least once a year. [27 CFR 555.122-.125]

69. When must ATF Form 5400.4, "Limited Permittee Transaction Report (LPTR)", be executed?

Before distribution of explosive materials to a limited permittee, the licensee or permittee must obtain an executed ATF F 5400.4 from the limited permittee with an original unaltered and unexpired Intrastate Purchase of Explosives Coupon (IPEC) attached. Except when delivery of explosive materials is made by a common or contract carrier who is an agent of the limited permittee, the licensee or permittee must verify the identity of the holder of the limited permit by examining an identification document (as defined in 555.11) and noting on the ATF F 5400.4 the type of document presented. The licensee or permittee must complete the appropriate section on ATF F 5400.4 to indicate the type and quantity of explosive materials distributed, the license or permit number of the seller, and the date of the transaction. The licensee or permittee must sign and date the form. [27 CFR 555.126(b)]

70. Do the ATF Forms 5400.4 have to be maintained by the licensee or permittee making the sale?

Yes. One copy of ATF F 5400.4 must be retained by the seller as part of his permanent records in chronological order by date of disposition, or in alphabetical order by name of limited permittee. They must be maintained for a period of five years. [27 CFR 555.126]

71. May I keep computerized records?

Yes. See ATF Ruling 2007-1.

Storage

72. Who must comply with the storage requirements?

Except for those items and activities given exempt status under 18 U.S.C. 845 (also see 27 CFR 555.141), or exempted under 27 CFR 555.32, Special Explosives Devices, all persons who store explosive materials must store them in conformity with the provisions of Subpart K of the regulations, unless the person or the materials are exempt from regulation. [18 U.S.C. 842(j); 27 CFR 555.29, 555.141, 555.201(a)]

73. What are the classes of explosive materials for storage purposes?

There are 3 classes of explosive materials:

(a) High explosives (for example, dynamite, flash powders, and bulk salutes);

(b) Low explosives (for example, black powder, safety fuses, igniters, igniter cords, fuse lighters, and "display fireworks", except for bulk salutes); and

(c) Blasting agents (for example, ammonium nitrate-fuel oil and certain water gels). [27 CFR 555.202]

74. May a person store explosive materials in a residence or dwelling?

No. Storage of explosive materials in a residence or dwelling is prohibited. [27 CFR 555.208(b), 555.210(b), 555.211(b)]

75. What is the "Table of Distances"?

This table lists the minimum acceptable distances separating explosives magazines from inhabited buildings, passenger railroads, public highways, and other explosives magazines. The table is contained in 27 CFR 555.218.

76. When low and high explosives are stored together, how is the distance determined to meet the table of distance requirements?

The table of distances at 27 CFR 555.218 would be applied using the total weight of explosive materials in the magazine. [27 CFR 555.218]

77. Is it necessary to inspect my explosives magazines on a regular basis?

Yes. Any person storing explosives must inspect the magazines at least once every 7 days to determine whether there has been unauthorized entry or attempted entry into the magazines or unauthorized removal of the contents of the magazines. [27 CFR 555.204]

78. What are the requirements for making changes or additions to an approved storage facility?

Making changes in construction to an approved explosives magazine or adding a magazine requires that ATF be notified. However, mobile or portable type 5 magazines and magazines used for the temporary (under 24 hours) storage of explosive materials are exempt from this requirement. See 27 CFR 555.63 for details.

79. Is any type of black powder fuse exempt from storage requirements?

Yes, 3/32-inch and other external burning pyrotechnic hobby fuses are exempt from the requirements of Federal explosives laws and regulations. [18 U.S.C. 845(a)(4-5); 27 CFR 555.11: definition of "ammunition", 555.141(a)(4), 555.141(b)]

80. With the exception of 3/32-inch pyrotechnic safety fuse for use in small arms, must black powder fuses generally be stored in approved explosives magazines?

Yes. Generally igniter fuses, time fuses, blasting fuses, safety fuses, or other black powder fuses by whatever name known, must be stored in approved magazines.

81. **Is smokeless powder designed for use in nall arms ammunition subject to the explosives orage requirements?**

Smokeless propellants designed for use in small arms ammuni-n are exempt from regulation under 18 U.S.C. Chapter 40 and regulations in 27 CFR Part 555. However, it should be noted it persons engaged in the business of importing or manufactur-g smokeless propellants must have a Federal explosives license. lditionally, smokeless propellant designed for use other than all arms ammunition is not exempt. Therefore, explosives oducts such as squibs, fireworks, theatrical special effects, or ier articles that may be utilizing smokeless propellants are gulated and must be stored accordingly.

82. **My office building, in which several com-ny employees work during the day in connec-n with my explosives business, is located in the neral area of my explosives magazine. Do the gulations and the Table of Distances apply to is building as an "inhabited building"?**

No. A building such as an office building or repair shop which part of the premises of an explosives business and is used by business in connection with the manufacture, transportation, rage, or use of explosive materials is not considered to be an habited building". [27 CFR 555.11: definition of "inhabited ilding", 555.218]

83. **Am I required to notify my State or local thorities about my explosives storage maga-nes?**

Yes. All persons who store explosive materials must notify the e department having jurisdiction over the site where explosive terials are manufactured or stored. Notification must be made lly by the end of the day on which storage begins and in iting within 48 hours from the time storage began. The tification must include the type of explosive materials, gazine capacity, and the location of each storage site. [27 CFR 5.11: Definition of "authority having jurisdiction for fire ety", 27 CFR 555.201(f)]

84. **What is the definition of a "case hardened ackle?"**

Case hardening involves putting carbon (or a combination of bon and nitrogen) into the surface of the steel to make it a h-carbon steel, which can be hardened by heat treatment. Only outer skin gets hard in this manner. The center is still tough l malleable. This makes for a strong lock with a tough surface.

85. **Can detonators be stored with detonating rd?**

No. However, products which are manufactured with a detona-attached to the detonating cord as an integral part need not be assembled and stored separately. [27 CFR 555.213]

86. **Are there storage requirements for oxidiz-ers, such as ammonium nitrate?**

In general, no. However, when a magazine or bin containing ammonium nitrate is located within the sympathetic detonation distance of other explosives or blasting agents, it must be stored in accordance with the table of distances in 27 CFR 555.220.

87. **Are State and local government agencies required to store their explosive materials in conformity with Federal storage regulations?**

Yes. There is no exemption in the law or regulations for the storage of explosive materials by any State or political subdivision thereof. [18 U.S.C. 842(j), 845(a)(6); 27 CFR 555.141(a)(3), (a)(5)]

Fireworks

Fireworks are defined in the Federal explosives regulations as any composition or device designed to produce a visible or an audible effect by combustion, deflagration, or detonation. Fireworks are further divided into two broad classifications, consumer fireworks or display fireworks as defined at 27 CFR Part 555.11.

88. **Are "consumer fireworks" subject to regula-tion under the Federal explosives laws?**

No. The importation, distribution, and storage of fireworks defined as consumer fireworks are exempted from the provisions of the Federal explosives laws. However, because they contain pyrotechnic compositions classed by ATF as explosive materials, the manufacture of consumer fireworks requires a manufacturer's license. In addition, pyrotechnic compositions used in the manu-facture of consumer fireworks must be stored in accordance with regulations in 27 CFR Subpart K. Consumer fireworks are defined as "any small firework device designed to produce visible effects by combustion and which must comply with the construc-tion, chemical composition, and labeling regulations of the U.S. Consumer Product Safety Commission, as set forth in title 16, Code of Federal Regulations, parts 1500 and 1507. Some small devices designed to produce audible effects are included, such as whistling devices, ground devices containing 50 mg or less of explosive materials, and aerial devices containing 130 mg or less of explosive materials. Consumer fireworks are classified as fireworks UN0336 and UN0337 by the U.S. Department of Transportation at 49 CFR 172.101. This term does not include fused set pieces containing components which together exceed 50 mg of salute powder." [27 CFR 555.11: definition of "consumer fireworks"; definition of "licensed manufacturer", 555.141(a)(7)]

89. Are "display fireworks" considered to be explosive materials subject to regulation under Federal explosives laws and regulations?

Yes. Display fireworks include, but are not limited to, salutes containing more than 2 grains (130 mg) of explosive materials, aerial shells containing more than 40 grams of pyrotechnic compositions, and other display pieces which exceed the limits of explosive materials for classification as "consumer fireworks". These fireworks are classified as fireworks UN0333, UN0334, or UN0335 by regulations of the U.S. Department of Transportation at 49 CFR 172.101. Display fireworks also include fused set pieces containing components which together exceed 50 mg of salute powder. [27 CFR 555.11: definition of "display fireworks"]

90. How must display fireworks be stored?

Display fireworks, with the exception of bulk salutes, are considered low explosives and, at a minimum, must be stored in type 4 storage magazines. They may also be stored in type 1 or type 2 magazines. Bulk salutes, which are defined as either salute components prior to final assembly into aerial shells, (or) finished salute shells held separately prior to being packed with other types of display fireworks, are classified as high explosives. As such, bulk salutes may only be stored in type 1 or type 2 magazines specifically constructed for the storage of high explosives. [27 CFR 555.11, 555.202(b), 555.203(d), 555.207, 555.208, 555.210]

91. Are "Articles Pyrotechnic" subject to the requirements of the Federal explosives regulations?

The importation, distribution, and storage of fireworks defined as "Articles Pyrotechnic", are exempt from the Federal explosives laws and regulations. However, because they contain pyrotechnic compositions classed by ATF as explosive materials, the manufacture of items defined as "articles pyrotechnic" requires an ATF manufacturer's license. In addition, pyrotechnic compositions used in the manufacture of articles pyrotechnic must be stored in accordance with regulations in 27 CFR Subpart K. [27 CFR 555.11: definitions of "articles pyrotechnic" and "consumer fireworks", 555.141(a)(7)]

92. Must partially assembled display fireworks be removed from a drying building for overnight storage?

Yes. At the end of a day's manufacturing operations, all dry explosive powders and mixtures and partially assembled and finished display fireworks must be removed from fireworks process buildings and stored in a magazine meeting the storage requirements in 27 CFR Part 555, Subpart K. [27 CFR 555.205, 555.221]

93. What areas of a fireworks manufacturing plant are considered to be "fireworks process buildings?"

Fireworks process buildings include any buildings in which pyrotechnic compositions or explosives materials are mixed, pressed, finished, or assembled. Fireworks process buildings do not include plant warehouses, office buildings, or other buildin[gs] and areas in which no fireworks, pyrotechnic compositions, or explosive materials are processed or stored. [27 CFR 555.11: definition of "fireworks process building"]

94. Under what conditions may I temporarily store display fireworks (including low explosive[s] for choreographed shows) on trucks?

See ATF Ruling 2007-2.

95. What types of fireworks require an ATF license or permit in order to be lawfully transpor[t]ed or received?

Any fireworks defined as "display fireworks" in 27 CFR 555.11 may be lawfully received or transported only by person[s] who hold a valid license or permit. No ATF license or permit is required to receive or transport "consumer fireworks" or "artic[les] pyrotechnic". [18 U.S.C. 842(a)(3); 27 CFR 555.26, 555.141(a)(7)]

Plastic Explosives

96. What is a plastic explosive?

A plastic explosive is defined as "an explosive material in flexible or elastic sheet form formulated with one or more high explosives which in their pure form has a vapor pressure less th[an] 10^{-4} Pa at a temperature of 250 C., is formulated with a binder material, and is as a mixture malleable or flexible at normal ro[om] temperature." [18 U.S.C. 841(q); 27 CFR 555.180(d)(4)]

97. What plastic explosives are required to contain detection agents?

All plastic explosives manufactured or imported on or after April 24, 1996, must contain a detection agent. Federal law enforcement agencies and the military may possess unmarked plastic explosives if they meet the requirements of the use-up period described in Question 103. [18 U.S.C. 841(q), 842(n); 27 CFR 555.180]

98. What are the permissible detection agent[s] for marking plastic explosives?

These agents are listed in the law and regulations at 18 U.S.C. 841(p) and 27 CFR 555.180(d)(3).

99. Is it lawful to manufacture plastic explosives that do not contain a detection agent[?]

No. [18 U.S.C. 842(l); 27 CFR 555.180(a)]

100. Is it lawful to import into the United States astic explosives that do not contain a detection ent?

No. The importation of plastic explosives into the United ates requires that the importer file ATF Form 6 certifying that imported plastic explosives contain the required detection ent, or is exempted from the marking requirements as provided the regulations. [18 U.S.C. 842(m); 27 CFR 555.180(b), 5.182, 555.183]

101. Is it lawful to ship, transport, transfer, ceive, or possess any plastic explosive that es not contain a detection agent?

No. However, a 15-year use-up period is provided for Federal v enforcement agencies and the military for unmarked plastic plosives imported into or manufactured in the U.S. prior to ril 24, 1996. [18 U.S.C. 842(n); 27 CFR 555.180(c)]

102. If a person acquired plastic explosives not ntaining a detection agent before April 24, 1996, ay he or she continue to lawfully possess the plosives?

No. With the exception of the use-up period provided by law Federal law enforcement agencies or the military, the time iod for lawful possession of unmarked plastic explosives minated on April 24, 1999. [18 U.S.C. 842(n); 27 CFR 5.180(c)]

103. Are police departments exempt from the ohibition against possessing unmarked plastic plosives after April 24, 1999?

No. Police departments and other State or local law enforce-nt agencies could lawfully possess unmarked plastic explo-es acquired on or before April 24, 1996, until April 24, 1999. ch agencies still possessing unmarked plastic explosives ould destroy them or abandon them to ATF. Contact the nearest F field office for information. [18 U.S.C. 842(n); 27 CFR 5.180(c)(1)]

U.S. Military Explosives

104. Would an ATF license or permit be needed to demilitarize (demil) U.S. military explosives?

As long as the demil operator has a valid Department of Defense contract to perform such operations, the operations would be exempt from 27 CFR Part 555 and no license or permit would be required. However, if title to the explosive materials has passed from the military to the demil operator and the operator intends to resell the explosives on the commercial market, then such operations may be regulated by ATF (e.g., storage, sales, manufacturing) and an ATF license or permit may be needed. Contact the nearest ATF field office for further information. [18 U.S.C. 845(a)(3), (a)(6) and 27 CFR 555.141(a)(3), (a)(5)]

105. Would a civilian contractor who is manufacturing explosive materials pursuant to a government contract for or on behalf of the United States military be entitled to the exemptions from the explosives laws and regulations?

Yes, provided that all the explosive materials in question are manufactured under a government contract. Any explosive materials manufactured in anticipation of receiving a government contract would not qualify for this exemption.

If the contractor manufactures any explosive materials not pursuant to a U.S. military contract, the manufacture and the explosive materials are subject to all requirements of the law and regulations. [18 U.S.C. 845(a)(3), (a)(6); 27 CFR 555.26, 555.41, 555.141(b)]

106. Is an ATF licensee or permittee, whose licensed premises are located on a U.S. military installation, subject to the regulations in 27 CFR Part 555?

All activities conducted outside the scope of a U.S. Government contract are subject to the requirements of Part 555, even if the activities are conducted on property owned by the military. [18 U.S.C. 845(a)(3), (a)(6); 27 CFR 555.26, 555.29, 555.41, 555.141(a)(3), (a)(5)]

Index to Questions and Answers

ATF Explosives Rulings (Revised 1/07)

Table of Contents

1. 27 CFR 55.11: Meaning of Terms (Also § 55.206)

An office or repair shop used in connection with the manufacture, etc. of explosive materials is not an "inhabited building."

ATF Ruling 75-20

ATF has held that a building, such as an office or repair shop, which is a part of the premises of an explosives manufacturer and is used in connection with the manufacture, transportation, storage, or use of explosive materials, is not an "inhabited building." Section 55.11 of 27 CFR defines inhabited building as "any building regularly occupied in whole or in part as a habitation for human beings, or any church, schoolhouse, railroad station, store, or other structure where people are accustomed to assemble, except any building occupied in connection with the manufacture, transportation, storage, or use of explosive materials."

Regulations in 27 CFR §§ 55.206 and 55.218 set forth provisions concerning the location of storage facilities and the minimum distances such storage facilities may be located from, among other things, "inhabited buildings." These provisions are intended to provide protection to persons who inhabit buildings located near premises where explosives are manufactured, stored, etc. However, it is the intent of § 55.11 to exempt buildings used by the explosives industry in connection with the manufacture, transportation, storage, or use of explosive materials from the table of distance requirements on "inhabited buildings." [75 CB 64]

2. 27 CFR 55.207: Construction of Type 1 Magazines (Also § 55.210)

Certain explosives storage facilities meeting standards of construction prescribed by the Department of Defense Explosives Safety Board for such storage are approved by the Bureau.

ATF Ruling 75-21

ATF has held that explosives storage facilities with smooth-finished concrete floors that were constructed under contract for the use of the Department of Defense (DOD) and that are presently being leased to licensees and permittees for the storage of commercial explosives are considered to be in compliance with the requirements for nonsparking floors, as set forth in 27 CFR §§ 55.207(a)(4), 55.207(b), and 55.210, for the storage of all types of fully packaged explosives, pyrotechnics and propellants, with the exception of black powder. Any other such magazines which have smooth finished concrete floors and which meet or exceed DOD construction specifications will also be considered to be in compliance with the requirements of Part 555 with respect to nonsparking floors. It is the responsibility of the licensee or permittee to provide verification that such facilities were manufactured under DOD specifications or that the facilities meet or exceed such specification standards. If the Division Director determines that the concrete floors of type 1 or type 4 magazines do not meet the preceding requirements, he will require such floors to be covered with a nonsparking material, such as epoxy paint or mastic. [75 CB 67]

3. 27 CFR 55.41: Licenses and Permits-General

Certain companies that manufacture explosive materials for use in their own operations are required to obtain licenses as manufacturers of explosive materials.

ATF Ruling 75-31

ATF has held that companies, such as public utility companies engaged in line and facility construction, which manufacture explosives on a regular or continual basis are considered to be engaged in the business of manufacturing explosive materials and must be appropriately licensed as required by 18 U.S.C. 842. The term "manufacturer" is defined in 18 U.S.C. 841(h) as "any person engaged in the business of manufacturing explosive materials for purposes of sale or distribution or for his own use." Although the term "engaged in the business" is not susceptible to a rigid definition within 18 U.S.C. §§ 841-848, it is interpreted to imply an element of continuity or habitual practice; an element clearly present in the operations of companies described herein. Therefore, these companies are considered to be "engaged in the business" and must be licensed as explosives manufacturers. [75 CB 65]

4. 27 CFR 555.109: Identification of Explosive Materials

Methods of marking containers of explosive materials are prescribed.

ATF Ruling 75-35

Editors note: ATF Ruling 75-35 was rendered obsolete pursuant to ATF 5F, 70 Federal Register 30626 (May 27, 2005), and effective July 26, 2005.

5. 27 CFR 55.11: Meaning of Terms- State of Residence

"State of residence" of business entities who use explosive materials; distribution of explosive materials by licensees to out-of-State business entities other than licensees and permittees; and distribution to nonresident employees of such entities are discussed.

Editor's Note: Provisions of ATF Ruling 76-4 were modified, in part, by the Safe Explosives Act.

Effective May 24, 2003, it is unlawful for any person to receive explosive materials unless such person holds an ATF license of permit. It is also unlawful for any licensee or permittee to knowingly distribute explosive materials to any person who does not hold a license or permit. The only relevance remaining in the term "State of residence" is for distribution of explosive materials to, and receipt by, limited permit holders. Pursuant to 18 U.S.C. 842(a)(3) and (a)(4), limited permit holders may, on not more than 3 separate occasions, lawfully receive explosive materials from a licensee or permittee whose premises are located within the state of residence of the limited permit holder. ATF Rule 76-4 continues to apply in determining whether a limited permit holder has acquired a "State of residence" for purposes of receipt of explosives under 18 U.S.C. 842(a)(4)(B).

6. 27 CFR 55.126: Explosives Transaction Record

Under certain conditions, a single Form 5400.4 may be used to cover a series of deliveries.

Editor's Note: The provisions of ATF Rule 76-10 were rendered obsolete by ATF No. 1, 68 FR 13791, Mar. 20, 2003.

7. 27 CFR 55.207: Construction of Type 1 Storage Facilities (Also § 55.208)

Alternate construction standards for storage facilities for explosive materials are prescribed.

ATF Ruling 76-18

Section 842(j) of 18 U.S.C. states: "It shall be unlawful for any person to store any explosive material in a manner not in conformity with regulations promulgated by the Secretary. In promulgating such regulations, the Secretary shall take into consideration the class, type, and quantity of explosive materials to be stored, as well as the standards of safety and security recognized in the explosives industry." The regulations in 27 CFR §§ 55.207 and 55.208 prescribe types of storage facilities for explosive materials and provide (among other things) that such storage facilities shall be bullet resistant.

Section 55.201(b) provides that alternate storage facilities may be authorized for the storage of explosive materials when it is shown that such alternate facilities are or will be constructed in a manner substantially equivalent to the standards of construction contained in the applicable regulations.

The term "bullet-resistant" means resistant to penetration of a bullet of 150 grain M2 ball ammunition having a nominal muzzle velocity of 2700 feet per second fired from a .30 caliber rifle from a distance of 100 feet perpendicular to the wall or door. It has been determined that a wide range of construction criteria meet the bullet-resistant requirements of regulations for construction of storage facilities for explosive materials. In order to promote standards of safety and security in the storage of explosive materials while allowing the industry a wide latitude in the selection of construction materials, it is held that storage facilities (magazines) that are constructed according to the following minimum specifications are bullet-resistant and meet the uirements of the regulations as set forth in 27 CFR Part 55 ll steel and wood dimensions are actual thicknesses. To meet concrete block and brick dimensions indicated, the manufacers' represented thicknesses may be used).

Exterior of ⅝ inch steel, lined with an interior of any type of nsparking material.

Exterior of ½ inch steel, lined with an interior of not less n ⅜ inch plywood.

Exterior of ⅜ inch steel, lined with an interior of two inches hardwood.

Exterior of ⅜ inch steel, lined with an interior of three hes of softwood or 2¼ inches of plywood.

Exterior of ¼ inch steel, lined with an interior of three hes of hardwood.

Exterior of ¼ inch steel, lined with an interior of five inches softwood or 5 ¼ inches of plywood.

Exterior of ¼ inch steel, lined with an intermediate layer of inches of hardwood and an interior lining of 1½ inches of wood.

Exterior of 3⁄16 inch steel, lined with an interior of four inches hardwood.

Exterior of 3⁄16 inch steel, lined with an interior of seven hes of softwood or 6¾ inches of plywood.

Exterior of 3⁄16 inch steel, lined with an intermediate layer of ee inches of hardwood and an interior lining of ¾ inch of wood.

Exterior of ⅛ inch steel, lined with an interior of five inches hardwood.

Exterior of ⅛ inch steel, lined with an interior of nine inches softwood.

Exterior of ⅛ inch steel, lined with an intermediate layer of r inches of hardwood and an interior lining of ¾ inch wood.

Exterior of any type of fire-resistant material which is ıcturally sound, lined with an intermediate layer of four inches solid concrete block, OR four inches of solid brick OR four hes of solid concrete; AND, an interior lining of ½ inch wood placed securely against the masonry lining.

Standard eight inch concrete block with voids filled with ll-tamped sand/cement mixture.

Standard eight inch solid brick.

Exterior of any type of fire-resistant material which is ıcturally sound, lined with an intermediate six inch space ed with well-tamped dry sand or well-tamped sand/cement xture.

(r) Exterior of 1/8 inch steel, lined with a first intermediate layer of ¾ inch plywood, a second intermediate layer of 3⅝ inches of well-tamped dry sand or sand/cement mixture and an interior lining of ¾ inch plywood.

(s) Exterior of any type of fire-resistant material, lined with a first intermediate layer of ¾ inch plywood, a second intermediate layer of 3⅝ inches well-tamped dry sand or sand/cement mixture, a third intermediate layer of ¾ inch plywood, and a fourth intermediate layer of two inches of hardwood OR 14 gauge steel AND an interior lining of ¾ inch plywood.

(t) Eight inch thick solid concrete. [76 CB 106]

8. 27 CFR 55.213: Quantity and Storage Restrictions(Also § 55.208)

Alternate magazine construction standards for storage of electric blasting caps with other explosive materials are prescribed.

ATF Ruling 77-24

Section 842(j) of 18 U.S.C. states: "It shall be unlawful for any person to store any explosive material in a manner not in conformity with regulations promulgated by the Secretary. In promulgating such regulations, the Secretary shall take into consideration the class, type, and quantity of explosive materials to be stored, as well as the standards of safety and security recognized in the explosives industry." The regulations in 27 CFR § 55.213 restrict the storage of blasting caps with other explosive materials. Section 55.201(b) provides that alternate storage magazines may be authorized for the storage of explosive materials when it is shown that such alternate magazines are or will be constructed in a manner substantially equivalent to the standards of construction contained in the applicable regulations.

ATF recognizes that the transportation and storage of explosive materials in the same

vehicle along with electric blasting caps is often desired. The Institute of Makers of Explosives established a recommended standard for such transport in their Safety Library Publication No. 22, dated November 5, 1971 [revised January 1985]. This standard prescribes the minimum construction criteria for:

(a) A container securely attached—

(1) Above the cab of the vehicle (see Figure 1, Appendix A), and

(2) To the vehicle frame under the cargo space (see Figure 2, Appendix A), or

(b) A built-in compartment in the cargo space of the vehicle (see Appendix B).

In addition to motorized vehicles, consideration was also given for the use of similar criteria on portable wheeled trailers being used as magazines under § 55.208(a) of the regulations (see

Appendix E). In order to insure standards of safety and security in the storage of explosive materials while allowing the industry a proper latitude in the construction of magazines, it is held that vehicles used for transporting and for storing explosive materials that are constructed in conformity with the standards listed below, and in compliance with all other safety and security provisions contained in Part 55 (e.g., effectively immobilized when unattended) will meet the requirements of ATF regulations. Even though constructed on the same vehicle, each compartment will be considered as a separate magazine. The two magazines on the vehicle will, however, be considered as one magazine when applying the American Table of Distances [see Table at § 55.218].

Construction Standards For Storage of Electric Blasting Caps (Non Mass-Detonating)

a. The container or compartment must provide for total enclosure of the electric blasting caps.

b. The partition between the explosives storage compartment and the electric blasting cap compartment must be of laminate construction consisting of A/C grade or better exterior plywood, gypsum board [sheetrock] and low carbon steel plates. In order of arrangement, the laminate must conform to the following, with minimum thickness of each lamination as indicated:
½ inch plywood
½ inch gypsum board [sheetrock],
⅛ inch low carbon steel, and,
¼ inch plywood, with the ¼ inch plywood facing the explosives storage compartment. See Appendix C for details of laminate construction. The door to the electric blasting cap compartment must be of metal construction or solid wood covered with metal; the outside walls and top must be of the same construction as the rest of the vehicle or trailer. If high explosives or bullet sensitive explosive materials are stored in the vehicle, then the storage compartment of the vehicle must be constructed so as to be bullet-resistant.

c. As an alternative to the construction requirements shown in paragraph b, a container for use only as illustrated in Appendix A may be used when constructed as follows:

1. The top, lid or door, and the sides and bottom of each container must be of laminate construction consisting of A/C grade or better exterior plywood, solid hardwood, gypsum board [sheetrock], and sheet metal. In order of arrangement, the laminate must conform to the following, with minimum thickness of each lamination as indicated:
¼ inch plywood,
1 inch solid hardwood,
½ inch plywood,
½ inch gypsum board [sheetrock] (OR ¼ inch particle board), and 22 gauge sheet metal, constructed inside to outside in that order.
See Appendix D for details of laminate construction.

2. The hardwood must be fastened together with wood screws, the ½ inch plywood must be fastened to the hardwood with wood screws, the inner ¼ inch plywood must be fastened to the hardwood with adhesive, and the 22 gauge sheet metal must be attached to the exterior of container with screws.

d. The laminate composite material must be securely bound together by waterproof adhesive or other equally effective means.

e. The steel plates at the joints of laminations must be secure by continuous fillet welds.

f. All interior surfaces of the container or compartment must constructed so as to prevent contact of contents with any sparking metal.

g. There must be direct access to the container or into a compartment from outside the vehicle.

h. Each container or compartment must have a snug fitting continuous piano-type hinged lid or door equipped with a locking device (or devices).

i. Without permitting direct access to contents under normal conditions, the locking or hinging mechanisms must perm at least one edge of the lid or door to rise or move outward least ½ inch when subjected to internal pressure.

j. The exterior of the container or compartment must be weather-resistant. [77 CB 191]

9. 18 U.S.C. 842(j): Storage of Explosives

27 CFR 55.208(b)(1), 55.210(b)(1), and 55.211(b)(1): Indoor Storage of Explosives in a Residence or Dwelling

ATF will approve variances to store explosives a residence or dwelling only upon certain conditions including, but not limited to, receipt of a certification of compliance with State and local law, and documentation that local fire safety officials have received a copy of the certificatior

ATF Ruling 2002-3

Section 842(j) of 18 U.S.C. states: "It shall be unlawful for a person to store any explosive material in a manner not in confo mity with regulations promulgated by the Secretary."

The regulations in 27 CFR 55.208(b)(1), 55.210(b)(1), and 55.211(b)(1) specify that no indoor magazine is to be located i residence or dwelling. Section 55.22 specifies that the Directo may allow alternate methods or procedures in lieu of a method procedure specifically prescribed in the regulations. Specifical section 55.22(a)(3) provides that such "variances" are permiss only in certain circumstances, including where "[t]he alternate method or procedure will not be contrary to any provision of la and will not . . . hinder the effective administration of this part

ATF has been advised that certain variances previously proved for storage of explosives in residences or dwellings are violation of State or local zoning law. ATF believes it is portant to ensure that approval of variances is in compliance th all State and local provisions.

To obtain a variance for indoor storage of explosives in a sidence or dwelling, ATF has determined that a person must bmit to ATF a certification signed under penalty of perjury ng with the request for the variance. The certification must:

1. State that the proposed alternative storage method will mply with all applicable State and local law;
2. Provide the name, title, address, and phone number of authority having jurisdiction for fire safety of the locality in ich the explosive materials are being stored; and,
3. Demonstrate that the person has mailed or delivered the rtification to the authority identified in (2).

When required by the Director, such persons must furnish her documentation as may be necessary to determine whether a riance should be approved.

Held, ATF will approve variances to store explosives in a sidence or dwelling only upon certain conditions including, but t limited to, receipt of a certification of compliance with State d local law, and documentation that local fire safety officials ve received a copy of the certification.

Date signed: August 23, 2002

. 18 U.S.C. 842(j): Storage of Explosives 27 FR 55.208(b)(1), 55.210(b)(1), and 55.211(b)(1): door Storage of Explosives in Business remises Directly Adjacent to a Residence or welling

ATF Ruling 2002-4

ATF requires approval of variances for indoor storage of plosives in business premises directly adjacent to a residence or elling.

Section 842(j) of 18 U.S.C. states: "It shall be unlawful for any rson to store any explosive material in a manner not in conforty with regulations promulgated by the Secretary."

The regulations in 27 CFR 55.208(b)(1), 55.210(b)(1), and .211(b)(1) specify that no indoor magazine is to be located in a sidence or dwelling. Section 55.22 specifies that the Director y allow alternate methods or procedures in lieu of a method or ocedure specifically prescribed in the regulations. Specifically, ction 55.22(a)(3) provides that such "variances" are permissible ly in certain circumstances, including where "[t]he alternate thod or procedure will not be contrary to any provision of law d will not … hinder the effective administration of this part."

ATF has been asked whether businesses that are directly acent to living quarters may lawfully store explosive materials the business premises. The issue presented is whether the emises amount to a "residence or dwelling" within the meaning the regulations cited above.

Even where the business premises are segregable from the living quarters by the existence of a door or a common wall, the business premises retain their character as a residence or dwelling. Accordingly, indoor storage of explosives in such premises is generally prohibited and can be allowed only pursuant to an approved variance.

Held, ATF requires approval of variances for indoor storage of explosives in business premises directly adjacent to a residence or dwelling. ATF may approve such variances upon receipt of all appropriate certification and other documentation as may be requested.

Date signed: August 23, 2002

11. 18 U.S.C. 842(f): Unlawful Acts, 27 CFR 555.105(b)(6)(iii): Distribution of Explosives to Limited Permittees

Distributors distributing explosive materials to holders of limited permits via common or contract carrier may verify receipt of the explosive materials by telephone, facsimile, e-mail, or other means within three business days of shipment in lieu of requiring the common or contract carrier to verify the identity of the person accepting delivery of the explosives. The distributor shall make a notation on ATF Form 5400.4 indicating whether the shipment was received and the date and time of the contact with the distributee.

ATF Ruling 2003-5

The Bureau of Alcohol, Tobacco, Firearms and Explosives (ATF) has received questions from the explosives industry regarding the requirement under 27 CFR 555.105(b)(6)(iii) that, effective May 24, 2003, a common or contract carrier hired by a Federal explosives licensee or permittee verify the identity of the distributee, note the type and number of the identification document, and provide this information to the distributor. The distributor is required to record this information on ATF Form 5400.4, Limited Permittee Transaction Report (LPTR). Industry members have informed ATF that this requirement places an undue burden on common and contract carriers. Drivers are concerned that verifying the identity of persons accepting delivery of explosive materials by examining an identification document and providing the identification information to the distributor will be overly time consuming. Drivers are also concerned that they could be held personally liable for delivering explosives to persons not authorized to receive them.

ATF imposed the verification requirements of section 555.105(b)(6)(iii) to ensure that when explosive materials are sold by a distributor to a holder of a limited permit and transported by a common or contract carrier hired by the distributor, the explosive materials are delivered only to a person authorized to receive them. ATF continues to believe it is important that sellers of explosive materials verify that such materials are delivered to persons authorized to receive them.

However, it was not ATF's intention to impose an undue burden on common or contract carriers.

Section 555.22, Title 27, CFR, provides that the Director may approve an alternate method or procedure in lieu of a method or procedure specifically prescribed in the regulations when he finds that:

(1) Good cause is shown for the use of the alternate method or procedure;

(2) The alternate method or procedure is within the purpose of, and consistent with the effect intended by, the specifically prescribed method or procedure and that the alternate method or procedure is substantially equivalent to that specifically prescribed method or procedure; and

(3) The alternate method or procedure will not be contrary to any provision of law and will not result in an increase in cost to the Government or hinder the effective administration of Part 555.

ATF finds that there is good cause to authorize a variance to the provisions of section 555.105(b)(6)(iii) due to the undue burden placed on common or contract carriers by the verification requirement. Accordingly, ATF authorizes the following alternate method or procedure to the identification verification requirements of section 555.105(b)(6)(iii):

The distributor shall, no later than three business days after shipment of the explosive materials, contact the distributee by telephone, facsimile, e-mail or any other means to ensure that the shipment has been received. The distributor shall make a notation on ATF Form 5400.4 indicating whether the shipment was received and the date and time of the contact with the distributee.

ATF finds that the above alternate method is consistent with the verification provisions of section 555.105(b)(6)(iii), because it will ensure that delivery has taken place and document the information in the distributor's records.

The alternate method is not contrary to any provision of law, will not increase the costs to ATF, and will not hinder the effective administration of the regulations in 27 CFR Part 555.

Held, pursuant to 27 CFR 555.22, ATF authorizes a variance from the requirements of 27 CFR 555.105(b)(6)(iii) for Federal explosives licensees and permittees making distributions of explosive materials to holders of limited permits via common or contract carrier. As an alternate method or procedure, distributors distributing explosive materials to holders of limited permits via common or contract carrier may verify receipt of the explosive materials by telephone, facsimile, e-mail or other means within three business days of shipment in lieu of requiring the common or contract carrier to verify the identity of the person accepting delivery of the explosives. The distributor shall make a notation on ATF Form 5400.4 indicating whether the shipment was received and the date and time of the contact with the distributee.

Date signed: May 23, 2003.

12. 27 CFR 555.11: Meaning of Terms

ATF provides guidance on three different priva roads and whether they are "highways" as defin in 27 CFR 555.11.

ATF Ruling 2005-2

The Federal explosives laws, 18 U.S.C. Chapter 40, require a persons to store explosive materials in a manner in conformity with regulations issued by the Attorney General. 18 U.S.C. 842 The Attorney General has delegated his authority to administer and enforce the Federal explosives laws to the Director, ATF. 28 CFR 0.130. Regulations in 27 CFR Part 555 implement the provisions of the Federal explosives laws. Storage regulations in 27 CFR Part 555, Subpart K, provide that outdoor magazines in which high explosives are stored must be located no closer to inhabited buildings, passenger railways, public highways, or ot magazines in which high explosives are stored than the minimu distances specified in the table of distances for storage of explosive materials in section 555.218 of the regulations. 27 CF 555.206.

Section 555.11 of the regulations defines the term "highway" as "any public street, public alley, or public road, including a privately financed, constructed, or maintained road that is regularly and openly traveled by the general public."

In Scenario A, a private road owned by a corporation is used the general public as an access road to a parking lot owned by corporation. The road is near an explosives magazine. The road does not have a gate, sign, or other means of restricting access the road. The road is also used by the general public on a daily basis to gain access to other public streets.

In Scenario B, a company that manufactures display firework a logging company, and an individual who owns buildings utili to store his collection of automobiles all occupy property to which the only access is a privately owned road. A separate par that leases to these three entities owns the property. The road is located on private property, and a locked gate at the entrance to the road prevents access by the general public. The display fireworks company, the logging company, and the individual storing automobiles all have keys to unlock the gate and travel the road when needed. The gate is locked at all times, and there no evidence that the road is open to anyone other than the two businesses and one individual who require access to their property.

In Scenario C, an explosives company maintains explosives magazines in a quarry area that has a roadway traversing throu the quarry. The quarry owns the property, and the road is maintained by the quarry. The road has a gate and there are signs advising no trespassing. However, when ATF officials visited th location on several occasions, the gate was left open and memb of the public regularly utilized the roadway as a shortcut betwe two major highways. There were no indications the owner of the property took any steps to prevent members of the public from utilizing the roadway.

Applying the regulatory definition of "highway" to the three scenarios, the road in Scenario A is clearly a highway that is subject to the tables of distance in Part 555. Although it is privately owned, it is regularly and openly traveled by members of the general public without restriction.

The roadway in Scenario B is not a "highway" as defined. Access is restricted at all times and there is no evidence the general public regularly travels on the roadway. Access to the road is provided to only a limited number of persons who have a legal right to travel the road. Accordingly, this road is not regularly and openly traveled by members of the general public.

ATF concludes that the roadway described in Scenario C is a "highway" as defined in 27 CFR 555.11. Although access to the roadway is restricted by a gate and "No trespassing" signs are posted, the gate is not closed at all times. Furthermore, ATF observation indicates that the roadway is regularly and openly traveled by members of the general public. Based on these facts, the roadway is a highway which is subject to the tables of distance in 27 CFR Part 555.

Held, a private road with no gate, signs, or other means of restricting access that is used by the general public as an access road to a parking lot and as access to other public streets is a "highway" as defined in 27 CFR 555.11.

Held further, a private road with a locked gate at the entrance that is locked at all times and used by a limited number of persons leasing or owning property accessed by the road is not a "highway" as defined in 27 CFR 555.11.

Held further, a private roadway traversing a quarry with a gate restricting access and a "no trespassing" sign is a "highway," as defined in 27 CFR 555.11, because the gate is not locked at all times and the general public regularly utilizes the roadway as a shortcut between two public highways.

Date approved: September 8, 2005

13. 27 CFR 555.11: Meaning of Terms

ATF provides guidance on two situations involving structures and whether they are "inhabited buildings" as defined in 27 CFR Part 555.

ATF Ruling 2005-3

The Federal explosives laws, 18 U.S.C. Chapter 40, require all persons to store explosive materials in a manner in conformity with regulations issued by the Attorney General. 18 U.S.C. 842(j). The Attorney General has delegated his authority to administer and enforce the Federal explosives laws to the Director, ATF. 28 CFR 0.130. Regulations in 27 CFR Part 555 implement the provisions of the Federal explosives laws. Storage regulations in 27 CFR Part 555, Subpart K, provide that outdoor magazines in which high explosives are stored must be located no closer to inhabited buildings, passenger railways, public highways, or other magazines in which high explosives are stored than the minimum distances specified in the table of distances for storage of explosive materials in section 555.218 of the regulations. 27 CFR 555.206.

The regulation at 27 CFR 555.11 defines the term "inhabited building" as "any building regularly occupied in whole or in part as a habitation for human beings, or any church, schoolhouse, railroad station, store, or other structure where people are accustomed to assemble, except any building occupied in connection with the manufacture, transportation, storage, or use of explosive materials."

In Scenario A, an explosives licensee leases explosives magazines to an individual who uses the magazines for storage of goods other than explosive materials. The magazines are located adjacent to magazines used by the explosives licensee for the storage of explosive materials. The magazines used by the lessee are not separated by the minimum distances required for the separation of magazines from "inhabited buildings" as required by the regulations in 27 CFR Part 555.

The magazines are visited regularly by the individual who stores property in the magazines, but no additional persons accompany the individual when he is present at the magazine. However, the individual hires contractors to repair equipment stored in the facilities from time to time, and 1-3 employees of the contractor may occasionally be present for short periods of time at the storage site. However, such visits occur no more than 3-5 times per year. In addition, an employee from the water company visits the storage site once a month to read the water meter, an employee from the power company reads the power meter once a month, and other vendors may be present at the site for short periods of time for other legitimate purposes.

In Scenario B, a licensed manufacturer of explosives X leases a unit in an industrial park that shares a common wall with a unit leased by licensed manufacturer Y. Both licensees store explosives in magazines located inside and outside the units. The magazines of Manufacturer X and the building used by Manufacturer Y are not separated by the minimum distance prescribed in 27 CFR Part 555 for the separation of magazines and inhabited buildings. Likewise, the magazines of Manufacturer Y and the building used by Manufacturer X are not separated by the minimum distance prescribed in 27 CFR Part 555 for the separation of magazines and inhabited buildings. In both structures, employees and contractors are regularly present during work hours for purposes of carrying on the manufacturing and distributing businesses of the two licensees. This includes personnel who work in the manufacturing plant, those who work on the loading dock to load and ship explosives products to customers, and those who work in the office taking orders, sending out invoices, and handling other clerical work for the businesses.

Applying the law and regulations to the facts of Scenario A, ATF concludes that the leased structures used by the individual to store items other than explosives are not being used as a habitation for human beings and are not buildings occupied in connection with the manufacture, transportation, storage, or use of explosive materials. Accordingly, the sole issue remaining is whether the structure is one where people are accustomed to assemble.

Noteworthy, the regulation uses the term "people," which is the plural version of "person." Thus, ATF believes that more than one person must "assemble" at the structure for it to be an "inhabited building." In addition, the word "assemble" is defined, in part, as "To bring or gather together in a group or whole." The word "assembly" is defined, in part, as "A group of persons gathered for a common purpose." The American Heritage Dictionary, Second College Edition, Houghton Mifflin Co., 1982. It is clear that the presence of one person at a structure or location cannot be an assembly of any sort. Accordingly, in situations where one person is present at a particular structure, whether on a regular or infrequent basis, such a structure is not an "inhabited building" as defined in 27 CFR 555.11.

Likewise, occasional visits to the storage facility by mail delivery persons or employees of public utility companies for brief periods of time would not be an "assembly" that would make the facility an inhabited building. However, if 2 or more repair persons are present at the facility to make repairs to equipment stored there, such persons would be there for a common purpose, and would have "assembled" at the structure. However, the structure would be an "inhabited building" only if it is a structure where people are accustomed to assemble. The word "accustom" is defined as "To familiarize, as by constant practice, use, or habit: accustomed himself to working long hours." The word "accustomed" is defined as "Usual, characteristic, or normal: worked with her accustomed thoroughness." The American Heritage Dictionary, Second College Edition, Houghton Mifflin Co., 1982. These definitions indicate that a structure will be one where people are accustomed to assemble only if there is some degree of continuity, regularity, or frequency to such assembly.

Infrequent, occasional visits to the storage site by 2 or more repair persons would not make the storage facility an "inhabited building," because such intermittent visits would not be "customary." Only where 2 or more persons are present at the site for a common purpose and on a regular basis would the building fit within the definition of "inhabited building."

To address Scenario B, it is apparent that the building leased by Manufacturer X is exempt from the definition of "inhabited building" as to the magazines of Manufacturer X, and the building leased by Manufacturer Y is likewise exempt as to the magazines of Manufacturer Y. This is because both buildings are occupied in connection with the manufacture, transportation, and storage of explosive materials. Clearly, the employees of both licensees are aware that explosive materials are present on the premises and they assume the risk of any such operation. A more difficult question is presented by the buildings of Manufacturer X and the magazines of Manufacturer Y and vice versa. ATF cannot assume that all employees are cognizant of the activities of their neighbors in the industrial park. Thus, it cannot be assumed that the employees are knowingly assuming the risk of explosive materials stored in magazines owned by the other licensee.

Given the plain language of the regulation, however, ATF does not believe it is appropriate to deny the coverage of the regulatory exemption to adjoining licensees on the basis of an assumption of the risk analysis. The current regulatory definition excludes from the definition of "inhabited building" any building occupied in connection with the manufacture, transportation, storage, or use of explosive materials, regardless of the knowledge of the building's occupants. Accordingly, ATF concludes that the industrial units occupied by Manufacturer X and Manufacturer Y are both exempted from the definition of "inhabited building" as to the magazines of each other as well as to their own magazines.

Held, a structure used to store items other than explosive materials that is visited on a regular basis by one individual is not an "inhabited building" as defined in 27 CFR 555.11, because it is not a structure where people are accustomed to assemble. Where 2 or more persons are present at the structure to repair equipment stored therein and such visits to the site are occasional and infrequent, the structure is not an "inhabited building" because the visits are not "customary." However, where 2 or more persons make regular visits to the structure for a common purpose, the structure is an "inhabited building," and explosives magazines may not be stored closer to the structure than the minimum distances specified in the regulations in 27 CFR Part 555.

Held further, buildings occupied by licensed explosives manufacturers in connection with the manufacture, transportation, storage, or use of explosive materials are not included within the definition of "inhabited building" as to magazines located on their own premises. In addition, buildings occupied by licensed explosives manufacturers in connection with the manufacture, transportation, storage, or use of explosives are not included within the definition of "inhabited building" as to magazines located on property owned by another licensee.

Date approved: November 25, 2005

4. 18 U.S.C. 842(f): Records Required for Explosives Licensees and Permittees

27 CFR 555, Subpart G: Records and Reports

27 CFR 555.22: Alternate Methods or Procedures; Emergency Variations from Requirements

Under specified conditions, approval is granted to utilize computerized records as required records under 27 CFR 555, Subpart G.

ATF Ruling 2007-1

Section 842(f), Title 18, United States Code, makes it unlawful for any licensee or permittee to willfully manufacture, import, purchase, distribute, or receive explosive materials without making such records as the Attorney General may by regulation require, including, but not limited to, a statement of intended use, the name, date, place of birth, social security number or taxpayer identification number, and place of residence of any natural person to whom explosive materials are distributed.

Regulations implementing section 842(f) are in 27 CFR Part 555, Subpart G. The regulations in this subpart specify the records required to be created and maintained by licensed importers (section 555.122), licensed manufacturers (section 555.123), licensed dealers (section 555.124), and permittees (section 555.125). The regulation in section 555.121 provides that licensees and permittees must keep records pertaining to explosive materials in permanent form (i.e., commercial invoices, record books) and in the manner required in Subpart G. In addition, sections 555.122-555.125 specifically allow licensees and permittees to use an alternate record to record the distribution of explosive materials when it is shown that the alternate records would accurately and readily disclose the information required by the regulations. These regulations require licensees and permittees who propose to use alternate records to submit a letter application to ATF describing the proposed alternate records and the need for them. Alternate records are not to be employed until approval from ATF is received.

Regulations at 27 CFR 555.22 allow for the approval and use of an alternate method or procedure in lieu of a method or procedure specifically prescribed in Part 555. ATF may approve an alternate method or procedure when it is found that—

(1) Good cause is shown for the use of the alternate method or procedure;

(2) The alternate method or procedure is within the purpose and consistent with the effect intended by, the specifically prescribed method or procedure and that the alternate method or procedure is substantially equivalent to that specifically prescribed method or procedure; and

(3) The alternate method or procedure will not be contrary to any provision of law and will not result in an increase in cost to the Government or hinder the effective administration of 27 CFR Part 555.

With advances in technology and the dramatic decrease in the cost of computers, many businesses rely upon computers to maintain records of their inventory, sales, customer lists, and other business information. Even the smallest home-based business utilizes computers to record and maintain business information. Creating and maintaining records in a computer database, rather than paper form, makes it easier to ensure accuracy of records and makes it less likely that records will be lost or misplaced. In addition, maintaining records via computer generally saves time and money in bookkeeping and auditing expenses. This utilization of computers has allowed companies to automate inventories, utilizing technology such as bar codes or RFID (radio frequency identification) chips. This facilitates better accountability of product overall, reducing the potential of everyday accounting errors. Over the years ATF has seen a significant increase in the number of requests from explosives licensees and permittees for authorization to utilize computerized records rather than paper records of acquisition and distribution and other required records, such as magazine transaction records. ATF routinely approves requests to utilize computerized records, with certain conditions, finding that the use of such records is substantially equivalent to methods of record keeping set forth in the regulations in 27 CFR Part 555, Subpart G.

Several explosives industry members have asked whether computerized records may be maintained without obtaining written approval from ATF if they contain all the required information specified in the regulations and are maintained in a permanent form. Additionally, industry members have questioned whether computer records in combination with paper records may be maintained if they are permanent and contain all the information required by the regulations.

ATF has determined that records of acquisition and disposition, magazine summary records, and the other records required by 27 CFR Part 555, Subpart G, satisfy the standard of permanency and are substantially equivalent to paper records if they meet the following criteria:

1. All data entered into the computer system must be recorded into the database and cannot be capable of being edited or modified at a later date. The software system must retain any correction of errors as an entirely new entry, without deleting or modifying the original entry. The system may allow for entries in a notes column to explain any correction.

2. The system must have a reliable daily memory backup capability to protect the data from accidental deletion or other system failure.

It is also acceptable for licensees/permittees to maintain required records using a combination of a computer program, commercial invoices, and other documents, provided that all of the required information is maintained in the records in permanent form. Any use of a computer for any portion of the required records must comply with the standards outlined above. However, each particular transaction must be self-contained with all the required information in the same recordkeeping medium. As one example, dispositions of explosives by a dealer cannot be separated by keeping the dates of disposition and the manufacturer's name or brand name in the computer, and all the other required information for that disposition on separate written documents.

ATF finds that good cause exists for authorizing the use of a computer to create and maintain the records required by 27 CFR Part 555, Subpart G, as the use of computers is accepted throughout the business community as a reliable, cost-efficient means of maintaining business records. ATF also finds that the use of a computer to maintain required records, contingent upon the requirements outlined above, is consistent with the effect intended by the requirements of Subpart G, as it will result in a permanent, reliable record that will accurately indicate acquisitions and dispositions of explosive materials.

Finally, ATF finds that the use of computer records properly containing all the required information should not hinder the effective administration of the Federal explosives laws or regulations – use of such records generally makes it easier for ATF to conduct inventories of product on hand and to audit required records. Accordingly, ATF concludes that the requirements for approval of an alternate method or procedure in accordance with 27 CFR Part 555, sections 555.22 and 555.122-555.125, are met.

Held, persons holding licenses and permits issued under 18 U.S.C., Chapter 40, may use computers to create and maintain all or any portion of the records required by 18 U.S.C. 842(f) and 27 CFR Part 555, Subpart G, if the following conditions are satisfied:

1. All data entered into the computer system must be recorded into the database and cannot be capable of being edited or modified at a later date. The software system must retain any correction of errors as an entirely new entry, without deleting or modifying the original entry. The system may allow for entries in a notes column to explain any correction.

2. The system must have a reliable daily memory backup capability to protect the data from accidental deletion or other system failure.

Held further, licensees and permittees who wish to use computers to create and maintain all or a portion of their required records in accordance with the requirements set forth in this ruling are not required to obtain advance approval in accordance with 27 CFR 555.22 or 555.121-125.

Held further, licensees and permittees utilizing a combination of a computer program, commercial invoices, and other paper documents as required records must ensure that the required information for a particular transaction is fully contained in the same recordkeeping medium.

Held further, this ATF approved alternate method or procedu for computerized records shall not be withdrawn unless the ho of said variance is so advised by ATF in writing or no longer holds a Federal explosives license or permit.

Date approved: January 18, 2007

15. 18 U.S.C. 842(j): Storage of Explosives

27 CFR 555.210: Construction of Type 4 Magazines

27 CFR 555.215: Housekeeping

27 CFR 555.201: Notification of Local Fire Officials

27 CFR 555.63: Explosives Magazine Changes

27 CFR 555.22: Alternative Methods or Procedures; Emergency Variations from Requirements

Under specified conditions, display fireworks may be temporarily stored in locked and attend motor vehicles at the explosives magazine site(and at fireworks display site(s) without meeting the locking requirements of 27 CFR 555.210 provided certain additional security measures are i place. Additionally, allowance per 27 CFR 555.21 is made for the fuel tanks containing volatile materials that may be on the temporary storage vehicles. Finally, slight variation is provided for notification requirements to ATF and local fire officials.

ATF Rule. 2007-2

The Federal explosives laws, 18 U.S.C. Chapter 40, require persons to store explosive materials in a manner in conformity with regulations issued by the Attorney General. 18 U.S.C. 84 The Attorney General has delegated the authority to administe and enforce the Federal explosives laws to the Director, ATF. CFR 0.130. Regulations in 27 CFR Part 555, implement the provisions of the Federal explosives laws.

The regulation at 27 CFR 555.210(a) states, in part, "[o]utd magazines are to be fire-resistant, weather-resistant, and theft-resistant." Partly to satisfy the theft-resistant requirement, this section requires that each door be equipped with two mortise locks; two padlocks fastened in separate hasps and staples; a combination of a mortise lock and a padlock; a mortise lock t requires two keys to open; or a three-point lock. Padlocks mus have at least five tumblers and a case-hardened shackle of at l

-inch diameter. Padlocks must be protected with not less than -inch steel hoods constructed so as to prevent sawing or lever tion on the locks, hasps, and staples.

The regulation at 27 CFR 555.215 states, in part, "[v]olatile aterials are to be kept a distance of not less than 50 feet from tdoor magazines."

The regulation at 27 CFR 555.201 requires, in part, that any rson storing explosive materials notify local fire authorities ally before the end of the day on which storage of the explosive aterials began and in writing within 48 hours from the time ch storage began.

The regulation at 27 CFR 555.63 requires that any licensee or rmittee who acquires (adds) a storage magazine must notify 'F at least five business days in advance of using any added plosives storage magazine.

Regulations at 27 CFR 555.22 allow for the approval and use an alternate method or procedure in lieu of a method or ocedure specifically prescribed in Part 555. ATF may approve alternate method or procedure when it is found that—

(1) Good cause is shown for the use of the alternate method procedure;

(2) The alternate method or procedure is within the rpose of, and consistent with the effect intended by, the ecifically prescribed method or procedure and that the alternate ethod or procedure is substantially equivalent to that specifi-lly prescribed method or procedure; and

(3) The alternate method or procedure will not be contrary any provision of law and will not result in an increase in cost to Government or hinder the effective administration of 27 CFR rt 555.

ATF has approved a significant number of variances for nporary storage for a specified amount of time before a display eworks event, as well as during and after the event until the naining explosive materials can be placed back into the propriate storage magazine.

Preparation of display fireworks shows and the transportation explosive materials to numerous show sites often take place er a period of several days. Preparing and temporarily storing fireworks for these shows ordinarily take place on delivery cks and trailers in one storage location where the proprietor eady maintains storage of explosives materials with a high gree of security and safety by complying with the provisions of CFR Part 555.

Many display fireworks shows also take several days to prepare the show site. During preparation and after the show is mpleted, explosive materials frequently must be temporarily red. This is often either extra product that was brought to the ow or misfires that have been maintained and must be returned permanent storage.

Allowing flexibility through alternate methods or procedures specific regulations increases both safety and security at these ow sites. These procedures are needed to increase public safety, as well as facilitate smooth operations for the display fireworks industry. The highest risk of incidents involving the accidental ignition of display fireworks is during handling, with the next highest risk being transportation. Providing no flexibility to allow storage in the delivery vehicles would require the industry to dangerously load and unload from storage magazines to vehicles and back into a storage magazine. Additionally, ATF believes that providing this guidance allows for preplanning by the proprietor and consistency of regulatory application nationwide.

One of the major dangers around explosives is fire. Therefore, the regulations require that volatile materials be maintained a distance of not less than 50 feet from outdoor explosives storage magazines. ATF believes that requiring attended storage for display fireworks temporarily stored in vehicles will ensure public safety, in lieu of the 50 foot separation requirement. The attendee should be able to alert the proper authorities if needed to ensure that a fire does not compromise this storage, or may relocate these temporary storage magazines to a safe location away from an identified fire.

Held, ATF will approve alternate methods or procedures for the temporary storage of display fireworks in locked and attended vehicles at explosives magazine site(s), as well as at the fireworks display site(s), under the following conditions:

1. The doors to each storage compartment containing explosive materials must be locked with at least one steel padlock having at least five tumblers and a casehardened shackle of at least ⅜-inch diameter. The padlock does not need to be protected by a steel hood. However, each temporary storage magazine must be attended at all times for security purposes. The vehicle is considered "attended" when an authorized individual is within 100 feet of all temporary storage and has an unobstructed view of the vehicle(s) containing the explosive materials. The individual must remain awake and observant of activities around the vehicle(s).

2. The person who temporarily stores the explosive materials must notify in writing the authority having jurisdiction for fire safety in the locality in which the explosive materials are stored no less than 3 Federal office business days prior to utilizing the additional temporary storage magazine(s).

3. The person who temporarily stores the explosive materials must notify ATF in writing of the location of this storage no less than 3 Federal office business days prior to utilizing the additional temporary storage magazine(s).

All other provisions of 27 CFR Part 555 must be complied with as prescribed.

Held further, this ATF-approved alternate method or procedure for the temporary storage of display fireworks in locked and attended vehicles shall not expire unless the holder of said variance is so advised by ATF or no longer holds a Federal explosives license or permit.

Date approved: January 18, 2007

16. 18 U.S.C. 842(j): Storage of Explosives

27 CFR 555.211(a): Immobilization of Outdoor Type 5 Mobile Storage Magazines

27 CFR 555.215: Housekeeping

27 CFR 555.22: Alternative Methods or Procedures; Emergency Variations from Requirements

Under specified conditions, blasting agents may be stored in mobile type 5 magazines (bulk delivery trucks) without meeting the prescribed immobilization requirements of 27 CFR 555.211.

ATF Rule 2007-3

The Federal explosives laws, 18 U.S.C. Chapter 40, require all persons to store explosive materials in a manner in conformity with regulations issued by the Attorney General. 18 U.S.C. 842(j). The Attorney General has delegated his authority to administer and enforce the Federal explosives laws to the Director, ATF. 28 CFR 0.130. Regulations in 27 CFR

Part 555 implement the provisions of the Federal explosives laws.

The regulation at 27 CFR 555.211(a)(1) states, in part, "[o]utdoor magazines are to be weather-resistant and theft-resistant." This section further states, "[w]hen unattended, vehicular magazines must have wheels removed or otherwise be effectively immobilized by kingpin locking devices or other methods approved by the Director."

The regulation at 27 CFR 555.215 states, in part, "[v]olatile materials are to be kept a distance of not less than 50 feet from outdoor magazines."

The regulation at 27 CFR 555.22 allows for the approval and use of an alternate method or procedure in lieu of a method or procedure specifically prescribed in Part 555. ATF may approve an alternate method or procedure when it is found that—

(1) Good cause is shown for the use of the alternate method or procedure;

(2) The alternate method or procedure is within the purpose of, and consistent with the effect intended by, the specifically prescribed method or procedure and that the alternate method or procedure is substantially equivalent to that specifically prescribed method or procedure; and

(3) The alternate method or procedure will not be contrary to any provision of law and will not result in an increase in cost to the Government or hinder the effective administration of 27 CFR Part 555.

ATF has approved a significant number of variances for an alternate means of immobilizing preloaded bulk delivery vehicles when some additional security measures were put in place.

Bulk delivery vehicles are routinely utilized for on-site delive of blasting services. Often these trucks contain a blasting agent defined under 27 CFR 555.11. Utilization of these bulk produc delivered on-site and used immediately has increased safety an security by reducing the number of remotely located storage trailers containing packaged blasting agent products. These delivery vehicles generally leave the explosives storage locatio at unusual times of day or night, and for safety reasons they are loaded the day before for next day's delivery. These trucks are incapable of being disabled by a kingpin locking device, and th requirement to remove the wheels for immobilization is obviou not feasible. Additionally, there are times when explosive mate remains on the vehicle when it returns from use. Most of these products degrade during handling, therefore, removing them fr the bulk vehicle would not be an option.

One of the major dangers surrounding explosives is fire. Therefore, the regulations at 27

CFR 555.215 require that volatile materials be maintained a distance of not less than 50 feet from outdoor explosive storage magazines. Because these preloaded storage vehicles contain a fuel tank filled with volatile materials, they must remain in an area protected from fire such as gravel, paved, or closely mowe designated parking area.

ATF believes that the following alternate method of operatio substantially equivalent to the prescribed methods. Increased safety for employees and the public provides good cause for thi alternate method. It is not contrary to law and will not result in any increased cost to the Government. Overall, ATF believes allowing for this flexibility assists with the effective administra tion of 27 CFR Part 555.

Held, ATF will approve alternate methods or procedures for t preloading and temporary storage of bulk blasting agents in delivery vehicles at explosive magazine site(s), when the secur and immobilization meets the following criteria:

1. All doors on the vehicle are locked, the ignition key i removed, and the key is secured away from the truck.

2. When the site is not in operation, outer perimeter security is established. This may be by a variety of means such a locked gate, security guards, fence, natural features, or a combination of these.

3. Each potential access point to explosive materials on storage vehicle will be secured with a minimum of one padlock that has at least five tumblers and casehardened shackle of at le ⅜-inch diameter.

4. Each vehicle shall be immobilized through the use of steering wheel locking device, lockable battery disconnect swit or both.

5. All vehicles preloaded with blasting agents shall be parked in a company designated area not susceptible to fire propagation such as bare dirt, gravel, rock, paving, or closely mowed parking lot.

All other provisions of 27 CFR Part 555 must be complied with as prescribed.

Held further, licensees and permittees who wish to use the alternate method or procedure set forth in this ruling are not required to obtain advance approval in accordance with 27 CFR 555.22.

Held further, this ATF-approved alternate method or procedure for the temporary storage of bulk blasting agent products in locked and properly secured vehicles shall not be withdrawn unless the holder of said variance is so advised by ATF in writing or no longer holds a Federal explosives license or permit.

Date approved: January 26, 2007

APPENDIX A

PERMANENTLY MOUNTED CONTAINERS

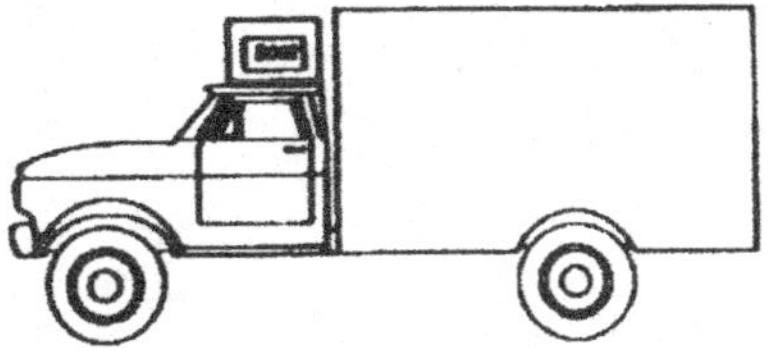

Figure 1

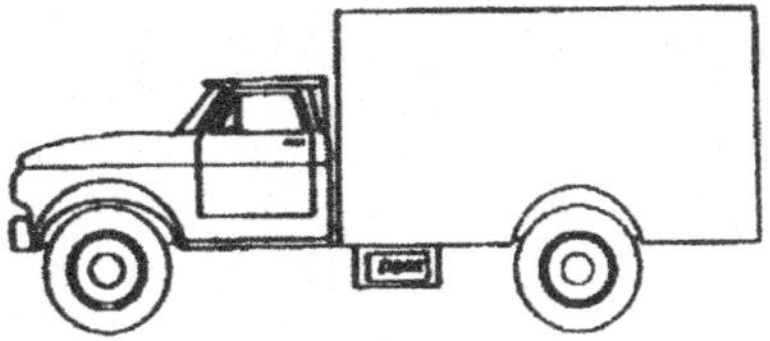

Figure 2

NOTE: The configurations shown in Figures 1 and 2 are equally applicable to multi-axle and "cab-over" vehicles.

[Diagrams: Courtesy of IME]

APPENDIX B

COMPARTMENTS

Figure 1

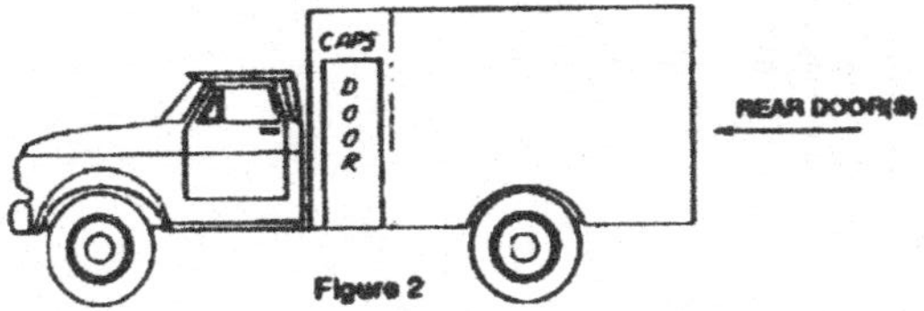

Figure 2

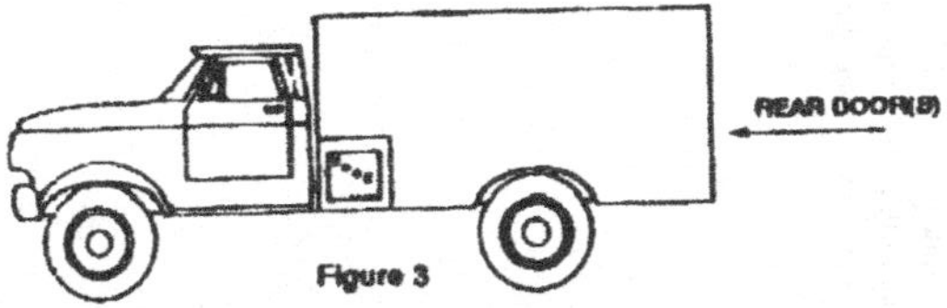

Figure 3

NOTE: The configurations shown in Figures 1 and 2 are equally applicable to multi-axle and "cab-over" vehicles.

[Diagrams: Courtesy of IME]

ATF Rul. 77–24, Appendix A and B

APPENDIX C

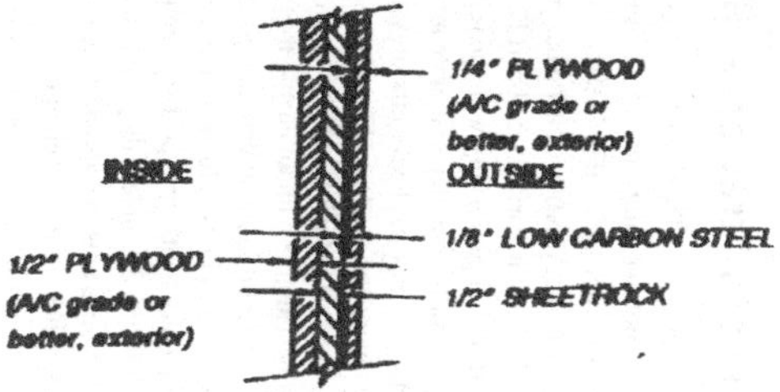

Sketch of laminate construction for container or compartment for electric blasting caps use, as illustrated in Appendix A, B, and E.

APPENDIX D

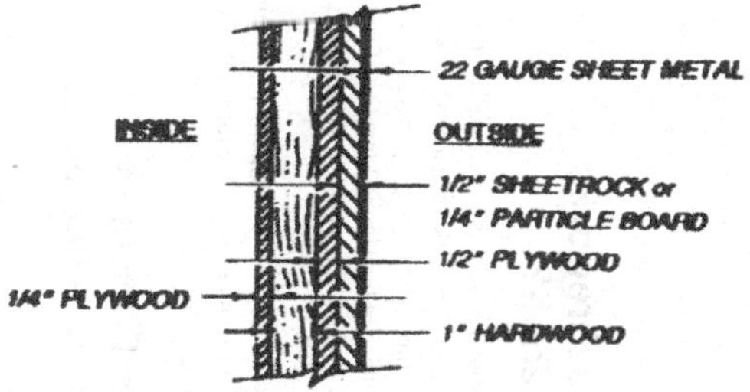

Sketch of laminate construction for container or compartment for electric blasting caps; restricted to use as illustrated in Appendix A.

[Diagrams: Courtesy of IME]

APPENDIX E

PORTABLE WHEELED TRAILERS

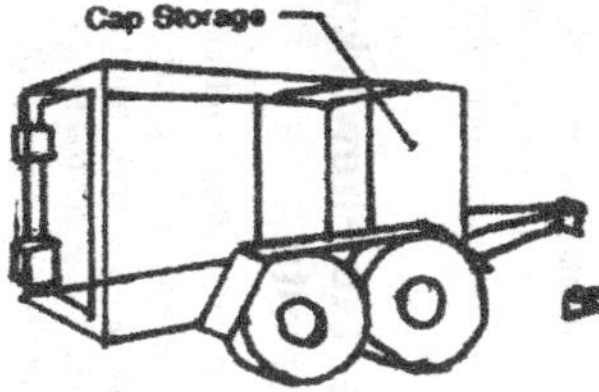

Figure 1

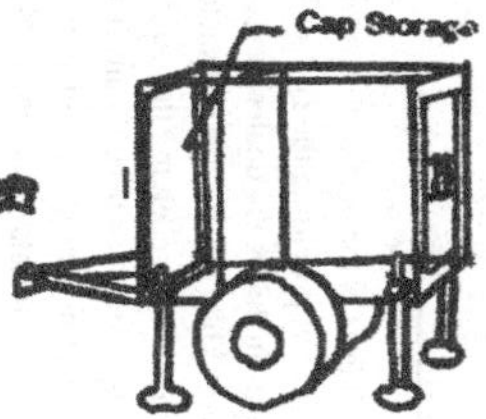

Figure 3

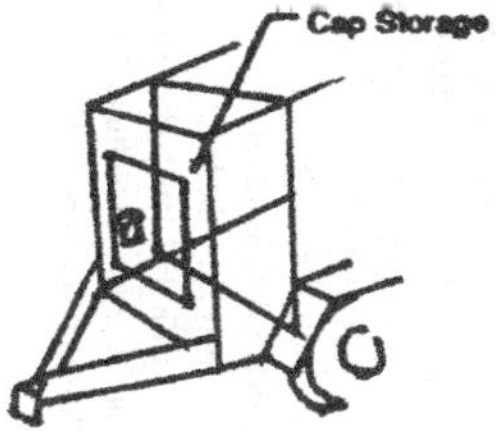

Figure 2

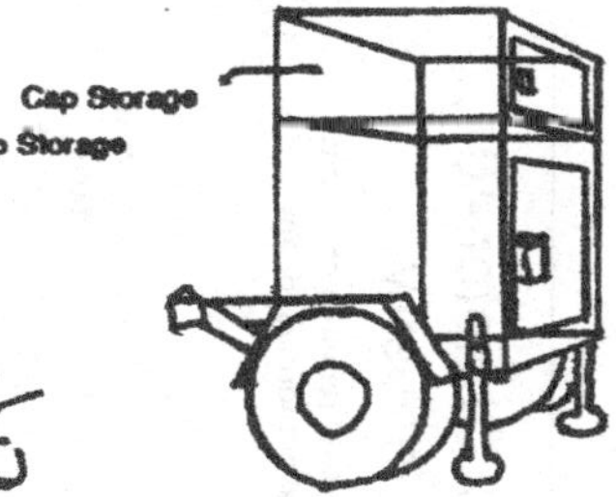

Figure 4

ATF Rul. 77–24, Appendix C – E

General Information (Revised 8/06)

Effect of 18 U.S.C. Chapter 40 On the Fireworks Industry

[Caution! This item discusses Federal requirements only. Please contact your State or local authorities for any additional requirements.]

Title XI of the Organized Crime Control Act of 1970 (18 U.S.C. Chapter 40) establishes controls over explosive materials, including black powder and other pyrotechnic compositions commonly used in fireworks. Part 555 of Title 27, Code of Federal Regulations (CFR), contains the regulations which implement Title XI. Section 555.141(a)(7) exempts "the importation, distribution, and storage of fireworks classified as UN0336, UN0337, UN0431, or UN0432 explosives by the U.S. Department of Transportation at 49 CFR 172.101 and generally known as 'consumer fireworks' or 'articles pyrotechnic.'" Section 555.141(a)(7) does not exempt "display fireworks," as defined in 555.11.

With Respect to Fireworks: Who needs a license?

1. Manufacturers of black powder;
2. Manufacturers of any other explosive material used in manufacturing consumer fireworks or display fireworks; and
3. Importers of, or dealers in, display fireworks.

With Respect to Fireworks: Who needs a permit?

1. All persons transporting, shipping, causing to be transported, or receiving display fireworks, regardless of whether for their own use or for commercial display purposes (Certain exemptions apply, e.g. agencies of the United States or of any State or political subdivisions thereof are exempt from permit requirements); and
2. A person, other than a licensee, transporting, shipping, causing to be transported, or receiving explosive materials for use in manufacturing display fireworks or consumer fireworks.

With Respect to Fireworks: Who may not need a license or permit?

Frequently, persons contracting for display fireworks (e.g., for Fourth of July observances) from a Federal explosives licensee or permittee receive a total service, including the services of a pyrotechnician who transports display fireworks in interstate or intrastate commerce to the site of the display and conducts and supervises the display. In these instances, the customers purchase and receive the contractor's services and not the explosive materials themselves (i.e. the cost of the services includes the contractor's expense in providing the fireworks utilized), and tl cost of the services includes the dealer's expense in providing t fireworks utilized. When business is transacted in this manner, customers purchasing and receiving the services need not obta Federal explosives licenses or permits under Part 555 as long a they are not transporting, shipping, causing to be transported, c receiving explosive materials. Note: the transportation of explc sive materials to the display sites would be authorized by the Federal explosives license or permit of the licensee or permitte providing the services.

With Respect to Fireworks: Types of permits

1. User permit: Allows the permit holder to transport, sl cause to be transported, and receive display fireworks in inters or foreign commerce for his or her own use and not for resale. This permit is issued at a cost of $100 for a 3-year period and i renewable at a cost of $50 for a 3-year period.
2. User-limited permit: Identical to the user permit but issued for a single purchase transaction, only. The fee is $75; t permit is nonrenewable.

With Respect to Fireworks: Storage

The law prohibits any person from storing any explosive materials in a manner not in conformity with the regulations promulgated by the Attorney General (18 U.S.C. 842(j)). Pursuant to this section, the Attorney General has prescribed storage regulations in 27 CFR Part 555, Subpart K. Display fireworks must be stored in conformity with the regulations. Display fireworks generally contain perchlorate mixture explo- sives, potassium chlorate base explosive mixtures, and black powder, which are entered on the List of Explosive Materials v numerous others. (The List, which is not all-inclusive, is annu compiled and readily available without charge from the addres set out in 27 CFR 555.23 or online at www.atf.gov.) Display fireworks must be stored as low explosives in magazines meet at a minimum, the requirements for type 4 storage magazines prescribed by 27 CFR 555.210 unless they contain other classe of explosives. Bulk salutes must be stored as high explosives i type 1 or type 2 magazines. The net weight of the explosive materials contained in the display fireworks may be used in determining compliance with table of distance requirements. T determine the actual weight of the materials, it may be necessa to contact their manufacturers. The manufacturer of exempt or nonexempt fireworks having stocks of explosive materials on hand to be used in the manufacture of fireworks must store the stocks in conformity with applicable storage requirements.

xplosives Dealer's and User's uide to Federal Explosives egulation

xplosives May Not be Distributed by censees (Or by Any Person) to Any Person ho:

1. Is under indictment for, or who has been convicted of a me punishable by imprisonment for a term exceeding one year.

2. Is an unlawful user of, or addicted to, marijuana or any pressant or stimulant drug or narcotic drug (as these terms are fined in section 102 of the Controlled Substances Act).

3. Has been adjudicated as a mental defective or has been mmitted to a mental institution.

4. Is a fugitive from justice.

5. Is an alien (with certain exceptions)

6. Has been discharged from the armed forces under shonorable conditions; or

7. Having been a citizen of the United States, has ounced citizenship.

8. Is less than 21 years of age.

ealers in Explosives Must:

Have a current and valid Federal explosives license.
Have proper storage facilities.
Keep accurate and complete records.
Verify that each buyer has a Federal explosives license or rmit.
Verify buyers' identities.

sers of Explosives

Federal permits are required of those who transport, ship, cause be transported, or receive explosive materials. Among other ngs, the permittee must keep complete and accurate records of acquisitions and dispositions of explosives materials. Unless ıerwise exempted by law, no person may receive or transport y explosive materials without a permit.

o person shall store any explosive material a matter not in conformity with applicable gulations.

ll persons must report to ATF and local thorities any loss or theft of their explosive aterials within 24 hours of discovery.

Federal license or permit does not confer y right or privilege to violate any state law local ordinance.

The above summary is general and does not purport to fully convey the Federal explosives law and regulations pertaining to dealers and users.

Black Powder Transactions

Public Law 93-639 (1975) allows nonlicensees/nonpermittees to purchase commercially manufactured black powder, in quantities of 50 pounds or less, solely for sporting, recreational or cultural purposes for use in antique firearms or antique devices. A nonlicensee or nonpermittee purchasing black powder under the exemption need not be a resident of the State in which the dealer is located. Also, the categories of persons to whom the distribution of explosive materials is prohibited do not apply to black powder transactions made under the exemption. Acquisitions of black powder not qualifying under this exemption are subject to the same regulatory requirements that govern any other low explosive.

All persons who distribute black powder, regardless of quantity, must be licensed as explosives dealers and, among other things, must provide adequate storage.

Explosives Security

Through prompt reporting of losses and thefts of explosives and increased emphasis on physical security, explosives licensees and permittees can contribute greatly to efforts by Federal, State and local authorities to reduce the incidence of bombings and other criminal misuse of explosives in the United States. The following actions are of prime importance and in some instances required:

Report . . . any thefts or losses of explosives within 24 hours of discovery, by telephone, to ATF (toll free: 1-800-800-3855) and to appropriate local authorities. Because the States and many municipalities have designated specific agencies to investigate the theft or loss of explosives, licensees and permittees are urged to be familiar with State and local reporting procedures and appropriate contact points.

Follow . . . telephone notification with a written report on ATF Form 5400.5, "Report of Theft or Loss—Explosive Materials," to the nearest ATF Division Office, and in accordance with the instructions on the form.

Observe . . . activity around magazines, within business premises, and on job sites, particularly if strangers appear to be loitering in the area in which explosives are being kept. On-site users should take special care to assure that explosives removed from storage for use on the job are either detonated or accounted for and unused items returned to storage.

Review . . . recordkeeping practices to assure that no discrepancies exist and that no figures in reported inventories have been manipulated, and correct any clerical errors promptly. Should any questions arise concerning explosives security procedures or any aspect of explosives regulation coming under the jurisdiction of ATF, do not hesitate to contact ATF.

Note: For Q&A's on regulatory requirements governing recordkeeping and storage, see "Questions and Answers" numbers 62-87.

Additional Information

The flow of useful information is an essential ingredient in the effective administration of regulatory programs. The Bureau of Alcohol, Tobacco, Firearms and Explosives is the Federal agency charged with the responsibility of administering laws impacting the explosives industry. We call your attention to the following publication distributed by ATF that merits your attention:

The Explosives Newsletter

During 1989 ATF developed the Explosives Newsletter, an information service for Federal explosives licensees and permittees which is intended to help explosives industry members better understand the Federal laws under which they must operate. It also includes other items of particular interest to the explosives industry. There is no charge for the Explosives Newsletter; licensees and permittees automatically receive copies when new editions are published.

Explosives industry members having questions on the Federal explosives laws and regulations may address their inquiries to:

Bureau of Alcohol, Tobacco, Firearms and Explosives
Explosives Industry Programs Branch
99 New York Avenue, N.E.
Mailstop 6E405
Washington, DC 20226

Direct e-mail inquiries on general questions or variance requests may be sent to the branch at EIPB@atf.gov.

List of Explosive Materials

Pursuant to the provisions of section 841(d) of title 18, U.S.C., and 27 CFR 555.23, the Director, Bureau of Alcohol, Tobacco, Firearms and Explosives, must revise and publish in the Federal Register at least annually a list of explosives determined to be within the coverage of 18 U.S.C. Chapter 40, Importation, Manufacture, Distribution and Storage of Explosive Materials. This chapter covers not only explosives, but also blasting agents and detonators, all of which are defined as explosive materials in section 841(c) of title 18, U.S.C. Accordingly, the following is the current List of Explosive Materials subject to regulation under 18 U.S.C. Chapter 40. Materials constituting blasting agents are marked by an asterisk. While the list is comprehensive, it is not all-inclusive. The fact that an explosive material may not be on the list does not mean that it is not within the coverage definitions in section 841 of title 18, U.S.C. Explosive materials are listed alphabetically by their common names, followed by chemical names and synonyms in brackets. This revised list is effective as of September 18, 2006.

List of Explosive Materials

A

Acetylides of heavy metals.
Aluminum containing polymeric propellant.
Aluminum ophorite explosive.
Amatex.
Amatol.
Ammonal.
Ammonium nitrate explosive mixtures (cap sensitive).
*Ammonium nitrate explosive mixtures (non-cap sensitive).
Ammonium perchlorate having particle size less than 15 microns.
Ammonium perchlorate composite propellant.
Ammonium perchlorate explosive mixtures.
Ammonium picrate [picrate of ammonia, Explosive D].
Ammonium salt lattice with isomorphously substituted inorganic salts.
*ANFO [ammonium nitrate-fuel oil].
Aromatic nitro-compound explosive mixtures.
Azide explosives.

B

Baranol.
Baratol.
BEAF [1,2-bis (2, 2-difluoro-2-nitroacetoxyethane)].
Black powder.
Black powder based explosive mixtures.
*Blasting agents, nitro-carbo-nitrates, including non-cap sensitive slurry and water gel explosives.
Blasting caps.
Blasting gelatin.
Blasting powder.
BTNEC [bis (trinitroethyl) carbonate].
BTNEN [bis (trinitroethyl) nitramine].
BTTN [1,2,4 butanetriol trinitrate].
Bulk salutes.
Butyl tetryl.

C

Calcium nitrate explosive mixture.
Cellulose hexanitrate explosive mixture.
Chlorate explosive mixtures.
Composition A and variations.
Composition B and variations.
Composition C and variations.
Copper acetylide.
Cyanuric triazide.
Cyclonite [RDX].
Cyclotetramethylenetetranitramine [HMX].
Cyclotol.
Cyclotrimethylenetrinitramine [RDX].

D

DATB [diaminotrinitrobenzene].
DDNP [diazodinitrophenol].
DEGDN [diethyleneglycol dinitrate].
Detonating cord.
Detonators.
Dimethylol dimethyl methane dinitrate composition.
Dinitroethyleneurea.
Dinitroglycerine [glycerol dinitrate].
Dinitrophenol.
Dinitrophenolates.
Dinitrophenyl hydrazine.
Dinitroresorcinol.
Dinitrotoluene-sodium nitrate explosive mixtures.
DIPAM [dipicramide; diaminohexanitrobiphenyl].
Dipicryl sulfone.
Dipicrylamine.
Display fireworks.
DNPA [2,2-dinitropropyl acrylate].
DNPD [dinitropentano nitrile].
Dynamite.

E

EDDN [ethylene diamine dinitrate].
EDNA [ethylenedinitramine].
Ednatol.
EDNP [ethyl 4,4-dinitropentanoate].
EGDN [ethylene glycol dinitrate].
Erythritol tetranitrate explosives.
Esters of nitro-substituted alcohols.
Ethyl-tetryl.
Explosive conitrates.
Explosive gelatins.
Explosive liquids.
Explosive mixtures containing oxygen-releasing inorganic salts and hydrocarbons.
Explosive mixtures containing oxygen-releasing inorganic salts and nitro bodies.
Explosive mixtures containing oxygen-releasing inorganic salts and water insoluble fuels.
Explosive mixtures containing oxygen-releasing inorganic salts and water soluble fuels.
Explosive mixtures containing sensitized nitromethane.

Explosive mixtures containing tetranitromethane (nitroform).
Explosive nitro compounds of aromatic hydrocarbons.
Explosive organic nitrate mixtures.
Explosive powders.

F

Flash powder.
Fulminate of mercury.
Fulminate of silver.
Fulminating gold.
Fulminating mercury.
Fulminating platinum.
Fulminating silver.

G

Gelatinized nitrocellulose.
Gem-dinitro aliphatic explosive mixtures.
Guanyl nitrosamino guanyl tetrazene.
Guanyl nitrosamino guanylidene hydrazine.
Guncotton.

H

Heavy metal azides.
Hexanite.
Hexanitrodiphenylamine.
Hexanitrostilbene.
Hexogen [RDX].
Hexogene or octogene and a nitrated Nmethylaniline.
Hexolites.
HMTD [hexamethylenetriperoxidediamine].
HMX [cyclo-1,3,5,7-tetramethylene 2,4,6,8-tetranitramine; Octogen].
Hydrazinium nitrate/hydrazine/aluminum explosive system.
Hydrazoic acid.

I

Igniter cord.
Igniters.
Initiating tube systems.

K

KDNBF [potassium dinitrobenzofuroxane].

L

Lead azide.
Lead mannite.
Lead mononitroresorcinate.
Lead picrate.
Lead salts, explosive.
Lead styphnate [styphnate of lead, lead trinitroresorcinate].
Liquid nitrated polyol and trimethylolethane.
Liquid oxygen explosives.

M

Magnesium ophorite explosives.
Mannitol hexanitrate.
MDNP [methyl 4,4-dinitropentanoate].
MEAN [monoethanolamine nitrate].
Mercuric fulminate.
Mercury oxalate.
Mercury tartrate.
Metriol trinitrate.
Minol-2 [40% TNT, 40% ammonium nitrate, 20% aluminum].
MMAN [monomethylamine nitrate]; methylamine nitrate.
Mononitrotoluene-nitroglycerin mixture.
Monopropellants.

N

NIBTN [nitroisobutametriol trinitrate].
Nitrate explosive mixtures.
Nitrate sensitized with gelled nitroparaffin.
Nitrated carbohydrate explosive.
Nitrated glucoside explosive.
Nitrated polyhydric alcohol explosives.
Nitric acid and a nitro aromatic compound explosive.
Nitric acid and carboxylic fuel explosive.
Nitric acid explosive mixtures.
Nitro aromatic explosive mixtures.
Nitro compounds of furane explosive mixtures.
Nitrocellulose explosive.
Nitroderivative of urea explosive mixture.
Nitrogelatin explosive.
Nitrogen trichloride.
Nitrogen tri-iodide.
Nitroglycerine [NG, RNG, nitro, glyceryl trinitrate, trinitroglycerine].
Nitroglycide.
Nitroglycol [ethylene glycol dinitrate, EGDN].
Nitroguanidine explosives.
Nitronium perchlorate propellant mixtures.
Nitroparaffins Explosive Grade and ammonium nitrate mixtures
Nitrostarch.
Nitro-substituted carboxylic acids.
Nitrourea.

O

Octogen [HMX].
Octol [75 percent HMX, 25 percent TNT].
Organic amine nitrates.
Organic nitramines.

P

PBX [plastic bonded explosives].
Pellet powder.
Penthrinite composition.
Pentolite.
Perchlorate explosive mixtures.
Peroxide based explosive mixtures.
PETN [nitropentaerythrite, pentaerythrite tetranitrate, pentaerythritol tetranitrate].
Picramic acid and its salts.
Picramide.
Picrate explosives.
Picrate of potassium explosive mixtures.
Picratol.
Picric acid (manufactured as an explosive).
Picryl chloride.

cryl fluoride.
.X [95% nitromethane, 5% ethylenediamine].
lynitro aliphatic compounds.
lyolpolynitrate-nitrocellulose explosive gels.
tassium chlorate and lead sulfocyanate explosive.
tassium nitrate explosive mixtures.
tassium nitroaminotetrazole.
rotechnic compositions.
YX [2,6-bis(picrylamino)] 3,5-dinitropyridine.

DX [cyclonite, hexogen, T4, cyclo-1,3,5,-trimethylene-2,4,6,-trinitramine; hexahydro-1,3,5-trinitro-S-triazine].

afety fuse.
alts of organic amino sulfonic acid explosive mixture.
alutes (bulk).
lver acetylide.
lver azide.
lver fulminate.
lver oxalate explosive mixtures.
lver styphnate.
lver tartrate explosive mixtures.
lver tetrazene.
urried explosive mixtures of water, inorganic oxidizing salt, lling agent, fuel, and sensitizer (cap sensitive).
nokeless powder.
datol.
dium amatol.
dium azide explosive mixture.
dium dinitro-ortho-cresolate.
dium nitrate explosive mixtures.
dium nitrate-potassium nitrate explosive mixture.
dium picramate.
ecial fireworks.
uibs.
yphnic acid explosives.

cot [tetranitro-2,3,5,6-dibenzo-1,3a,4,6a tetrazapentalene].
TB [triaminotrinitrobenzene].
TP [triacetonetriperoxide].
EGDN [triethylene glycol dinitrate].
tranitrocarbazole.
trazene [tetracene, tetrazine, 1(5-tetrazolyl)-4-guanyl tetrazene hydrate].
trazole explosives.
tryl [2,4,6 tetranitro-N-methylaniline].
trytol.
ickened inorganic oxidizer salt slurried explosive mixture.
METN [trimethylolethane trinitrate].
NEF [trinitroethyl formal].
NEOC [trinitroethylorthocarbonate].
NEOF [trinitroethylorthoformate].
NT [trinitrotoluene, trotyl, trilite, triton].
rpex.
idite.

Trimethylol ethyl methane trinitrate composition.
Trimethylolthane trinitratenitrocellulose.
Trimonite.
Trinitroanisole.
Trinitrobenzene.
Trinitrobenzoic acid.
Trinitrocresol.
Trinitro-meta-cresol.
Trinitronaphthalene.
Trinitrophenetol.
Trinitrophloroglucinol.
Trinitroresorcinol.
Tritonal.

U

Urea nitrate.

W

Water-bearing explosives having salts of oxidizing acids and nitrogen bases, sulfates, or sulfamates (cap sensitive).
Water-in-oil emulsion explosive compositions.

X

Xanthamonas hydrophilic colloid explosive mixture.

Approved: September 18, 2006.
Michael J. Sullivan, Acting Director.

[FR Doc. E6–15850 Filed 9–26–06; 8:45 am]

PREFACE

This handbook is primarily for the use of persons in the business of importing, manufacturing, and dealing in firearms defined by the National Firearms Act (NFA) or persons intending to go into an NFA firearms business. It should also be helpful to collectors of NFA firearms and other persons having questions about the application of the NFA.

This publication is not a law book. Rather, it is intended as a "user friendly" reference book enabling the user to quickly find answers to questions concerning the NFA. Nevertheless, it should also be useful to attorneys seeking basic information about the NFA and how the law has been interpreted by ATF. The book's Table of Contents will be helpful to the user in locating needed information.

Although the principal focus of the handbook is the NFA, the book necessarily covers provisions of the Gun Control Act of 1968 and the Arms Export Control Act impacting NFA firearms businesses and collectors.

The book is the product of a joint effort between ATF and the National Firearms Act Trade and Collectors Association. ATF takes this opportunity to express its appreciation to the Association for its assistance in writing and making this publication possible.

TABLE OF CONTENTS

CHAPTER 1. INTRODUCTION

CHAPTER 2. WHAT ARE "FIREARMS" UNDER THE NFA?

CHAPTER 3. REGISTRATION OF NFA FIREARMS

CHAPTER 4. TAXES IMPOSED BY THE NFA

CHAPTER 5. QUALIFYING TO DO BUSINESS IN NFA FIREARMS

CHAPTER 6. MAKING NFA FIREARMS BY NONLICENSEES

CHAPTER 7. MANUFACTURING NFA FIREARMS

CHAPTER 8. IMPORTING NFA FIREARMS

CHAPTER 9. TRANSFERS OF NFA FIREARMS

CHAPTER 10. COLLECTORS OF NFA FIREARMS

CHAPTER 11. EXPORTATION OF NFA FIREARMS

CHAPTER 12. RECORDKEEPING

CHAPTER 13. REQUIRED REPORTS AND NOTIFICATIONS TO ATF

CHAPTER 14. GOING OUT OF BUSINESS

CHAPTER 15. PENALTIES AND SANCTIONS

CHAPTER 1. INTRODUCTION

Section 1.1 History of the National Firearms Act (NFA)

1.1.1 The NFA of 1934. The NFA was originally enacted in 1934.[1] Similar to the current NFA, the original Act imposed a tax on the making and transfer of firearms defined by the Act, as well as a special (occupational) tax on persons and entities engaged in the business of importing, manufacturing, and dealing in NFA firearms. The law also required the registration of all NFA firearms with the Secretary of the Treasury. Firearms subject to the 1934 Act included shotguns and rifles having barrels less than 18 inches in length, certain firearms described as "any other weapons," machineguns, and firearm mufflers and silencers.

While the NFA was enacted by Congress as an exercise of its authority to tax, the NFA had an underlying purpose unrelated to revenue collection. As the legislative history of the law discloses, its underlying purpose was to curtail, if not prohibit, transactions in NFA firearms. Congress found these firearms to pose a significant crime problem because of their frequent use in crime, particularly the gangland crimes of that era such as the St. Valentine's Day Massacre. The $200 making and transfer taxes on most NFA firearms were considered quite severe and adequate to carry out Congress' purpose to discourage or eliminate transactions in these firearms. The $200 tax has not changed since 1934.

As structured in 1934, the NFA imposed a duty on persons transferring NFA firearms, as well as mere possessors of unregistered firearms, to register them with the Secretary of the Treasury. If the possessor of an unregistered firearm applied to register the firearm as required by the NFA, the Treasury Department could supply information to State authorities about the registrant's possession of the firearm. State authorities could then use the information to prosecute the person whose possession violated State laws. For these reasons, the Supreme Court in 1968 held in the *Haynes* case that a person prosecuted for possessing an unregistered NFA firearm had a valid defense to the prosecution - the registration requirement imposed on the possessor of an unregistered firearm violated the possessor's privilege from self-incrimination under the Fifth Amendment of the U.S. Constitution.[2] The *Haynes* decision made the 1934 Act virtually unenforceable.

1.1.2 Title II of the Gun Control Act of 1968. Title II amended the NFA to cure the constitutional flaw pointed out in *Haynes*.[3] First, the requirement for possessors of unregistered firearms to register was removed. Indeed, under the amended law, there is no mechanism for a possessor to register an unregistered NFA firearm already possessed by the person. Second, a provision was added to the law prohibiting the use of any information from an NFA application or registration as evidence against the person in a criminal proceeding with respect to a violation of law occurring prior to or concurrently with the filing of the application or registration.[4] In 1971, the Supreme Court reexamined the NFA in the *Freed* case and found that the 1968 amendments cured the constitutional defect in the original NFA.[5]

---

[1] National Firearms Act, Public Law 474, approved June 26, 1934.
[2] Haynes v. U.S., 390 U.S. 85 (1968).
[3] Gun Control Act of 1968, Public Law 90-618, approved October 22, 1968.
[4] 26 U.S.C. 5848
[5] U.S. v. Freed, 401 U.S. 601 (1971)

Title II also amended the NFA definitions of "firearm" by adding "destructive devices" and expanding the definition of "machinegun."

1.1.3 Firearm Owners' Protection Act. In 1986, this Act amended the NFA definition of "silencer" by adding combinations of parts for silencers and any part intended for use in the assembly or fabrication of a silencer.[6] The Act also amended the GCA to prohibit the transfer or possession of machineguns.[7] Exceptions were made for transfers of machineguns to, or possession of machineguns by, government agencies, and those lawfully possessed before the effective date of the prohibition, May 19, 1986.

Section 1.2 Meaning of terms. Certain terms and abbreviations used in this book are defined as follows:

1.2.1 "AECA" means the Arms Export Control Act, 22 U.S.C. 2778.

1.2.2 "ATF" means the Bureau of Alcohol, Tobacco, Firearms and Explosives, U.S. Department of Justice.

1.2.3 "ATF Ruling" means a formal ruling published by ATF stating its interpretation of the law and regulations as applied to a specific set of facts.

1.2.4 "CFR" means the Code of Federal Regulations in which Federal firearms regulations are published.

1.2.5 "DIO" means an ATF Director of Industry Operations responsible for regulating the firearms industry within an ATF field division.

1.2.6 "FFL" means a Federal firearms licensee, person or entity having a license to import, manufacture, or deal in firearms under the GCA.

1.2.7 "FTB" means ATF's Firearms Technology Branch.

1.2.8 "GCA" means the Gun Control Act of 1968, 18 U.S.C. Chapter 44.

1.2.9 "NFA" means the National Firearms Act, 26 U.S.C. Chapter 53.

1.2.10 "NFRTR" means the National Firearms Registration and Transfer Record containing the registration of NFA firearms.

1.2.11 "SOT" means a special occupational taxpayer, a person or entity qualified to import, manufacture, or deal in NFA firearms by having paid the special (occupational) tax to do so under the NFA.

---

6 Firearm Owners' Protection Act, Public Law 99-308, approved May 19, 1986.
7 18 U.S.C. 922(o)

1.2.12 "U.S.C." means the United States Code in which Federal firearms laws are codified.

Section 1.3 Administration and enforcement of Federal firearms laws

Until January 24, 2003, authority to administer and enforce Federal firearms laws was the responsibility of the Bureau of ATF within the U.S. Department of the Treasury. As a result of enactment of Section 1111 of the Homeland Security Act of 2002, ATF and its firearms authorities were transferred to the U.S. Department of Justice, effective January 24, 2003. ATF's name was also changed to the Bureau of Alcohol, Tobacco, Firearms and Explosives. ATF continues to have the authority to administer and enforce Federal firearms laws. The Department of State retained its authority over the enforcement of the export provisions of the AECA that relate to firearms.

Section 1.4 What are regulations and rulings?

1.4.1 Regulations. Regulations interpret the statutes (the law) and explain the procedures for compliance. The Administrative Procedure Act (APA) generally requires agencies to publish proposed regulations in the Federal Register as a notice of proposed rulemaking, giving the public the opportunity to comment on the proposals before they may be issued as final regulations. The APA provides no specific comment period for proposed rules under the NFA.[8] As specifically provided for in the GCA, GCA regulations require a comment period of at least 90 days.[9] An exception in the APA eliminates the need to provide any notice or comment period with respect to AECA regulations. Regulations have the force and effect of law. Courts will uphold a regulation if they find reasonable legal basis for it and if it generally is within the scope of the statute.

1.4.2 Rulings. ATF publishes rulings in its periodic bulletins and posts them on the ATF website. These contain ATF's interpretation of the law and regulations as they pertain to a particular fact situation. Rulings do not have the force and effect of law but may be cited as precedent with respect to substantially similar fact situations. Courts will recognize and apply such rulings if they are determined to correctly interpret the law and regulations.

Section 1.5 Other ATF publications

1.5.1 ATF's internet website. This is the best source for up to date information on ATF's firearms administration and enforcement activities, including amendments to the law, rulings, regulations, and open letters to firearms industry members. The website address is http://www.atf.gov.

1.5.2 ATF P 5300.4. This is ATF's publication, "Federal Firearms Regulations Reference Guide." ATF supplies the publication to all FFLs and SOTs. It contains all Federal firearms laws and regulation (except those pertaining to the firearms and ammunition excise tax and State Department export regulations), ATF firearms rulings, articles on various firearms issues, and questions and answers. The publication can be found on ATF's website, http://www.atf.gov, and downloaded.

---

[8] 5 U.S.C. 552

[9] 18 U.S.C. 926(b)

Section 1.6 ATF points of contact

Chief, National Firearms Act Branch
Bureau of Alcohol, Tobacco, Firearms and Explosives
244 Needy Road
Martinsburg, WV 25405
Phone: (304) 616-4500
Fax: (304) 616-4501

Chief, Federal Firearms Licensing Center
Bureau of Alcohol, Tobacco, Firearms and Explosives
244 Needy Road
Martinsburg, WV 25405

Phone: (304) 616-4600 or 1-866-662-2750
Fax: (304) 616-4501 or 1-866-257-2749

Chief, Firearms & Explosives Imports Branch
Bureau of Alcohol, Tobacco, Firearms and Explosives
244 Needy Road
Martinsburg, WV 25405
Phone: (304) 616-4550
Fax: (304) 616-4551

Chief, Firearms Technology Branch
Bureau of Alcohol, Tobacco, Firearms and Explosives
244 Needy Road
Martinsburg, WV 25405
Phone: (304) 260-5476
Fax: (304) 260-1701

Section 1.7 ATF Forms

ATF forms may be ordered from ATF's Distribution Center by use of the Center's order form on ATF's website at http://www.atf.gov. After entering the website, click on "Forms." They may also be obtained by calling the Center at (301) 583-4696 or writing the ATF Distribution Center at 1519 Cabin Branch Drive, Landover, MD 20785-3816.

CHAPTER 2. WHAT ARE "FIREARMS" UNDER THE NFA?

Section 2.1 Types of NFA firearms

The NFA defines the specific types of firearms subject to the provisions of the Act. These definitions describe the function, design, configuration and/or dimensions that weapons must have to be NFA firearms. In addition to describing the weapon, some definitions (machinegun, rifle, shotgun, any other weapon) state that the firearm described also includes a weapon that can be readily restored to fire. A firearm that can be readily restored to fire is a firearm that in its present condition is incapable of expelling a projectile by the action of an explosive (or, in the case of a machinegun, will not in its present condition shoot automatically) but which can be restored to a functional condition by the replacement of missing or defective component parts. Please be aware that case law is not specific but courts have held that the "readily restorable" test is satisfied where a firearm can be made capable of renewed automatic operation, even if it requires some degree of skill and the use of tools and parts.

2.1.1 Shotgun A shotgun is a firearm designed to be fired from the shoulder and designed to use the energy of the explosive in a fixed shotgun shell to fire through a smooth bore either a number of projectiles or a single projectile for each pull of the trigger.[10] A shotgun subject to the NFA has a barrel or barrels of less than 18 inches in length.

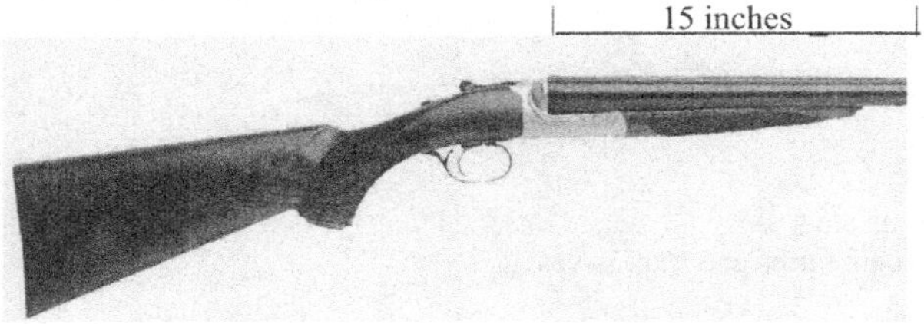

The ATF procedure for measuring barrel length is to measure from the closed bolt (or breech-face) to the furthermost end of the barrel or permanently attached muzzle device. Permanent methods of attachment include full-fusion gas or electric steel-seam welding, high-temperature (1100°F) silver soldering, or blind pinning with the pin head welded over. Barrels are measured by inserting a dowel rod into the barrel until the rod stops against the bolt or breech-face. The rod is then marked at the furthermost end of the barrel or permanently attached muzzle device, withdrawn from the barrel, and measured.

[10] 26 U.S.C. 5845(d)

.1.2 Weapon made from a shotgun. A weapon made from a shotgun is a shotgun type weapon that as an overall length of less than 26 inches or a barrel or barrels of less than 18 inches in length.

25 inches

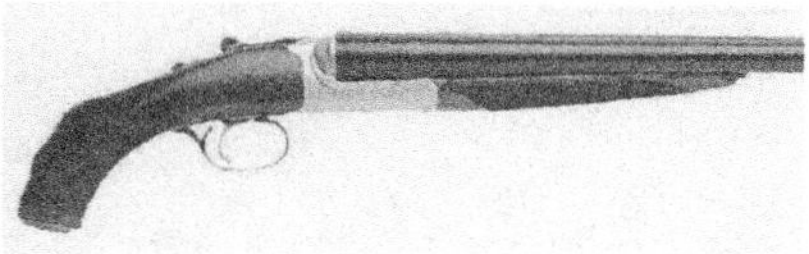

he overall length of a firearm is the distance between the muzzle of the barrel and the rearmost portion f the weapon measured on a line parallel to the axis of the bore.

.1.3 Rifle. A rifle is a firearm designed to be fired from the shoulder and designed to use the energy of n explosive in a fixed cartridge to fire only a single projectile through a rifled barrel for each single pull f the trigger.[11] A rifle subject to the NFA has a barrel or barrels of less than 16 inches in length.

12 inches

he ATF procedure for measuring barrel length is to measure from the closed bolt (or breech-face) to ne furthermost end of the barrel or permanently attached muzzle device. Permanent methods of ttachment include full-fusion gas or electric steel-seam welding, high-temperature (1100°F) silver oldering, or blind pinning with the pin head welded over. Barrels are measured by inserting a dowel rod nto the barrel until the rod stops against the bolt or breech-face. The rod is then marked at the urthermost end of the barrel or permanently attached muzzle device, withdrawn from the barrel, and neasured.

26 U.S.C. 5845(c)

2.1.4 Weapon made from a rifle. A weapon made from a rifle is a rifle type weapon that has an overall length of less than 26 inches or a barrel or barrels of less than 16 inches in length.

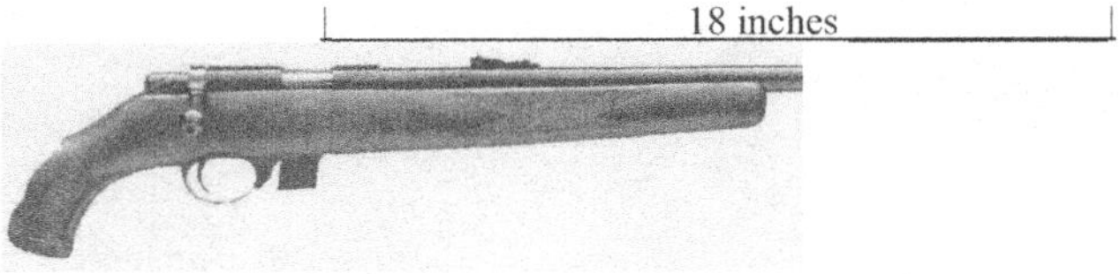

The overall length of a firearm is the distance between the muzzle of the barrel and the rearmost portion of the weapon measured on a line parallel to the axis of the bore.

2.1.5 Any other weapon. Firearms meeting the definition of "any other weapon" are weapons or devices capable of being concealed on the person from which a shot can be discharged through the energy of an explosive. Many "any other weapons" are disguised devices such as penguns, cigarette lighter guns, knife guns, cane guns and umbrella guns.

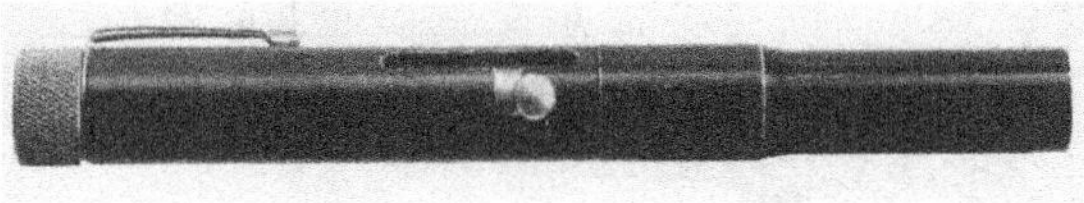

pengun

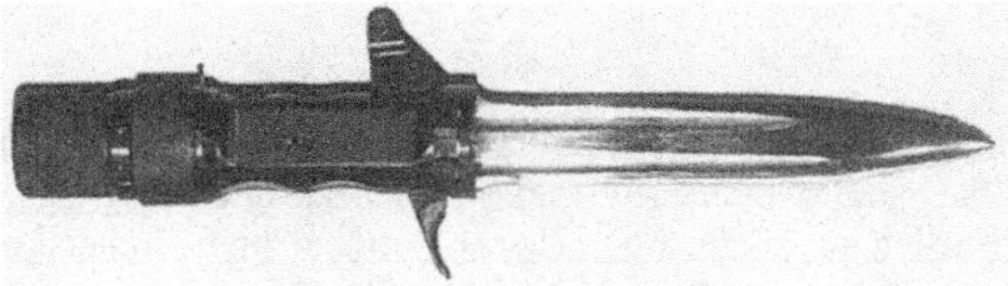

knife gun

umbrella gun

Also included in the "any other weapon" definition are pistols and revolvers having smooth bore barrel designed or redesigned to fire a fixed shotgun shell.

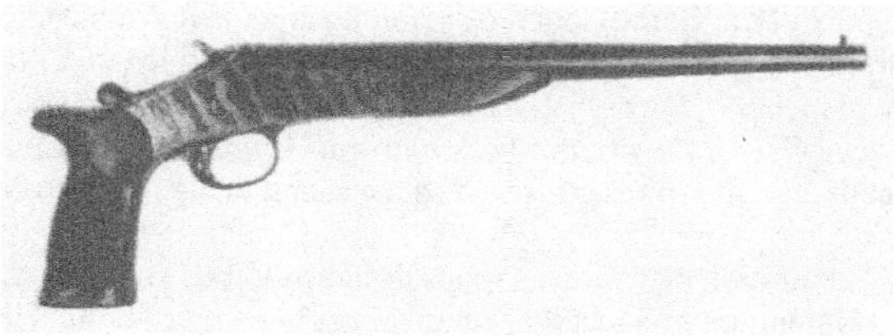

H&R Handy Gun

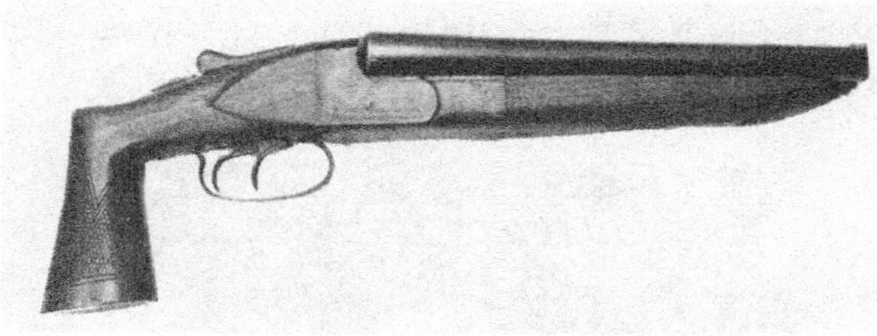

Ithaca Auto & Burglar Gun

Vhile the above weapons are similar in appearance to weapons made from shotguns, they were riginally manufactured in the illustrated configuration and are not modified from existing shotguns. As result, these weapons do not fit within the definition of shotgun[12] or weapons made from a shotgun[13].

he "any other weapon" definition also includes specifically described weapons with combination hotgun and rifle barrels 12 inches or more but less than 18 inches in length from which only a single ischarge can be made from either barrel without manual reloading. The firearm most commonly ssociated with this portion of the definition is the Marble's Game Getter.

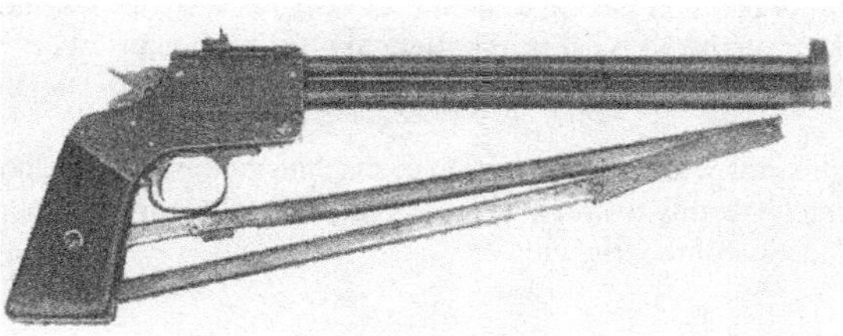

Marble's Game Getter

26 U.S.C. 5845(d)
26 U.S.C. 5845(a)(2)

NOTE: One version of the Marble's Game Getter was produced with 18-inch barrels and a folding shoulder stock. This model of the Game Getter, as manufactured, is not subject to the provisions of the NFA because it has barrels that are 18 inches in length and the overall length of the firearm, with stock extended, is more than 26 inches. *However, if the shoulder stock has been removed from the 18-inch barrel version of the Game Getter, the firearm has an overall length of less than 26 inches and is an NFA weapon.* Specifically, the firearm is classified as a weapon made from a rifle/shotgun.

The "any other weapon" definition excludes weapons designed to be fired from the shoulder that are no capable of firing fixed ammunition or a pistol or revolver having a rifled bore. However, certain alterations to a pistol or revolver, such as the addition of a second vertical handgrip, create a weapon th no longer meets the definition of pistol or revolver.[14] A pistol or revolver modified as described is an "any other weapon" subject to the NFA because the weapon is not designed to be fired when held in on hand.

semiautomatic pistol with second vertical handgrip

As stated above, a pistol or revolver having a rifled bore does not meet the definition of "any other weapon" and is not subject to the NFA. It is important to note that any pistol or revolver having a barre without a rifled bore does not fit within the exclusion and is an "any other weapon" subject to the NFA.

2.1.6 Machinegun. Firearms within the definition of machinegun include weapons that shoot, are designed to shoot, or can be readily restored to shoot, automatically more than one shot without manual reloading by a single function of the trigger.

---

[14] 27 CFR 479.11

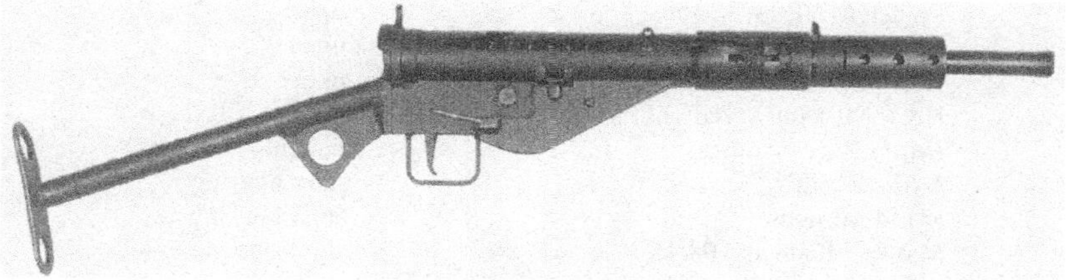

STEN MK II submachinegun

The definition of machinegun also includes the frame or receiver of a machinegun.

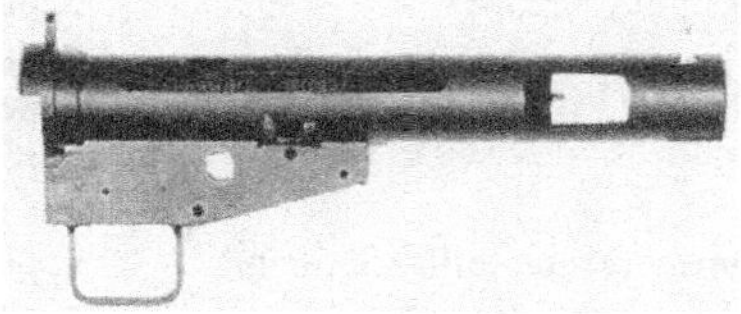

STEN MK II submachinegun receiver

Of all the different firearms defined as NFA weapons, machineguns are the only type where the receiver of the weapon by itself is an NFA firearm. As a result, it is important that the receiver of a machinegun be properly identified. Many machineguns incorporate a "split" or "hinged" receiver design so the main portion of the weapon can be easily separated into upper and lower sections. Additionally, some machineguns utilize a construction method where the receiver is composed of a number of subassemblies that are riveted together to form the complete receiver.

The following table lists specific models of machineguns incorporating the above designs and the portion of the weapon that has been held to be the receiver. This list is not all- inclusive. For information concerning a split or hinged receiver type machinegun not listed below, contact FTB at (304) 260-1699.

Model	Receiver
Armalite AR10	lower
Armalite AR15 (all variations)	lower
Armalite AR18	lower
Beretta AR70	lower
British L1A1	upper
Browning M1917	right side plate
Browning M1919 (all variations)	right side plate
Browning M2 & M2HB	right side plate
Colt M16 (all variations)	lower
Czech Vz 61	lower
FN FNC	lower
Model	**Receiver**
FN CAL	upper
FN FAL	upper

French MAT 49	upper
German MP38 & MP40	upper
H&K G3 (all variations)	upper
H&K MP5 (all variations)	upper
IMI UZI	upper
M61 Vulcan	outer housing
M134 Minigun	outer housing
Maxim MG08 and 08/15	right side plate
SIG AMT	upper
SIG STG 57	upper
SIG 550 Series (all variations)	upper
Soviet PPsH 41	upper
Soviet PPS 43	upper
Steyr MPi 69	upper
Steyr MPi 81	upper
Thompson submachinegun (all variations)	upper
Vickers water cooled machineguns	right side plate

The "designed to shoot automatically more than one shot without manual reloading by a single function of the trigger" portion of the definition relates to the characteristics of the weapon that permit full automatic fire. ATF has also held that the "designed" definition includes those weapons which have not previously functioned as machineguns but possess design features which facilitate full automatic fire by simple modification or elimination of existing component parts. ATF has published rulings concerning specific firearms classified as machineguns based on this interpretation of the term "designed."[15]

Included within the definition of machinegun is any part designed and intended solely and exclusively, or combination of parts designed and intended, for use in converting a weapon into a machinegun. This portion of the machinegun definition addresses what are commonly referred to as conversion kits. The "any part designed and intended solely and exclusively" language refers to a part that was produced for no other reason than to convert a weapon into a machinegun. Illustrated below are examples of such parts.

conversion sear for H&K semiautomatic firearms

[15] Appendix B (ATF Rulings 82-2, 82-8, 83-5)

Drop in Auto Sear for AR15 type semiautomatic firearms

he above parts are designed solely and exclusively for use in converting a weapon into a machinegun nd are classified as machineguns.

he "combination of parts designed and intended for use in converting a weapon into a machinegun" anguage refers to a group of parts designed and intended to be used in converting a weapon into a nachinegun. A typical example is those M2 carbine parts that are only used to permit fully automatic ire in a US Carbine M1 or M2.

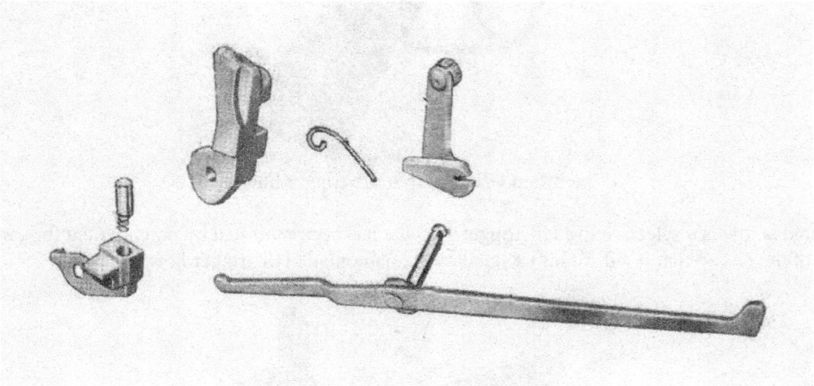

M2 Carbine conversion kit

he above parts consisting of an M2 selector lever, selector lever spring, disconnector lever assembly, 12 disconnector, disconnector spring, disconnector plunger and M2 hammer are classified as a achinegun. These parts are used specifically for fully automatic fire and have no application in a emiautomatic carbine. While other parts such as an M2 sear, operating slide, trigger housing and stock re used in the fully automatic carbine, these parts are also appropriate for use in semiautomatic M1 arbines.[16]

herefore, the M2 sear, operating slide, trigger housing and stock are not a combination of parts esigned and intended for use in converting a weapon into a machinegun. Other commonly encountered

---

TM9-1267, Cal. .30 Carbines M1, M1A1, M2, and M3, United States Government Printing Office, 1953

conversion kits include modified trigger housings and/or trigger paks for Heckler & Koch (HK) type semiautomatic firearms. As originally manufactured, semiautomatic HK firearms (HK, 41, 43, 91, 93 and SP89) were specifically designed such that they will not accept fully automatic trigger housings or trigger paks for HK selective fire weapons such as the G3 and MP5. If selective fire trigger paks or trigger housings are modified so that they will function with semiautomatic HK firearms, the modified components are classified as parts designed and intended solely and exclusively, or combination of part designed and intended for use in converting a weapon into a machinegun. These modified parts are also machineguns as defined.

The following illustration shows a selective fire HK trigger pak with a selective fire trigger housing tha has been modified to function with a HK semiautomatic firearm by removing the forward pivot point o "ears" from the trigger housing.

modified HK selective fire trigger housing

Illustrated below is a selective fire HK trigger pak that has been modified by notching the forward lowe corner of the pak so that it will fit into a standard semiautomatic HK trigger housing.

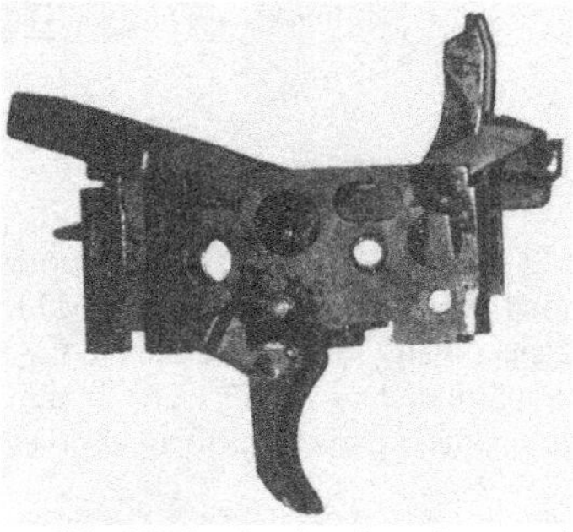

modified HK selective fire trigger pak

NOTE: standard selective fire HK trigger housings and trigger paks as originally manufactured are component parts for machineguns. These unmodified parts, in and of themselves, are not subject to the NFA. However, when adapted to function with a semiautomatic HK firearm the modified parts have been redesigned and are intended for use in converting a weapon into a machinegun.

'he following illustration shows a semiautomatic HK trigger pak with HK conversion sear installed.

HK semiautomatic trigger pak with conversion sear installed

'or the conversion sear to function the trigger or the trigger pak must be modified to increase the earward travel of the trigger. When the trigger is modified a notch is cut into the trailing leg to provide nore travel before the trigger contacts the upper trigger stop. When the trigger pak is modified, the pper trigger stop is either removed or relocated. *IMPORTANT NOTE: should the conversion sear be emoved from the trigger pak and the modified pak left in the firearm, the weapon will still be capable of illy automatic fire. Therefore, it is important that registered HK conversion sears be kept with their espective trigger paks. This is particularly important in instances where HK type firearms are sold as eing "sear ready" or "sear host guns". If these weapons contain semiautomatic trigger paks modified o function with conversion sears the firearms are capable of fully automatic fire (without the onversion sear) and as such are machineguns as defined.*

'oncerning the installation of conversion kits in semiautomatic firearms, it must be pointed out that the eceiver of the firearm may not be modified to permit fully automatic fire. Such modification results in he making of a machinegun which is prohibited by 18 U.S.C. 922(o).

'he definition of machinegun also includes a combination of parts from which a machinegun can be ssembled if such parts are in the possession or under the control of a person. An example of a firearm neeting this section of the definition is a semiautomatic AR15 rifle possessed with an M16 bolt carrier, ammer, trigger, disconnector and selector. If the semiautomatic AR15 is assembled with the described Л16 parts and the rifle is capable of fully automatic fire, the weapon possessed in conjunction with the Л16 parts, whether assembled or not, is a machinegun as defined.[17]

' ATF P 5300.4 (9/05), Federal Firearms Regulations Reference Guide – 2005, p. 155

An additional example of a combination of parts from which a machinegun can be assembled is a STEN submachinegun "parts kit" possessed with a length of metal tube to be used as a replacement receiver and instructions for assembling the parts into a functional machinegun. The parts kit as sold does not contain a firearm receiver although remnants of the destroyed receiver may be present. A machinegun parts kit in this condition is not subject to the GCA or the NFA.

Unfinished receiver tubes with instructions and/or templates for use in the assembly of a functional machinegun are also commercially available. These tubes with instructions/templates, in and of themselves, are not subject to the GCA or NFA.

When the parts kit is possessed in conjunction with the above described unfinished receiver tube, a combination of parts from which a machinegun can be assembled exists and is a machinegun as defined

2.1.7 Silencer. A firearm silencer and a firearm muffler are defined as any device for silencing, muffling, or diminishing the report of a portable firearm.[18] Firearm silencers are generally composed of an outer tube, internal baffles, a front end cap, and a rear end cap.

complete firearm silencer

The definition of a silencer also includes any combination of parts, designed or redesigned, and intended for use in assembling or fabricating a firearm silencer or firearm muffler.

The following illustration depicts parts that are designed and intended for use in assembling a firearm silencer. Another example of parts redesigned and intended for use in assembling or fabricating a firearm silencer are automotive engine freeze plugs that have been modified by drilling a hole through their center to permit passage of a bullet.

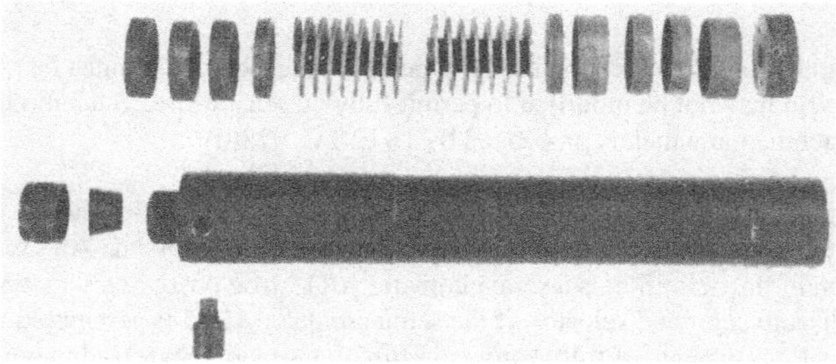

silencer parts

[18] 18 U.S.C. 921(a)(24)

Also included within the silencer definition is any part intended only for use in the assembly or fabrication of a firearm silencer.

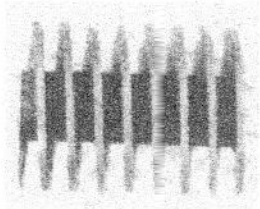

silencer baffle

Any of the above illustrated components meet the definition of a firearm silencer and are subject to the NFA. *NOTE: the language in the definition of silencer contains no provisions that permit an owner of a registered silencer to possess spare or replacement components for the silencer. However, licensed manufacturers who are SOTs may possess spare silencer components in conjunction with their manufacturing operations.*

2.1.8 Destructive device. The destructive device definition contains different categories that address specific types of munitions. Each category describes the devices subject to the definition based on the material contained in the item, the dimensions of the bore of certain weapons, and a combination of parts for use in converting the described items into destructive devices.

2.1.8.1 Explosive devices. The first portion of the definition deals with explosive, incendiary and poison gas munitions. The definition specifies that any explosive, incendiary or poison gas bomb, grenade, mine or similar device is a destructive device.

explosive bomb

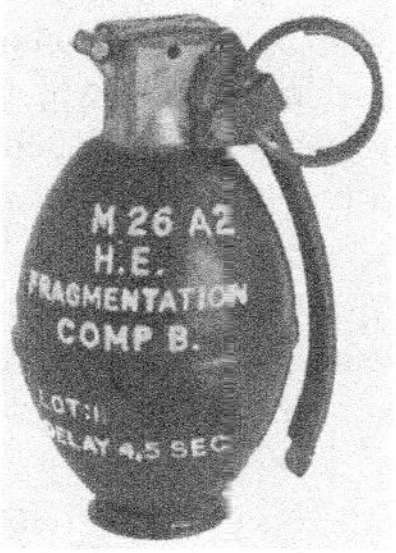

explosive grenade

This portion of the definition includes a rocket having a propellant charge of more than four ounces and a missile (projectile) having an explosive or incendiary charge of more than one-quarter ounce.

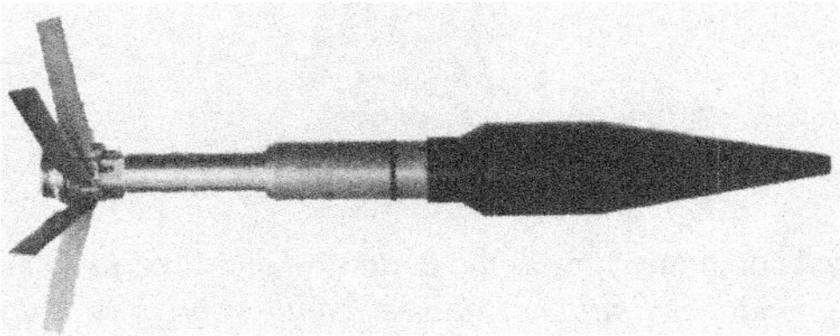

rocket with more than 4 ounces of propellant

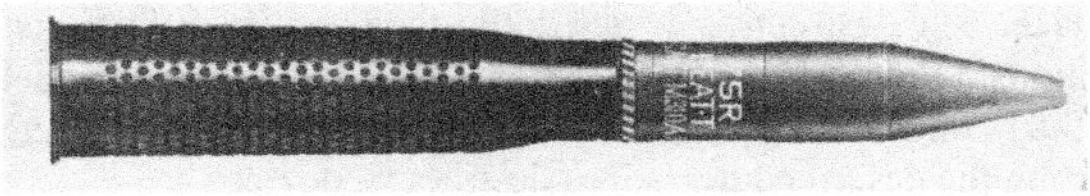

projectile

NOTE: Missiles (projectiles) less than caliber 20mm generally are not large enough to accommodate more than one-quarter ounce of explosive or incendiary material. In the case of 20mm high explosive (HE) or high explosive incendiary (HEI) projectiles, it is imperative to determine the model designation of the specific item as some 20mm HE and HEI projectiles contain more than one-quarter ounce of explosive or incendiary material and are destructive devices. Other 20mm HE and HEI projectiles do not contain more than one-quarter ounce of explosive and are not destructive devices. Therefore, it is incumbent upon persons interested in 20mm HE and HEI ammunition to determine the amount of explosives contained in a specific projectile. HE and HEI missiles (projectiles) larger than 20mm generally contain more than one-quarter ounce of explosive or incendiary material and are destructive devices.

2.1.8.2 Large caliber weapons. The second section of the definition states that any type of weapon by whatever name known which will, or which may be readily converted to, expel a projectile by the action of an explosive or other propellant, the barrel or barrels of which have a bore diameter of more than one-half inch in diameter is a destructive device. This portion of the definition specifically excludes a shotgun or shotgun shell which the Attorney General finds is generally recognized as particularly suitable for sporting purposes. ATF has issued rulings classifying specific shotguns as destructive devices because they have a bore of more than one half inch in diameter and were found to not be particularly suitable forfor sporting purposes.[19]

The majority of weapons covered by this portion of the destructive device definition are large caliber military weapons such as rocket launchers, mortars and cannons.

[19] Appendix B (ATF Rulings 94-1, 94-2)

RPG 7 launcher (bore diameter 1.57 inches)

120mm mortar (bore diameter 4.7 inches)

It is important to note that the large caliber firearms covered by this section are defined as weapons that expel a projectile by the action of an explosive *or other propellant*. This is the only place in the GCA and NFA where a propellant other than an explosive must be considered when classifying a weapon. Examples of weapons having a bore diameter of more than one-half inch in diameter and that expel a projectile by means other than an explosive are mortars that utilize compressed air as a propellant and some rocket launchers.

Certain destructive devices may also meet the definition of machinegun because in addition to having a bore diameter of more than one-half inch the weapons are capable of fully automatic fire. ATF treats NFA firearms of this type as both machineguns and destructive devices. The weapons are coded as machineguns in the NFRTR with an annotation that they are also destructive devices. Any such weapons manufactured on or after May 19, 1986 are subject to 18 U.S.C. 922(o). In instances where a weapon of this type is being transferred, it is imperative that State and local laws where the weapon is being transferred do not prohibit possession of destructive devices or machineguns.

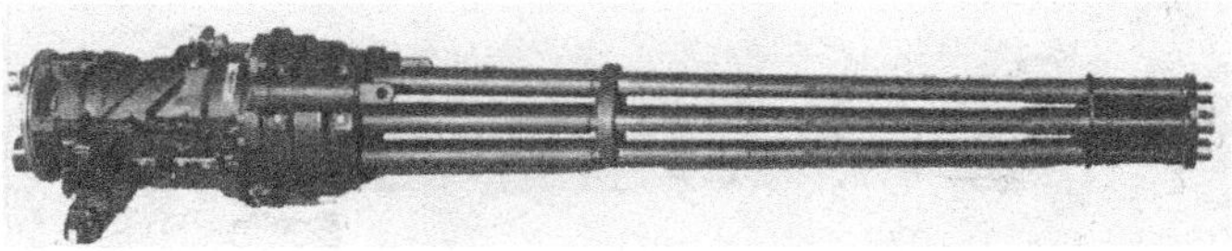

M61 20mm full automatic cannon

In addition to defining destructive devices, the definition also specifically excludes certain items from that classification. As previously stated, any shotgun or shotgun shell which the Attorney General finds is generally recognized as particularly suitable for sporting purposes is not a destructive device. Additionally, the following items are also excluded from the definition:

- Any device which is neither designed nor redesigned for use as a weapon.
- Any device, although originally designed for use as a weapon, which is redesigned for use as a signaling, pyrotechnic, line throwing, safety or similar device.
- Surplus ordnance sold, loaned or given by the Secretary of the Army pursuant to the provisions of 10 U.S.C. 4684(2), 4685, or 4686.
- Any other device which the Attorney General finds is not likely to be used as a weapon, or is an antique, or is a rifle which the owner intends to use solely for sporting purposes.

It should not be assumed that any device meeting the above descriptions is automatically excluded from the definition of a destructive device. ATF has ruled that certain pyrotechnic devices are destructive devices.[20] ATF should be contacted to confirm the classification of any items that appear to meet the above exclusions. Additionally, many of the items excluded from the definition of destructive device may contain a firearm receiver and would still be a firearm as defined in the GCA.

2.1.9 Unserviceable firearm. An unserviceable firearm is a firearm that is incapable of discharging a shot by the action of an explosive and is incapable of being readily restored to a firing condition. The most common method for rendering a firearm unserviceable, and that recommended by ATF, is to weld the chamber of the barrel closed and weld the barrel to the receiver.[21] The chamber of the barrel should be plug welded closed and all welds should be full fusion, deep penetrating, and gas or electric steel welds. In instances where the above procedure cannot be employed to render a firearm unserviceable, FTB should be contacted for alternate methods.

It is important to remember that rendering a firearm unserviceable does not remove it from the definition of an NFA firearm. An unserviceable NFA firearm is still subject to the import, registration, and transfer provisions of the NFA. However, there is no tax imposed on the transfer of an unserviceable

[20] Appendix B (ATF Ruling 95-3)
[21] ATF Form 5 (5320.5), Instruction 6a

irearm as a "curio or ornament." See 26 U.S.C. 5852(e). *NOTE: "curio or ornament" is only lescriptive of unserviceable firearms transferred exempt from transfer tax. An unserviceable firearm ransferred as a "curio or ornament" is not necessarily a "curio or relic" firearm for purposes of the GCA unless the weapon is classified as a curio or relic under the GCA.* For further information on curio r relic classification see section 2.2.

ection 2.2 Antique firearm. Firearms defined by the NFA as "antique firearms" are not subject to ny controls under the NFA.[22] The NFA defines antique firearms based on their date of manufacture nd the type of ignition system used to fire a projectile. Any firearm manufactured in or before 1898 hat is not designed or redesigned for using rimfire or conventional center fire ignition with fixed mmunition is an antique firearm. Additionally, any firearm using a matchlock, flintlock, percussion ap or similar type ignition system, irrespective of the actual date of manufacture of the firearm, is also n antique firearm.

NFA firearms using fixed ammunition are antique firearms only if the weapon was actually nanufactured in or before 1898 *and* the ammunition for the firearm is no longer manufactured in the Jnited States and is not readily available in the ordinary channels of commercial trade. To qualify as an ntique firearm, a fixed cartridge firing NFA weapon must meet both the age and ammunition vailability standards of the definition.

Concerning ammunition availability, it is important to note that a specific type of fixed ammunition that as been out of production for many years may again become available due to increasing interest in lder firearms. Therefore, the classification of a specific NFA firearm as an antique can change if mmunition for the weapon becomes readily available in the ordinary channels of commercial trade.

ection 2.3 Curios or relics. Curios or relics are firearms that are of special interest to collectors.[23] NFA firearms can be classified as curios or relics under the same criteria used to classify conventional irearms as curios or relics.[24]

An NFA firearm that is recognized as a curio or relic is still an NFA "firearm" and is still subject to the egistration and transfer provisions of the NFA. The primary impact of a curio or relic classification is hat a properly registered NFA firearm classified as a curio or relic may be lawfully transferred interstate o, or received interstate by, a person licensed as a collector of curios or relics under the GCA.

ection 2.4 Applications to remove firearms from the scope of the NFA as collector's items.

Certain NFA weapons can be removed from the provisions of the NFA as collector's items.[25] The rocedures for requesting removal of an NFA firearm are the same as used for requesting a destructive levice determination.[26]

---

26 U.S.C. 5845(a), (g)
27 CFR 478.11
27 CFR 478.26
26 U.S.C. 5845(a)
27 CFR 479.24 - 479.25

An NFA firearm removed from the NFA as a collector's item is no longer subject to any of the provisions of the NFA. In most cases, the weapon will still be a firearm as defined in the GCA and subject to regulation under the GCA. In some situations, the weapon that is removed from the NFA as collector's item will be an antique firearm as defined in the GCA.[27] In these instances, the weapon would no longer be a firearm as defined in Federal law.

The Attorney General does not have the authority to remove a machinegun or a destructive device from the provisions of the NFA as collector's items.[28] Therefore, applications to remove machineguns or destructive devices from the NFA as collector's items cannot be approved.

Section 2.5 Removal of firearms from the scope of the NFA by modification/elimination of components.

Firearms, except machineguns and silencers, that are subject to the NFA fall within the various definitions due to specific features. If the particular feature that causes a firearm to be regulated by the NFA is eliminated or modified, the resulting weapon is no longer an NFA weapon.

For example, a shotgun with a barrel length of 15 inches is an NFA weapon. If the 15- inch barrel is removed and disposed of, the remaining firearm is not subject to the NFA because it has no barrel. Likewise, if the 15 inch barrel is modified by permanently attaching an extension such that the barrel length is at least 18 inches and the overall length of the weapon is at least 26 inches, the modified firearm is not subject to the NFA. *NOTE: an acceptable method for permanently installing a barrel extension is by gas or electric steel seam welding or the use of high temperature silver solder having a flow point of 1100 degrees Fahrenheit.*

A shot pistol ("any other weapon") such as an H&R Handy Gun may be removed from the NFA by either disposing of the smooth bore barrel or permanently installing a rifled sleeve chambered to accept a standard pistol cartridge into the smooth bore barrel. Modified by sleeving the barrel, an H&R Handy Gun is no longer an NFA weapon because it now has a rifled bore.

Large caliber destructive devices that are not also machineguns can be removed from the NFA by disposing of the barrel. If the barrel of a 37mm cannon is removed and disposed of, the remaining weapon has no barrel or bore diameter. As an alternative, the barrel of a destructive device may be functionally destroyed. To destroy the barrel of a destructive device the following operations must be performed:

- Cut a hole, equal to the diameter of the bore, on a 90-degree angle to the axis of the bore, through one side of the barrel in the high pressure (chamber) area.
- Weld the barrel to the receiver of the weapon.
- Weld an obstruction into the barrel to prevent the introduction of a round of ammunition.

2.5.1 Removal of machineguns and silencers from the scope of the NFA. Machineguns are defined to include the receiver of a machinegun and the definition of silencer includes each component of a

[27] 18 U.S.C. 921(a)(16)
[28] 26 U.S.C. 5845(a)

silencer. Therefore, to remove these weapons from the provisions of the NFA, the receiver of a machinegun or all the components of a silencer must be destroyed.

The preferred method for destroying a machinegun receiver is to completely sever the receiver in specified locations by means of a cutting torch that displaces at least one-quarter inch of material at each cut location. ATF has published rulings concerning the preferred destruction of specific machineguns.[29]

A machinegun receiver may also be properly destroyed by means of saw cutting and disposing of certain removed portions of the receiver. To ensure that the proposed saw cutting of a particular machinegun receiver is acceptable, FTB should be contacted for guidance and approval of any alternative destruction proposal. *Note: a machinegun receiver that is not properly destroyed may still be classified as a machinegun, particularly in instances where the improperly destroyed receiver is possessed in conjunction with other component parts for the weapon.*

A silencer may be destroyed by completely severing each component by means of a cutting torch that has a tip of sufficient size to displace at least one-quarter inch of material at each cut location.

Concerning the outer tube(s) of a silencer, these components may be destroyed by crushing them flat in lieu of cutting with a torch.

Anyone interested in destroying an NFA weapon by means other than described above should contact FTB to discuss possible alternatives.

---

[29] Appendix B (ATF Rulings 2003-1, 2003-2, 2003-3, 2003-4)

CHAPTER 3. REGISTRATION OF NFA FIREARMS

Section 3.1 The National Firearm Registration and Transfer Record (NFRTR)

The NFRTR is the central registry of all NFA firearms in the U.S. which are not in the possession or under the control of the U.S. Government. The registry includes (1) the identification of the firearm, (2) date of registration, and (3) identification and address of the person entitled to possession of the firearm (the person to whom the firearm is registered).[30]

Section 3.2 Who may register NFA firearms

3.2.1 Amnesty registration. When the NFA was amended in 1968, a 30-day amnesty period immediately following the law's effective date was established during which persons possessing unregistered firearms could register them in the NFRTR.[31]

The 1968 amendments also provided for the establishment of additional amnesty periods not exceeding 90 days per period.[32] To date, no additional amnesty periods have been declared. Requests for further amnesty periods have been denied, principally because additional periods could jeopardize pending ATF investigations and prosecutions of NFA violations.

3.2.2 Registration by State and local agencies. To be lawfully possessed by States and political subdivisions of the States (for example, local police departments), NFA firearms must be registered in the NFRTR. The regulations permit State and local police organizations acquiring unregistered NFA firearms for official use, by seizure, forfeiture, or abandonment, to register them in the NFRTR by filing ATF Forms 10. Appendix C contains a copy of the form. Firearms registered on Forms 10 are for official use only and subsequent transfers will be approved only to other government agencies for official use.[33] For example, they may not be traded to an FFL/SOT in exchange for other firearms or police equipment.

3.2.3 Registration by makers. Persons other than FFLs and SOTs desiring to make an NFA firearm are required to first register the firearm by filing Form 1 with ATF and obtaining approval of the form and registration of the firearm. Appendix C contains a copy of the form. ATF will approve a making application on Form 1 if the maker pays the $200 making tax required by the NFA, identifies the firearm as the form requires, includes his/her fingerprints and photographs if the maker is an individual, and if the making and the maker's possession of the firearm would not place the maker in violation of any Federal, State or local law.[34] A law enforcement certification is also required if the maker is an individual. Note also that ATF will not approve the making of a machinegun it determines would violate 18 U.S.C. 922(o). Section 922(o) generally prohibits the possession of machineguns manufactured on or after May 19, 1986.

---

[30] 26 U.S.C. 5841(a)
[31] Section 207(b) of the Gun Control Act of 1968, Public Law 90-618, approved October 22, 1968
[32] Section 207(d), ibid
[33] 27 CFR 479.104
[34] 26 U.S.C 5822; 27 CFR 479.62, 479.63, 479.64

.2.4 **Registration by importers.** FFLs/SOTs qualified as importers must register imported firearms y filing ATF Forms 2, Notice of Firearms Manufactured or Imported, no later than 15 days from the ate the imported firearms were released by Customs. Upon timely receipt by ATF of a Form 2 and a opy of Form 6A showing Customs' release of an imported firearm, ATF will register the firearm to the nporter.[35] Appendix C contains copies of Forms 2 and 6A. Note that the NFA prohibits importation of IFA firearms unless they are being imported for the use of the United States or a State agency, for cientific or research purposes, or for testing or use as a model by a registered importer or solely for use s a sales sample by a registered importer or registered dealer.[36] In the case of an imported machinegun, ection 922(o) of the GCA would also apply.

.2.5 **Registration by manufacturers**. FFLs/SOTs qualified as manufacturers must register ıanufactured firearms by filing ATF Form 2, Notice of Firearms Manufactured or Imported. All rearms manufactured during a single day must be listed on one Form 2. The form must be filed no ıter than the close of the next business day. Receipt of the form by ATF will serve to register the listed rearms to the manufacturer.[37] Appendix C contains a copy of Form 2.

.2.6 **Registration to transferees**. Registered firearms may be transferred by their registered wners/possessors to transferees. Other than Form 10 registration, there is no mechanism in the NFA to ıwfully transfer unregistered NFA firearms.

.2.6.1 **Transfers by persons other than FFLs/SOTs to other such persons**. Transferors of registered rearms must file ATF Forms 4, Application for Tax Paid Transfer and Registration of a Firearm, to egister the firearm to the transferee and pay the applicable transfer tax.[38] Appendix C contains a copy f the form. The form must be approved by ATF before the transfer may be made.[39] ATF will not pprove the form if the transfer, receipt, or possession of the firearm would place the transferee in iolation of any Federal, State, or local law.[40] A law enforcement certification is also required on ATF orm 4.

.2.6.2 **Transfers by FFLs/SOTs to persons other than FFLs/SOTs**. Transferors of registered rearms must file ATF Forms 4, Application for Tax Paid Transfer and Registration of a Firearm, to egister the firearm to the transferee and pay the applicable transfer tax. Appendix C contains a copy of ıe form. The form must be approved by ATF before the transfer may bemade.[41] ATF will not approve ıe form if the transfer, receipt, or possession of the firearm would place the transferee in violation of ıy Federal, State, or local law.[42]

.2.6.3 **Transfers by non-FFLs/SOTs to FFLs/SOTs**. Transferors of registered firearms must file TF Forms 4, Application for Tax Paid Transfer and Registration of a Firearm, to register the firearm to

---

27 CFR 479.112
26 U.S.C. 5844
27 CFR 479.103
26 U.S.C. 5812(a); 27 CFR 479.84
26 U.S.C. 5812(b); 27 CFR 479.86
26 U.S.C. 5812(a); 27 CFR 479.85
26 U.S.C. 5812(b); 27 CFR 479.86
26 U.S.C. 5812(a); 27 CFR 479.85

the transferee and pay the applicable transfer tax.[43] Appendix C contains a copy of the form. The form must be approved by ATF before the transfer may be made.[44] ATF will not approve the form if the transfer, receipt, or possession of the firearm would place the transferee in violation of any Federal, State, or local law.

3.2.6.4 Transfers by FFLs/SOTs to other FFLs/SOTs. Transferors must file ATF Forms 3, Application for Tax-Exempt Transfer of Firearm and Registration to Special Occupational Taxpayer, to register the firearm to the transferee.[45] Appendix C contains a copy of the form. In these transactions, the transferor has no liability for the transfer tax. The form must be approved by ATF before the transfer may be made.[46] ATF will not approve the form if the transfer, receipt, or possession of the firearm would place the transferee in violation of any Federal, State, or local law.

3.2.6.5 Transfers to State and local government agencies. Transferors must file ATF Forms 5, Application for Tax Exempt Transfer and Registration of a Firearm, to register the firearm to such agency.[47] Appendix C contains a copy of the form. In these transactions, the transferor has no liability for the transfer tax. The Form must be approved by ATF before the transfer may be made.[48]

Section 3.3 Status of unregistered firearms

Firearms not lawfully registered as required by the NFA may not be registered and legitimized by their possessors. They are contraband and unlawful to possess.[49] However, see Section 2.4 for information on removing NFA firearms from the scope of the NFA because of their status as collectors' items, modification, or elimination of certain component parts.

Section 3.4 ATF disclosure of NFA registration information

3.4.1 Restrictive use of information. The NFA provides that no information or evidence obtained from an application, registration, or records required to be submitted or retained by a natural person (individual) in order to comply with the NFA or the NFA regulations shall be used directly or indirectly as evidence against the person in a criminal proceeding with respect to a violation of law occurring prior to or concurrently with the filing of the application or registration.[50] Filing false information is an exception to this prohibition.[51]

3.4.2 Prohibition on ATF's disclosure of tax returns or tax return information. NFA forms are treated as tax returns and registration information in the NFRTR is considered to be tax return

---

[43] 26 U.S.C. 5812(a); 27 CFR 479.84
[44] 26 U.S.C. 5812(b); 27 CFR 479.86
[45] 26 U.S.C. 5812(a); 27 CFR 479.88
[46] 26 U.S.C. 5812(b); 27 CFR 479.88(b)
[47] 26 U.S.C. 5812(a); 27 CFR 479.90
[48] 26 U.S.C. 5812(b); 27 CFR 479.90(b)
[49] U.S. v. Freed, 401 U.S. 601 (1971)
[50] 26 U.S.C. 5848(a)
[51] 26 U.S.C. 5848(b)

nformation. ATF is generally prohibited from disclosing tax returns and tax return information.[52] Iowever, firearms registration information may be disclosed to registered owners/possessors of the irearms.

.4.3 Determining the registration status of an NFA firearm. The situation may arise when a person inds in his or her possession an NFA firearm and is uncertain whether the firearm is lawfully registered. Naturally, the person will want to query the NFA Branch to determine the registration status of the irearm. Because of the restriction on disclosure of NFA registration information discussed in section .4.2, ATF will not respond to the person's telephone request for the registration status of the firearm. 'o communicate with the person, the NFA Branch will respond to the request if the person verifies his r her identity to the Branch in writing. If the firearm is registered to the person in the NFRTR, the Branch will so advise the person and, if the circumstances warrant, provide the person with a copy of the egistration. See also Section 3.5 on lost or stolen registration documents.

Section 3.5 Lost or stolen registration documents. A person possessing a firearm registered as equired by the NFA must retain proof of registration, that is, the registration form showing registration f the firearm to the person, which must be made available to ATF upon request.[53] If a registrant liscovers that a Form 1, 2, 3, 4, 5, 6A, or 10 is stolen, lost or destroyed, the registrant must immediately eport the theft, loss, or destruction in writing to the NFA Branch.[54] The report must contain the details f the situation. ATF will issue a duplicate copy of the registration document as the circumstances varrant.

Section 3.6 Correcting incorrect registration documents. Occasionally, the registered possessor of n NFA firearm may notice that the registered firearm does not match the registration document. Perhaps the serial number is slightly different. In this situation, the registrant should take a photograph f the markings on the firearm (or a rubbing) and send it to the NFA Branch with a written request to orrect the serial number as documented on the NFRTR. ATF will respond to the request by letter tating that the NFRTR has been corrected and advising the registrant to keep the letter with the egistration document as evidence of proper registration.

Section 3.7 Maintaining registration documents. A person possessing an NFA firearm registered as equired by law must retain proof of registration, that is, the document showing the person's registration, which must be made available to ATF upon request.[55]

---

[52] 26 U.S.C. 6103
[53] 26 U.S.C. 5841(e)
[54] 27 CFR 479.142
[55] 26 U.S.C. 5841(e)

CHAPTER 4. TAXES IMPOSED BY THE NFA

Section 4.1 Taxes

The NFA imposes tax on the making and transfer of firearms. The Act also requires annual payment of special (occupational) tax (SOT) by licensees engaged in the business of manufacturing, dealing in, or importing NFA firearms. The NFA also provides for exemptions from the making, transfer, and special (occupational) taxes in specific situations.

4.1.1 Making tax. A tax of $200 is imposed on the making of all NFA firearms made by other than qualified manufacturers of firearms.[56] The tax must be paid prior to making the firearm. Payment of th making tax is submitted with the Application to Make and Register a Firearm (ATF Form 1). See Appendix C for a copy of Form 1. The making tax is also imposed when a registered, unserviceable NFA firearm is reactivated.[57] See the exemptions from tax discussed in Section 4.2.1.

4.1.2 Transfer tax. Transfer of a serviceable NFA firearm is subject to the transfer tax.[58] The tax mu be paid prior to the transfer. Payment of the tax is submitted with the Application to Transfer and Register a Firearm (ATF Form 4). See Appendix C for a copy of Form 4. ATF will accept payment of the tax from either the transferor or transferee. The tax on the transfer of short barrel shotguns, short barrel rifles, machineguns, silencers and destructive devices is $200. The transfer tax for firearms classified as "any other weapon" is $5. See the exemptions from tax discussed in Section 4.2.2.

NOTE: there is often confusion concerning the tax on "any other weapons." The majority of NFA weapons are subject to a making tax of $200 and a transfer tax of $200. Many individuals have the mistaken belief that the rate of tax for making an "any other weapon" is $5 because the transfer tax on "any other weapons" is $5. As discussed in Section 4.1.1, the making tax on all types of NFA firearms is $200.

4.1.3 Special (occupational) tax. On first engaging in business, each importer, manufacturer, and dealer in NFA firearms must pay a special (occupational) tax for each place of business.[59] Special tax i paid by return, specifically ATF Form 5630.7, Special Tax Registration and Return. Appendix C contains a copy of the return. Subsequent to the initial payment of the SOT, the tax is due on or before July 1 of each year. The rate of tax for importers and manufacturers is $1000 per year. The rate of tax for dealers is $500 per year. The NFA provides a reduced rate of SOT for importers and manufacturers whose gross receipts for the most recent taxable year are less than $500,000.[60] See the exemptions fror tax discussed in Section 4.2.3.

---

[56] 26 U.S.C. 5821
[57] ATF Form 1 (5320.1), Instruction k
[58] 26 U.S.C. 5811
[59] 26 U.S.C. 5801
[60] 26 U.S.C. 5801(b)(1)

Section 4.2 Exemptions from tax

4.2.1 Making tax. A manufacturer who has paid the SOT is exempt from payment of the making tax.[61] Likewise, there is no making tax imposed on the making of an NFA firearm by or on behalf of a Federal or State agency.[62]

4.2.2 Transfer tax. Transfers of registered firearms between SOTs are exempt from the transfer tax.[63] Likewise, there is no transfer tax imposed on the transfer of firearms to a Federal or State agency.[64]

4.2.2.1 Unserviceable firearms. An unserviceable firearm may be transferred as a curio or ornament without payment of the transfer tax.[65]

4.2.3 Special (occupational) tax. A person required to pay SOT is exempt from the tax if it is established that the business is conducted exclusively with, or on behalf of, an agency of the United States.[66] This exemption must be obtained by filing a letter application addressed to the NFA Branch setting out the manner in which the applicant conducts business, the type of firearms to be manufactured, and satisfactory proof of the existence of the applicant's contract with the Government.[67] This exemption must be renewed by letter on or before July of each year. Approval of the application entitles the applicant to the exemption.

4.2.4 Exportation of firearms. An NFA firearm may be exported without payment of the transfer tax provided that proof of the exportation is furnished in such form and manner as the regulations prescribe.[68] The form to be filed is ATF Form 9 (Firearms), Application and Permit for Exportation of Firearms.[69] See Appendix C for a copy of the form. Approval of the form by ATF is required before exportation. *NOTE*: *exportation of NFA firearms without an approved Form 9 will subject their registered owner to NFA transfer tax.*

---

[61] 26 U.S. 5852(c)
[62] 26 U.S.C. 5852(b), 5853(b)
[63] 26 U.S.C. 5852(d)
[64] 26 U.S.C. 5852(a), 5853(a)
[65] 26 U.S.C. 5852(e)
[66] 26 U.S.C. 5851
[67] 27 CFR 479.33
[68] 26 U.S.C. 5854
[69] 27 CFR 479.114

CHAPTER 5. QUALIFYING TO DO BUSINESS IN NFA FIREARMS

Section 5.1 Licensing under the GCA

5.1.1 Application for GCA license. Any "person" intending to engage in the business of importing, manufacturing, or dealing in NFA firearms, which are also defined as "firearms" under the GCA, must first apply for and obtain a license.[70] The license application is ATF Form 7 (5310.12). Appendix C contains a copy of the form. Licenses are issued for a period of 3 years.

5.1.1.1 Definition of "person" for GCA purposes. The GCA defines "person" to include "any individual, corporation, company, association, firm, partnership, society, or joint stock company."[71] ATF recognizes the term to include a limited liability company ("LLC") organized under State law. Thus, an LLC must obtain a GCA license to engage in a GCA firearms business and ATF will issue a license to an LLC meeting the licensing standards.

5.1.2 License fees.[72]

5.1.2.1 Manufacturers. If the applicant is a manufacturer of destructive devices, the fee is $1,000 per year. For firearms other than destructive devices, the fee is $50 per year.

5.1.2.2 Importers. If the applicant is an importer of destructive devices, the fee is $1,000 per year. For firearms other than destructive devices, the fee is $50 per year.

5.1.2.3 Dealers. If the applicant is a dealer in destructive devices, the fee is $1,000 per year. For firearms other than destructive devices, the fee is $200 for 3 years, except that the fee for renewal is $90 for 3 years.

5.1.3 Licensing standards under the GCA.[73]

5.1.3.1 Age. If the applicant is an individual, he or she must be at least 21 years of age.

5.1.3.2 Prohibited persons. Persons who fall within the prohibited persons categories set forth in 18 U.S.C. 922(g) and (n) do not qualify for licensing. The same is true of business entities having "responsible persons" under any one of the disabilities.

5.1.3.3 "Responsible persons." If the applicant is a corporation, partnership, or association (including an LLC), the applicant would not qualify for licensing if the entity itself has GCA disabilities under section 922(g) or (n) or has any individual possessing directly or indirectly the power to direct or cause the direction of the management and policies of the firm under disabilities. These management and policy-making individuals include only those having responsibilities with respect to firearms. For example, they would not include individuals only

---

[70] 18 U.S.C. 923(a); 27 CFR 478.44
[71] 18 U.S.C. 921(a)(1)
[72] 18 U.S.C. 923(a)(1); 27 CFR 478.42
[73] 18 U.S.C. 923(d)(1)

having duties related to administration or personnel. Those having management and policy-making responsibilities with respect to firearms are referred to as "responsible persons" and must be listed on the firm's license application and their fingerprints and photographs must accompany the application. If the firm has a "responsible person" under disabilities, the firm would become eligible for licensing if the person is removed from his or her firearms management or policy-making position.

5.1.3.4 Removing disabilities. The GCA provides a mechanism for ATF to grant relief to persons under Federal firearms disabilities who apply for relief.[74] However, Congress has denied ATF's use of funds to process relief applications or grant relief. An exception was made for corporations. Thus, a corporation convicted of a "crime punishable by imprisonment for a term exceeding one year" may apply for and obtain relief from ATF. Not all disabilities are permanent. *Examples*: a person convicted of a crime in Federal court could get relief by applying to the U.S. Pardon Attorney for a presidential pardon and obtaining a pardon; a person convicted of a State crime may, depending on State law, get relief by a State pardon, expunction or set-aside of the conviction, or restoration of rights if the person would have no remaining firearms disabilities under State law[75]; and a person under a restraining order relating to domestic violence would be relieved from disabilities if the order is terminated.

5.1.3.5 Prior willful violations. The applicant must not have willfully violated any provision of the GCA or the regulations.

5.1.3.6 False statements. The applicant has not willfully failed to disclose any material information required or made any material false statements in connection with the application.

5.1.3.7 Business premises. The applicant must have a permanent premises from which business is to be conducted. The applicant must certify that conducting business from that location is not prohibited by State or local law; that within 30 days after license approval the business will comply with State and local laws applicable to the business and business will not be conducted until State and local requirements have been met; and that the chief law enforcement officer at the proposed business location has been advised of the intent to apply for a license.

5.1.3.8 May a license be obtained to do business solely away from the licensed premises at gun shows? Under the law, a licensee is entitled to do business at gun shows within the State in which the licensed premises is located. However, a person seeking a license to deal in firearms solely away from the licensed premises at gun shows does not qualify for licensing. Licensing in this instance would not meet the requirement discussed in Section 5.1.3.8 that the applicant must have a permanent premises from which business is conducted.

5.1.3.9 Gun storage and safety devices. Applicants for dealers licenses must certify that secure gun storage or safety devices will be available to nonlicensees with certain exceptions: storage or safety devices are temporarily unavailable because of theft, casualty loss, consumer sales, backorders, or similar reasons beyond the licensee's control.

---

18 U.S.C. 925(c)
18 U.S.C. 921(a)(20)

5.1.4 Multiple business locations. A separate license must be obtained for each location at which business will be conducted.[76]

5.1.4.1 Locations solely for storage. No license is required to cover a separate warehouse use by the FFL solely for storage of firearms if required records are maintained at the licensed premises served by the warehouse.[77]

5.1.5 Establishing a common expiration date for licenses at multiple locations. For the convenien of FFLs, ATF Ruling 73-9 (see Appendix C) provides that FFLs holding licenses at more than one location may establish a common expiration date for the licenses issued to their several locations. FFL wishing to establish such a date for all licenses issued to them may apply in writing to the Federal Firearms Licensing Center.

5.1.6 Engaging in business as both an importer and a manufacturer. Persons intending to engage in business as firearms importers and manufacturers, even if the businesses will be conducted from the same premises, must have both an importer's license and a manufacturer's license.[78]

5.1.7 Do importers and manufacturers need a dealer's license to deal in the firearms they import or manufacture? Licensed importers and manufacturers are not required to have a separate dealer's license to deal in firearms they import or manufacture. A license as an importer or manufacturer authorizes the FFL to deal in the types of firearms authorized by the license to be imported or manufactured.[79]

5.1.8 License renewal. Licenses are issued for a 3-year period. Prior to the expiration of a license, th FFL will receive from ATF a license renewal application, ATF Form 8, that should be filed with ATF prior to the expiration date shown on the current license. If a Form 8 is not received 30 days before the expiration date, the FFL should contact ATF's Federal Firearms Licensing Center. It is the responsibility of the FFL to timely obtain and file a renewal application. Appendix C contains a copy the form. ATF has 60 days within which to act on an original or renewal license application.[80]

5.1.8.1 Right to operate while renewal application is pending; "letters of continuing operations." ATF's approval of a license renewal application may occur after the expiration date shown on the current license. If a license renewal application is timely filed, but not acted upon until after the expiration date of the applicant's current license, may the applicant lawfully conduct the business after that date? The law, 5 U.S.C. 558, allows an FFL to continue to lawfully conduct business under a current license until a timely filed renewal application is acte upon. To show the FFL's current suppliers that the license is still valid while the renewal application is pending, the FFL may obtain a "letter of continuing operations" from the Federal Firearms Licensing Center, showing that the transferee's license is still valid even though it appears on its face to have expired.

---

[76] 27 CFR 478.50
[77] Ibid
[78] 27CFR 478.41(b)
[79] Ibid
[80] 27 CFR 478.45

ection 5.2 Payment of special (occupational) tax to do business in NFA firearms. See also Section .1.3.

.2.1 Every "person" who engages in the business of importing, manufacturing, or dealing in irearms (including pawnbrokers) shall pay a special tax.[81] The tax must be paid on or before the ate of commencing the taxable business and every year thereafter on or before July 1. The tax is not rorated and is computed for the entire tax year (July 1 through June 30), regardless of the portion of the ear during which the taxpayer engages in business. Persons commencing business at any time after uly 1 in any year are liable for the tax for the entire tax year.

> **5.2.1.1 Definition of "person" for NFA purposes**. For purposes of the NFA, "person" is defined as a "partnership, company, association, trust, estate, or corporation, as well as a natural person."[82] ATF recognizes the term to include a limited liability company ("LLC") organized under State law. Thus, an LLC engaged in an NFA firearms business must pay the NFA special tax.

.2.2 Tax must be paid for each business location. The special tax must be paid for each premises where business will be conducted.[83]

.2.3 Rate of tax.

> **5.2.3.1 Importers and manufacturers.** The tax is generally $1,000 a year or fraction thereof.[84] See 26 U.S.C. 5801(b) and 27 CFR 479.32a for the reduced tax rate for certain "small importers and manufacturers." Those entitled to the reduced tax rate are subject only to a $500 tax per year.

> **5.2.3.2 Dealers**. The tax is $500 a year or fraction thereof.[85]

.2.4 How to pay special tax. Special tax must be paid by filing ATF Form 5630.7, Special Tax Registration and Return, together with a check or money order for the amount of the tax. The form and ax payment must be sent to the Bureau of ATF, Attention NFA, P.O. Box 403269, Atlanta, Ga. 30384-269. Upon filing a properly completed Form 5630.7 and payment of the special tax, the taxpayer will e issued a special tax stamp as evidence of tax payment [86]

> **5.2.4.1 Employer identification number (EIN)**. The tax return, Form 5630.7, must contain a valid EIN number issued by the Internal Revenue Service (IRS). If the taxpayer does not have an EIN number, the IRS in the taxpayer's locality should be contacted to obtain a number. The number may also be obtained by contacting the IRS on-line.

---

26 U.S.C. 5801
27 CFR 479.11
27 CFR 479.38
26 U.S.C 5801(a)(1)
26 U.S.C. 5801(a)(2)
27 CFR 479.36

5.2.4.2 Renewal of special tax. In May of each year, ATF will send a renewal tax return, ATF Form 5630.5R (NFA Special Tax Renewal Registration and Return) and ATF Form 5630.5RC (NFA Special Tax Location Registration Listing) to each person who paid the special tax for the current tax year. The taxpayer should make any necessary changes to the information pre-printed on the forms and return the forms with remittance to ATF as directed on the forms.

5.2.5 Exemption from special tax. Any person required to pay special tax shall be relieved from tax payment if it is established that the business is conducted exclusively on behalf of the U.S. Government.[87] The exemption may be obtained by filing with the NFA Branch a letter stating the manner in which business is conducted, the type of firearms to be manufactured, and proof of the person's contract with the U.S.[88] The exemption must be renewed on or before July 1 of each year.

5.2.6 Collectors acquiring NFA firearms for their personal collections by acquiring dealers' licenses and paying NFA special tax. Some NFA firearms collectors, who are not engaged in any firearms business, have been known to acquire a GCA license to deal in firearms and pay the NFA special tax to acquire NFA firearms for their personal firearms collections. They do so for a number of reasons: (1) to acquire firearms from nonlicensees residing out-of-state; (2) to circumvent requirements imposed on individuals to provide their fingerprints and photographs in order to receive NFA firearms and law enforcement certifications authorizing their receipt of such firearms; and (3) to avoid NFA transfer tax on firearms they receive from FFLs/SOTs.

Warning: These transactions violate the law and can only lead to trouble for the collector. In these instances, the collector has committed Federal felonies by falsely stating on a license application and special tax return that the collector intends to conduct a firearms business. Any NFA firearms received tax free by the collector are subject to transfer tax and the collector's receipt of the firearms tax free violated the NFA. As held in ATF Ruling 76-22, these transfers are unlawful and the firearms received are subject to seizure and forfeiture.

Section 5.3 Registration by firearms manufacturers and exporters with the U.S. Department of State (DOS). DOS has the responsibility of enforcing the provisions of the Arms Export Control Act (AECA) relating to the export of firearms and other defense articles on the U.S. Munitions List. Nevertheless, any person engaged in the U.S. in the business of manufacturing or exporting firearms or other defense articles is required to register with DOS's Directorate of Defense Trade Controls.[89] The regulations expressly state that "Manufacturers who do not engage in exporting must nevertheless register."

5.3.1 How manufacturers should register. DSP-9 (Registration Statement) and a transmittal letter required by DOS regulations at 22 CFR 122.2(b) must be submitted to DOS by the registrant with a check or money order payable to DOS of one of the fees prescribed by 22 CFR 122.3(a).[90] Appendix C contains a copy of Form DSP-9. The registration and transmittal letter must be signed by a senior officer who has been empowered by the registrant to sign the documents. The registrant must also

[87] 26 U.S.C. 5851
[88] 27 CFR 479.33
[89] 22 CFR 122.1
[90] 22 CFR 122.2(a)

submit documentation demonstrating that it is incorporated or otherwise authorized to do business in the U.S.

Section 5.4 Registration by firearms importers with ATF. Persons engaged in the business in the U.S. of importing firearms and other articles enumerated on the U.S. Munitions Import List must register with ATF's Director.[91]

5.4.1 How importers should register. An application to register must be filed on ATF Form 4587, in duplicate, with the Director and be accompanied with the registration fee at the rate prescribed by the regulations at 27 CFR 447.32(b).[92] Appendix C contains a copy ATF Form 4587. Fees paid in advance for whole future years of a multiple year registration will be refunded upon request if the registrant ceases to engage in importing such firearms or other articles on the List. A request for refund must be submitted to ATF's Director, attention Firearms and Explosives Imports Branch, prior to the beginning of any year for which refund is claimed.

Sec. 5.5 Filing by Facsimile Transmission. The NFA Branch allows an FFL who is qualified to import, manufacture, or deal in NFA to file certain registration forms by facsimile transmission (fax) directly to the NFA Branch. The forms are: Form 2, Form 3, Form 5 (where the application is not accompanied by fingerprint cards), and Form 9. However, before the FFL can file by fax, the FFL must first submit an affidavit, in original, to the Chief, NFA Branch. The affidavit concerns who files the forms, the treatment of the signature on the fax as original, and that the copies are to be treated as originals. The format for the affidavit and a discussion of the procedure is in an ATF memo to Federal Firearms Licensees and NFA Special (Occupational) Taxpayers, dated January 31, 2006, and can be found in Appendix B.

---

[91] 27 CFR 447.31

[92] 27 CFR 447.32

CHAPTER 6. MAKING NFA FIREARMS BY NONLICENSEE

Section 6.1 Requirements for making NFA firearms. Persons not otherwise prohibited from possessing firearms may submit an application to make an NFA firearm, other than a machinegun.[93] The application process requires submission of ATF Form 1, Application to Make and Register a Firearm, in duplicate, along with FBI FD-258, Fingerprint Card, in duplicate, and payment of the $200 making tax. Appendix C contains a copy of Form 1.

Section 6.2 Preparation of Form 1. Every person (other than a licensed manufacturer who has also paid the required SOT to manufacture NFA weapons) must complete the Form 1.[94] Two identical copies of the application must be prepared. All entries must be made in ink. All required signatures must be original and entered in ink. Photocopies or other facsimile or carbon copy signatures are not acceptable. Under no circumstances will a form filled in by use of a lead pencil be accepted. All changes made on the form must be initialed and dated by the applicant.

6.2.1 Description of firearm. If an existing firearm or firearm receiver is being used, the name and location of the original manufacturer of the weapon should be entered in Block 4(a). If the applicant is making a completely new firearm, the applicant's name and location should be entered in Block 4(a). The type of firearm being made, i.e., short barrel rifle, short barrel shotgun, any other weapon, silencer or destructive device, is to be entered in Block 4(b). The caliber or gauge of the firearm is to be entered in Block 4(c). If a model designation has been assigned to the firearm, that designation is to be placed in Block 4(d). If the weapon has no model designation, enter "none" in Block 4(d). The length of the barrel is to be entered, in inches, in Block 4(e) and the overall length of the firearm is to be entered, in inches, in Block 4(f).

All NFA firearms must be identified by a serial number and other specified markings[95]. If an existing firearm is being used in the making of the NFA weapon, and that firearm is serialized, the existing serial number should be used (unless it duplicates a serial number already used by the maker on Form 1) and entered in Block 4(g). If the weapon is of new manufacture, the applicant must assign a unique serial number and enter it in Block 4(g). For example, a unique serial number could be composed of at least 4 digits preceded by the initials of the maker. *NOTE*: alpha characters, e.g., a name, will not be accepted as a serial number. If a name is to be used, there must be at least one numeric character in addition to the alpha characters.

The serial number must be engraved or stamped on the receiver of the firearm and the caliber, model, and identification of the maker must be engraved on the barrel or frame or receiver of the weapon.[96] The marking and identification requirements for a maker are the same as for a manufacturer. Refer to section 7.4 for a detailed discussion of the requirements.

---

[93] 18 U.S.C. 922(o)
[94] 26 U.S.C. 5822
[95] 27 CFR 479.102
[96] Ibid

•.2.2 **Photograph of applicant.** An unmounted photograph of the applicant approximately 2 x 2 inches nd taken within the past year must be affixed to the indicated space on each copy of the application. 'he photograph must be original. Photocopies of photographs are not accepted.

•.2.3 **FBI Form FD-258, fingerprint card.** A completed Form FD-258 containing the fingerprints of he applicant must be submitted in duplicate with the Form 1. The fingerprints should be taken by omeone qualified to do so and must be clear, unsmudged and classifiable. The person taking the ingerprints must also enter the identification data regarding the applicant and must complete the ingerprint cards by signing as the person taking the fingerprints. *NOTE*: If the fingerprint cards are not roperly prepared or the fingerprints are not legible, the application cannot be acted upon. If this ituation occurs, ATF will contact the applicant for a new set of fingerprint cards.

•.2.4 **Law enforcement certification.** As provided by the regulations, the law enforcement ertification located on the back of Form 1 must be completed and signed by the local chief of police or ounty sheriff, the head of the State police, the State or local district attorney or prosecutor, "or such ther person whose certificate may in a particular case be acceptable to the Director."[97] Examples of ertifying officials, other than those specifically mentioned in the regulations, who have been found to e acceptable are State attorneys general, heads of district State police offices and certain State court udges. Judges' certifications have been accepted if the judges preside over courts of general urisdiction having original jurisdiction in all civil and criminal cases or the authority to conduct riminal jury trials in felony cases. Generally, State magistrates and constables do not have such uthority. *NOTE: no official's certification will be accepted unless the official has jurisdiction over the lace where the applicant resides.* The signature on each copy of the certification must be an original ignature in ink. *NOTE*: *if the Form 1 being submitted is to reactivate a properly registered nserviceable firearm, the law enforcement certification is not required.* See Section 6.6.

> **6.2.4.1 What if the proposed maker is unable to find any official in his or her jurisdiction willing to sign the law enforcement certification?** The inability to find any official to sign the certification will not excuse the requirement for the certification. In that event, the proposed maker will not be able to obtain an approved Form 1 to lawfully make an NFA firearm.

ection 6.3 Submission of Form 1. The completed Form 1, in duplicate, with fingerprint cards, hotographs of the applicant, and payment of the $200 making tax should be mailed to:

Iational Firearms Act Branch
Bureau of Alcohol, Tobacco, Firearms and Explosives
44 Needy Road
Aartinsburg, West Virginia 25405

'ayment of the making tax is to be in the form of a check or money order payable to the Bureau of Alcohol, Tobacco, Firearms and Explosives.

.3.1 **State permit or license for possession of an NFA weapon.** If the applicant's state of residence

---

[97] 27 CFR 479.63

requires a permit or license for the possession of an NFA weapon before the applicant takes possession a certified copy of the license or permit should be submitted with the Form 1 application.

Section 6.4 Approval of Form 1. Non-FFL/SOT's may seek approval to manufacture an NFA firear (e.g., short-barreled rifles, short-barreled, shotguns, wallet guns, etc.) via submission of an ATF Form 1 Upon receipt of the completed Form 1, ATF will process the application and, if approved, a tax stamp will be affixed to the original of the form and the approved application will be returned to the applicant Approval by ATF will effect registration of the firearm to the applicant. Upon receipt of the approved application, the applicant may make the firearm described on the approved Form 1. The approved form must be retained by the applicant and made available at all times for inspection by ATF officers or investigators. *Note: Under no circumstances may the firearm in question be made prior to receipt of the approved Form 1.*

The approval of the Form 1 application authorizes the applicant to make the firearm. The approval doe not authorize the applicant to convey or ship the firearm to another person to manufacture the NFA firearm. If another person will manufacture the NFA firearm, the other person would be the maker and the application must be submitted by that person. Subsequent to the making, the firearm could then be transferred, subsequent to an approved Form 4 application, to the person who wanted the modification be made.

If the applicant on the Form 1 lacks the skill, ability, and/or equipment to manufacture the NFA firearm the applicant, after receipt of the approved Form 1, can have the firearm created or modified at a premises other than shown on the approved Form 1 as long as the creation or modification was done under the direct oversight of the applicant, thus having the applicant retain custody and control of the firearm. If the location is outside the applicant's State and the firearm being made is a short barreled rifle, short barreled shotgun, destructive device, or an unserviceable machinegun which is being reactivated, the applicant will also need to request permission to transport the firearm interstate as required by 27 CFR 478.28.

Section 6.5 Disapproval of Form 1. If the submitted application is disapproved, ATF will advise the applicant, provide a reason(s) for the disapproval, and return the making tax payment.

Section 6.6 Reactivation of a registered unserviceable NFA firearm. The Form 1 may be used to reactivate a properly registered unserviceable firearm, including registered unserviceable machineguns. Block 4(i) of the form should indicate that a registered unserviceable weapon is being reactivated. The remainder of the form should be prepared and submitted with the making tax in the same manner as described above except that a law enforcement certification is not required. The existing serial number on the unserviceable firearm should be used. If the reactivation will be performed by a Class 2 manufacturer, see Chapter 7 for additional information.

Section 6.7 Incorrect Description of Firearm. If the original registration document for the unserviceable firearm contains incorrect descriptive information for the weapon, a letter should be written to the NFA Branch providing the proper description and/or indicating what portions of the description need to be changed. The letter should contain photographs of the actual markings on the firearm. If a correction of the recorded serial number is needed, a photograph or pencil rubbing of the serial number is required. ATF will provide a response indicating that the NFRTR has been amended t eflect the correct information. This confirmation from ATF should be retained with the registration document for the firearm.

Section 6.8 Withdrawal or Cancellation of an ATF Form 1 and Refund of Making Tax. The Form 1 applicant may withdraw the application by submitting a request, in writing, to the Chief, NFA Branch. The applicant may also cancel an approved Form 1 application by submitting a request, in writing, to the Chief, NFA Branch. The request for cancellation must state that the firearm was not made and must include the approved Form 1 with the stamp affixed. The NFA Branch will amend the NFRTR to reflect the withdrawal or cancellation of the application and arrange for the refund of the making tax. A request for cancellation may only be done if the firearm has not already been manufactured. If the firearm is made, the tax liability is incurred and the applicant cannot seek a cancellation.

CHAPTER 7. MANUFACTURING NFA FIREARMS

Section 7.1 Qualifying to manufacture NFA firearms

7.1.1 Licensing under the GCA. Persons intending to engage in the business of manufacturing NFA firearms that also meet the definition of "firearm" in the GCA must first apply for and obtain a GCA manufacturer's license.[98] The license application is ATF Form 7 (5310.12).[99] Appendix C contains a copy of the form. Licenses are issued for a period of 3 years. See Section 5.1 for a discussion of the license fees, licensing standards, and other provisions of the GCA relating to licensing.

7.1.1.1 Engaging in business at multiple locations. A separate license must be obtained for each location where business will be conducted.[100]

7.1.1.2 Engaging in business as both an importer and a manufacturer. Persons intending to engage in business as firearms importers and manufacturers, even if the business will be conducted from the same premises, must have both an importer's license and a manufacturer's license.[101]

7.1.1.3 Do importers and manufacturers need a dealer's license to deal in the firearms they import or manufacture? Licensed importers and manufacturers are not required to have a separate dealer's license to deal in firearms they import or manufacture.[102] A license as an importer or manufacturer authorizes the FFL to deal in the types of firearms authorized by the license to be imported or manufactured.

7.1.1.4 Manufacturers of destructive devices. Manufacturers are reminded that if the firearms they manufacture include destructive devices, or if they solely manufacture destructive devices, they must apply for and obtain a license as a manufacturer of destructive devices. They should also note that the fee for such license is generally $1,000 per year, rather than $50 per year for a license to manufacture firearms other than destructive devices. A license is issued for a period of 3 years. Qualifying to manufacture destructive devices also entitles the FFL to manufacture and deal in firearms other than destructive devices.[103]

7.1.2 Payment of special (occupational) tax to do business in NFA firearms. Every person who engages in the business of manufacturing NFA firearms must pay a special tax.[104] The tax must be paid on or before the date of commencing the taxable business and every year thereafter on or before July 1. The tax is paid by filing ATF Form 5630.7, Special Tax Registration and Return, together with a check or money order for the amount of the tax.[105] Appendix C contains a copy of the form. The special tax

---

[98] 18 U.S.C. 923(a)
[99] 27 CFR 478.44(a)
[100] 27 CFR 478.50
[101] 27 CFR 478.41(b)
[102] ibid
[103] ibid
[104] 26 U.S.C 5801
[105] 27 CFR 479.34(a)

must be paid for each premises where business will be conducted.[106] See Section 5.2 for a more thorough discussion of NFA's requirements relating to the special tax.

7.1.3 Registration by firearms manufacturers with the U.S. Department of State. The State Department has the responsibility of enforcing the provisions of the Arms Export Control Act (AECA) relating to the export of firearms and other defense articles on the U.S. Munitions List. Nevertheless, any person engaged in the U.S. in the business of manufacturing or exporting firearms or other defense articles is required to register with the State Department's Directorate of Defense Trade Controls (DDTC).[107] The regulations expressly state that "Manufacturers who do not engage in exporting must nevertheless register." Form DSP-9 (Registration Statement) and a transmittal letter must be submitted to DDTC by the registrant with a check or money order payable to the State Department of one of the fees prescribed by State Department regulations.[108] Appendix C contains a copy of Form DSP-9. See Section 5.3 for a more detailed discussion of the registration requirement.

> **7.1.3.1 Registration exemption for fabricating articles for "research and development."** State Department regulations in 22 CFR 122.1(b)(4) exempt from the registration requirement persons who fabricate articles for experimental or scientific purpose, including research and development. The regulations do not define "experimental or scientific purpose" or "research and development." Consequently, manufacturers who believe they may be entitled to the exemption should contact DDTC to determine if they are entitled to the exemption.

Section 7.2 What is a "manufacturer"?; What is "manufacturing"?

7.2.1 "Manufacturer". As defined by the GCA, a "manufacturer" is any person engaged in the business of manufacturing firearms or ammunition for purposes of sale or distribution.[109] Similarly, NFA defines the term to mean any person "who is engaged in the business of manufacturing firearms", that is, firearms subject to the NFA.[110] To determine who is a "manufacturer" of firearms, we must look to see whether the person manufactures firearms as discussed in Section 7.2.2.

7.2.2 "Manufacturing". "Manufacturing" is not defined by the law, regulations, or any formal ATF ruling. Nevertheless, the term has been interpreted by ATF to cover activities other than producing a firearm from scratch. As interpreted by ATF, the term covers virtually any work performed on a firearm during the process of preparing the firearm for subsequent sale. For example, a person having a contract with a manufacturer to apply finishing or other work on firearms, or firearms frames or receivers, to prepare them for subsequent sale by the manufacturer would be a "manufacturer" required to qualify as such. Of course, if the person produced firearms parts other than frames or receivers for the manufacturer or performed work on firearms parts not defined as "firearms," the person would not be a "manufacturer."

---

[106] 27 CFR 479.31(b)
[107] 22 CFR 122.1
[108] 22 CFR 122.2(a)
[109] 18 U.S.C 921(a)(10)
[110] 26 U.S.C. 5845(m)

7.2.3 What is the difference between manufacturing and gunsmithing? Performing the work of a gunsmith requires a dealer's license under the GCA, not a manufacturer's license.[111] Nevertheless, a license as an importer or manufacturer also entitles the licensee to conduct business as a gunsmith. The term "dealer" under the GCA includes a gunsmith, that is, "any person engaged in the business of repairing firearms or fitting special barrels, stocks, or trigger mechanisms to firearms." However, gunsmiths occasionally perform the work of a manufacturer and do so without the required manufacturer's license. Indeed, there is no distinction between the activities of a gunsmith and a manufacturer in terms of the physical things done to a firearm. What distinguishes gunsmithing from manufacturing is the purpose for which the work is done. If a gunsmith performs work on a customer's personal gun for the customer's personal use, the function is lawfully performed pursuant to the gunsmith's license as a dealer under the GCA. However, if the gunsmith performs work on guns as a step in the process of preparing them for subsequent sale, the work is "manufacturing" requiring a manufacturer's license and, if the firearm is an NFA firearm, a special tax stamp under the NFA. Here are some examples:

(1) John Doe has a personal firearm and takes it to a gunsmith, a licensed dealer, for modification. The work performed in this instance is the legitimate work of a gunsmith and may be performed pursuant to the gunsmith's dealer's license. The gunsmith need not be licensed as a manufacturer, or hold a special tax stamp in the case of an NFA firearm, to perform the work.

(2) Company A is a licensed manufacturer, but contracts with other licensees to perform finishing work on NFA firearms it manufactures. One such contractor is a gunsmith, a licensed dealer. After receiving the finished firearms, Company A offers the firearms for sale. In this instance, the gunsmith, as well as Company A, is engaged in business as an NFA firearms manufacturer and needs a manufacturer's license and special tax stamp to do so.

7.2.4 Do you know how ATF would classify your product? There is no requirement in the law or regulations for a manufacturer to seek an ATF classification of its product prior to manufacture. Nevertheless, a firearms manufacturer is well advised to seek an ATF classification before going to the trouble and expense of producing it. Perhaps the manufacturer intends to produce a GCA firearm but not an NFA firearm. Submitting a prototype of the item to ATF's Firearms Technology Branch (FTB) for classification in advance of manufacture is a good business practice to avoid an unintended classification and violations of the law.

7.2.4.1 ATF classification letters. ATF letter rulings classifying firearms may generally be relied upon by their recipients as the agency's official position concerning the status of the firearms under Federal firearms laws. Nevertheless, classifications are subject to change if later determined to be erroneous or impacted by subsequent changes in the law or regulations. To make sure their classifications are current, FFLs/SOTs should stay informed by periodically checking the information published on ATF's website, particularly amendments to the law or regulations, published ATF rulings, and "open letters" to industry members.

---

[111] 18 U.S.C. 921(a)(11)(B)

ection 7.3 Registering the manufacture of NFA firearms. A manufacturer qualified to engage in usiness under the GCA and NFA may make NFA firearms without payment of the making tax.[112] Iowever, the manufacturer must report and register each firearm made by filing with the NFA Branch n accurate notice of the manufacture on ATF Form 2, Notice of Firearms Manufactured or Imported, xecuted under the penalties of perjury.[113] Appendix C contains a copy of Form 2. The NFA Branch's eceipt of the form effectuates the registration of the firearms listed on the form.[114] See Section 7.4.6 on he manufacture of silencer parts.

.3.1 Preparation of ATF Forms 2. The form and its contents are prescribed by ATF regulation.[115] he regulation states that the form must set forth the name and address of the manufacturer, identify the nanufacturer's special tax stamp and license, and show the date of manufacture, the type, model, length f barrel, overall length, caliber, gauge or size, serial numbers and other marks of identification, and lace where the firearms are kept.

.3.2 Filing ATF Forms 2. All firearms manufactured during a single day must be included on one orm 2 and must be filed by the manufacturer no later than the close of the next business day. The nanufacturer must prepare the form in duplicate, file the original with the NFA Branch, and keep the opy with the records required to be kept.

> **7.3.2.1 Existing firearms modified into NFA firearms or reactivated.** If an existing firearm is modified into an NFA firearm or reactivated, the existing serial number of the firearm must be entered into Block 8g on Form 2 and the name and address of the original manufacturer or importer must be entered into Block 8b. It is unlawful for any person knowingly to transport, ship, or receive in interstate or foreign commerce any firearm which has had the importer's or manufacturer's serial number removed, obliterated, or altered, or to receive or possess any such firearm that has at any time been shipped or transported in interstate or foreign commerce.[116]

.3.3 Reactivation of a registered unserviceable NFA firearm. Although unserviceable NFA irearms are "firearms" required to be registered under the NFA, the reactivation of such firearm is onsidered to be the making or manufacture of an NFA firearm. Consequently, a qualified NFA nanufacturer who reactivates an unserviceable firearm must file a Form 2 with ATF showing the eactivation.[117] The existing serial number on the unserviceable firearm should be used.

> **7.3.3.1 Incorrect description of firearm.** If the original registration document for the unserviceable firearm contains incorrect descriptive information for the weapon, a letter should be written to the NFA Branch providing the proper description and/or indicating what portions of the description need to be changed. The letter should contain photographs of the actual markings on the firearm. If a correction of the recorded serial number is needed, a photograph or pencil rubbing of the serial number is required. ATF will provide a response indicating that the

---

[2] 27 CFR 479.68
[3] 27 CFR 479.103
[4] ibid
[5] ibid
[6] 18 U.S.C. 922(k)
[7] See Instruction "c" on Form 2.

NFRTR has been amended to reflect the correct information. This confirmation from ATF should be retained with the registration document for the firearm.

7.3.3.2 Reactivation of a registered unserviceable NFA firearm. From time to time, a qualified manufacturer may receive a registered, unserviceable NFA firearm from a nonlicensee for reactivation. To lawfully do so, the firearm should be received pursuant to an approved ATF Form 5 for tax-exempt transfer (the transfer of a registered unserviceable firearm is not subject to the NFA transfer tax).[118] Upon reactivating the firearm, the manufacturer must file Form 2 with ATF to register the manufacture or reactivation. To return the firearm to its owner, the manufacturer must obtain an approved Form 4 from ATF to transfer the firearm.[119] Since the firearm is no longer unserviceable, the return of the firearm pursuant to an approved Form 4 would be subject to the $200 NFA transfer tax, except that the transfer tax for a firearm defined as "any other weapon" is $5. In these instances, Forms 4 need not contain law enforcement certifications.

Manufacturers may also receive unserviceable NFA firearms from FFLs/SOTs for reactivation. Receipt of the firearms must be pursuant to approved Form 3. Upon reactivation, the manufacturer must file Form 2 to register the manufacture or reactivation. Approved Forms 3 must also be used to return reactivated firearms to the FFLs/SOTs.

Section 7.4 The identification of firearms.

7.4.1 Serial numbers. Each manufacturer of a firearm must legibly identify it by engraving, stamping (impressing), or otherwise conspicuously placing on the firearm's frame or receiver an individual serial number not duplicating any serial number placed by the manufacturer on any other firearm.[120] The requirement that the marking be "conspicuously" placed on the firearm means that the marking must be wholly unobstructed from plain view. For firearms manufactured on or after January 30, 2002, the serial number must be to a minimum depth of .003 inch and in a print size no smaller than 1/16 inch.

7.4.1.1 What is an acceptable serial number? Alpha characters (letters), for example a name, are not acceptable as a serial number. A proper serial number may contain such characters or letters, but it must have at least one numeric character (number). ATF takes the view that marking "legibly" means using exclusively Roman letters (A, B, C, and so forth) and Arabic numerals (1, 2, 3, and so forth).[121] Deviations from this requirement have been found to seriously impair ATF's ability to trace firearms involved in crime.

7.4.2 Additional information. Certain additional information must also be conspicuously placed on the frame, receiver, or barrel of the firearm by engraving, casting, stamping (impressing), that is, they must be placed in such a manner that they are wholly unobstructed from plain view. For firearms manufactured on or after January 30, 2002, this information must be to a minimum depth of .003 inch. The additional information includes:

---

[118] 27 CFR 479.91
[119] 27 CFR 479.84
[120] 27 CFR 479.102
[121] ATF Ruling 2002-6

(1) The model, if such designation has been made;

(2) The caliber or gauge;

(3) The manufacturer's name (or recognized abbreviation); and
(4) The city and State (or recognized abbreviation) where the manufacturer maintains its place of business.[122]

7.4.3 Measuring the depth of markings. The depth of all markings is measured from the flat surface of the metal and not the peaks or ridges. The height of serial numbers is measured as the distance between the latitudinal ends of the character impression bottoms (bases).

7.4.4 Obtaining variances to the marking requirements. Requests for variances from the marking requirements of 27 CFR 478.92 and 27 CFR 479.102 should be submitted by letter to ATF's Firearms Technology Branch (FTB). The letter can be sent via mail to Chief, Firearms Technology Branch, 244 Needy Road, Martinsburg, WV 25405. The letter can also be sent to the marking variance e-mail address at: marking_variances@atf.gov. The marking variance request may be submitted by any of the parties involved in the variance. However, if the primary manufacturer is in possession of all the information including the names of the identity of the secondary manufacturers and the manufacturing processes they may be performing on the firearm, it is preferred that the primary manufacturer submit the request to FTB.

The marking variance letter of request should clearly state the following information:

- manufacturer, importer, or maker of the firearm(s),
- recipient of the firearm(s),
- identify the name, city and State that will be displayed on the firearm(s),
- model designation, if designated,
- identify the type/style of firearm (pistol, machinegun, short-barreled rifle, etc.),
- caliber or gauge if assigned, and
- serial number scheme.

In identifying the serial number scheme to be used, you must supply a different serial scheme for each model and you must state the exact beginning serial number of the serial scheme you wish to use. Although letters and characters may be used, the serial number must use at least one number it the scheme. Please note, using the letter X, or the use of characters (#, *, etc.) as digit/character holders is unacceptable. For example, an incorrectly submitted serial scheme would be ALZXXXX. A correctly submitted serial scheme would be ALZ0001. You do not need to provide an ending serial number when submitting your serial number scheme.

7.4.4.1 Variances in the name and location of the manufacturer. As stated above, the regulations require firearms to be marked with the manufacturer's name, city and State or recognized abbreviation of the information. FTB will only grant marking variances for

---

[122] 27 CFR 479.102

abbreviations regarding city and State names that are commonly recognized by the United State Postal Service. If you intend to use a name or abbreviation other than your licensed name or recognized abbreviation, you must contact the Federal Firearms Licensing Center and complete ATF Form 5300.38 to have your Federal Firearms License amended to reflect the addition of a trade name or a "doing business as" name to your license. You may not use a name or abbreviation until it is approved.

7.4.4.2 Variances for manufacturers' contractors. As pointed out in Section 7.2.2, some manufacturers contract with other entities to perform certain work on their firearms prior to thei ultimate sale. In those instances the contractors are also "manufacturers" who must be licensed as a Type 07 Manufacturer or as a Type 10 Manufacturer of Destructive Devices in order to perform any manufacturing function on the firearm. Additionally, the regulations require that the secondary manufacturer mark the firearm with their identifying information to include name city and State. A qualified, secondary manufacturer may request a variance to adopt the markings of the initial manufacturer. If a manufacturer is working with a secondary manufacturer, either the manufacturer or the secondary manufacturer can submit a letter of request to FTB for a marking variance.

You should be aware that marking variances for the manufacturer of machineguns is limited to the making of the receiver from one special (occupational) taxpayer manufacturer to another. The machinegun can be made on behalf of a manufacturer who intends on making the machineguns for stockpile for future sale to other dealers as sales samples, or for sale to law enforcement and the military. A machinegun cannot be transferred to a secondary manufacturer such as a bluer, Parkerizer, heat treater etc. In order to have a machinegun receiver blued, Parkerized or heat treated, etc., the possessor of the machinegun must transport the machinegun and remain in possession of the machinegun while it is being blued, Parkerized or heated treated by the secondary manufacturer.

7.4.4.3 Marking Destructive Devices. In the case of a destructive device, FTB may authorize an alternate means of identification upon receipt of a manufacturer's letter of request. The letter of request should indicate that the engraving, casting or stamping the weapons would be dangerous or impracticable[123] The variance would allow an alternate *method* of marking such devices. For example, ATF may permit the required markings to be placed on the device by paint or stencil. A variance in this format will contain the information that is proscribed by the regulations. For example, lot numbers of ammunition classified as destructive devices would be acceptable in lieu of the information required by the regulations.

7.4.4.4 Marking parts, other than frames or receivers, defined as NFA firearms. FTB may authorize alternate means of identifying such parts upon receipt of the manufacturer's letter application showing that such other identification is reasonable and will not hinder the effective administration of the regulations.[124] See Section 7.4.6 for information on marking silencer parts

[123] ibid
[124] ibid

7.4.5 Marking frames or receivers that are not complete weapons at the time of disposition. Firearms frames or receivers that are not components of complete firearms at the time of disposition must be identified with all the required markings, that is, serial numbers and all the additional markings discussed in Sections 7.4.1 and 7.4.2.[124]

7.4.6 Marking silencer parts. Some FFLs/SOTs assemble silencers, for subsequent sale, from parts acquired from their contractors (*NOTE*: this activity is the "manufacture" of NFA firearms requiring the assembler and the contractor to qualify as manufacturers under the GCA and the NFA). Under these circumstances, ATF takes the position that contractors are not required to place identifying markings on silencer parts. They may, however, place an assembler's markings on these parts if the assembler so desires. It should also be noted that these contractors are not required to register the parts they produce by filing Forms 2, nor are they required to obtain approved Forms 3 to transfer the parts to assemblers.

Section 7.5 Manufacturing machineguns. Section 922(o), Title 18, U.S.C., makes it unlawful to possess or transfer a machinegun, except for transfers to or possession by Federal and State agencies or the transfer or possession of a machinegun lawfully possessed before the effective date of the statute, May 19, 1986. So, machineguns "lawfully possessed" before the effective date are those manufactured before May 19, 1986 and registered in the NFRTR. See also Section 7.6 on the manufacture of machineguns by contractors of FFLs/SOTs.

7.5.1 Manufacture of machineguns for sale to government agencies or as "sales samples". Notwithstanding the prohibition in Section 922(o), qualified manufacturers may manufacture machineguns on or after May 19, 1986 for sale to Federal and State agencies or to FFLs/SOTs as "sales samples" for demonstration to prospective governmental customers.[125] The weapons may be manufactured and stockpiled in contemplation of future sales, but their registration and subsequent transfer are conditioned upon and restricted to sales only to government agencies or as sales samples. See Chapter 9 for a discussion of the required documentation supporting these sales. See also Section 7.6.1 on the manufacture of machineguns by contractors of FFLs/SOTs.

7.5.2 Manufacture of machineguns for exportation. Notwithstanding the prohibition in Section 922(o), qualified manufacturers may manufacture machineguns on or after May 19, 1986 for exportation in compliance with the Arms Export Control Act and Department of State regulations.[126] The weapons may be manufactured and stockpiled for future exportation.

7.5.3 May machineguns be manufactured for distribution to U.S. Government contractors? As interpreted by ATF, Section 922(o) contains no exception that would permit the lawful manufacture and stockpiling of machineguns for transfer to Government contractors. The only exceptions to the machinegun transfer and possession prohibitions in the statute are for machineguns manufactured for sale to Federal and State agencies, for distribution and use as sales samples, or for exportation.

---

125 ibid
125 27 CFR 479.105(c)
126 ibid

7.5.4 May machinegun receivers be manufactured and used as replacement parts for machineguns lawfully registered and possessed prior to May 19, 1986? As previously stated, 18 U.S.C. 922(o) generally makes it unlawful to possess or transfer any machinegun, including a machinegun frame or receiver, manufactured after May 18, 1986. Exceptions are provided for weapons produced by a qualified manufacturer for sale to government entities, as dealer sales samples, or for exportation. There is no exception allowing for the lawful production, transfer, possession, or use of a post-May 18, 1986 machinegun receiver as a replacement receiver on a weapon produced prior to May 19, 1986.

Section 7.6 Manufacture of NFA firearms by contractors. Qualified manufacturers of NFA firearms may contract with other persons to manufacture their NFA firearms. However, contractors who manufacture NFA firearms, or perform work on existing firearms as part of the manufacturing process, must also be qualified to manufacture NFA firearms. They are also subject to all NFA requirements imposed on NFA manufacturers relative to the manufacture, registration, transfer, and marking of NFA firearms. See Section 7.4 on the marking of firearms by manufacturers and variances that may be obtained from marking requirements.

7.6.1 Contractors' manufacture of machineguns. As previously stated, machineguns manufactured in the U.S. on or after May 19, 1986 must be solely for sale to U.S. or State governmental entities, for distribution as "sales samples," or for exportation (see Sections 7.5.1 and 7.5.2). However, a qualified NFA manufacturer may contract with other qualified firms to produce machineguns it intends to distribute only for those purposes. Since the number of machineguns that may be transferred between licensees is limited by 27 CFR 479.105, the parties involved will need to seek an alternate method or procedure under 479.26. For example, manufacturer B contracts with manufacturer A to build B's machinegun receivers and applies for and receives a variance that allows A to identify the receivers with B's identifying marks. A is to provide the receivers for assembly and distribution. However, since A is manufacturing the machineguns and must register them on a Form 2, the transfer is limited due to the provisions of 922(o) and 479. 105. Both parties would need to apply for an alternate procedure or method that would allow A to transfer these machineguns to B.

7.6.2 Manufacture of machineguns solely for purposes of testing. The manufacture of machineguns solely for testing or research purposes is not recognized as a legitimate exception to the ban on possession or transfer of firearms under 18 U.S.C. 922(o). As previously stated, manufacturers may only manufacture machineguns on or after May 19, 1986 and stockpile the same if they are manufactured and held for sale to Federal or State agencies, for distribution as "sales samples," or for exportation.

Section 7.7 Manufacturing NFA firearms exclusively for the United States. A person or entity engaged in the business of manufacturing NFA firearms exclusively for the United States or any agency of the U.S. may be relieved from compliance with any provision of the NFA regulations, 27 CFR Part 479, with respect to such firearms.[127] This exemption may be obtained by filing with the NFA Branch a letter application setting out the manner in which the applicant conducts business, the type of firearm to be manufactured, and satisfactory proof of the existence of the manufacturer's Government contract

[127] 26 U.S.C. 5851(a)

nder which the applicant intends to under which the applicant intends to operate.[128] Approval of the pplication entitles the applicant to the exemption. The manufacture of weapons for the U.S. *and* other ustomers (for example, manufacture for exportation to overseas customers) would not qualify for the xemption. The exemption must be renewed with another approved letter application on or before July of each year.[129]

ection 7.8 Locations. Your licensed premises is where your manufacturing must be done. ATF is ften asked by a manufacturer whether he or she can conduct a manufacturing operation elsewhere, such s a nearby machine shop. While, in general, component parts can be made at a site other than the censed premises, if the part being made is a receiver, silencer part, or a procedure is being performed such as shortening of the barrel of a rifle or shotgun) where the complete firearm is there, the operation ıust be done on the licensed manufacturing premises. If components are modified or fabricated at a ›cation where all other parts for a complete firearm are present, the location must be licensed as a ıanufacturer.

[8] 27 CFR 479.33(b)
[9] 26 U.S.C 5851(b)

CHAPTER 8. IMPORTING NFA FIREARMS

Section 8.1 Qualifying to import NFA firearms

8.1.1 Licensing under the GCA. Persons intending to engage in the business of importing NFA firearms that also meet the definition of "firearm" in the GCA must first apply for and obtain a GCA importer's license.[130] The license application is ATF Form 7. [131] Appendix C contains a copy of the form. Licenses are issued for a period of 3 years. See Section 5.1 for a discussion of the license fees, licensing standards, and other provisions of the GCA related to licensing.

8.1.1.1 Engaging in business at multiple locations. A separate license must be obtained for each location where business will be conducted.[132]

8.1.1.2 Engaging in business as both an importer and a manufacturer. Persons intending t engage in business as firearms importers and manufacturers, even if the business will be conducted from the same premises, must have both an importer's license and a manufacturer's license.[133]

8.1.1.3 Do importers and manufacturers need a dealer's license to deal in the firearms th import or manufacture? Licensed importers and manufacturers are not required to have a separate dealer's license to deal in firearms they import or manufacture. A license as an import or manufacturer authorizes the FFL to deal in the types of firearms authorized by the license to be imported or manufactured.[134]

8.1.1.4 Importers of destructive devices. Importers are reminded that if the firearms they import include destructive devices, or if they solely import destructive devices, they must apply for and obtain a license as an importer of destructive devices. They should also note that the fe for such license is $1,000 per year, rather than $50 per year for a license to import firearms othe than destructive devices. Qualifying to import destructive devices also entitles the FFL to impo and deal in firearms other than destructive devices.[135]

8.1.2 Payment of special (occupational) tax to do business in NFA firearms. Every person who engages in the business of importing NFA firearms must pay a special tax.[136] The tax must be paid on or before the date of commencing the taxable business and every year thereafter on or before July 1. The tax is paid by filing ATF Form 5630.7, Special Tax Registration and Return, together with a check or money order for the amount of the tax.[137] Appendix C contains a copy of the form. The special tax

---

[130] 18 U.S.C. 923(a)
[131] 27 CFR 478.44(a)
[132] 27 CFR 478.50
[133] 27 CFR 478.41(b)
[134] ibid
[135] ibid
[136] 26 U.S.C. 5801
[137] 27 CFR 479.34(a)

nust be paid for each premises where business will be conducted.[138] See Section 5.2 for a more horough discussion of NFA's requirements relative to the special tax.

.1.3 Registration by importers of U.S. Munitions Import List articles with ATF. ATF has the esponsibility of enforcing the provisions of the Arms Export Control Act (AECA) relating to the nportation of firearms and other defense articles on the U.S. Munitions Import List. Any person ngaged in the U.S. in the business of importing firearms or other defense articles is required to register vith ATF. Form 4587, Application to Register as an Importer of U.S. Munitions Import List Articles, nust be submitted to ATF by the registrant with a check or money order payable to the Bureau of Alcohol, Tobacco, Firearms and Explosives of one of the fees prescribed by 27 CFR 447.32. Appendix C contains a copy of Form 4587. See Section 5.4 for a more detailed discussion of the registration equirement.

ection 8.2 What is an "importer?"; What is "importation?"

.2.1 "Importer." As defined by the GCA, an "importer" is any person engaged in the business of nporting or bringing firearms or ammunition into the United States for purposes of sale or istribution."[139] Similarly, the NFA defines the term to mean any person "who is engaged in the usiness of importing firearms into the United States, that is, firearms subject to the NFA.[140]

.2.2 "Importation." Importation is defined in the regulations as the bringing of a firearm within the imits of the United States or any territory under its control or jurisdiction, from a place outside thereof whether such place be a foreign country or territory subject to the jurisdiction of the United States), vith the intent to unlade.[141] However, bringing a firearm from a foreign country or a territory subject to he jurisdiction of the United States into a Foreign Trade Zone (FTZ) for storage pending shipment to a oreign country or subsequent importation into this country shall not be deemed importation.

ection 8.3 Importation of NFA firearms.

With certain exceptions, NFA firearms may only be imported for the use of the United States or any tate or possession or political subdivision thereof such as a law enforcement agency; for scientific or esearch purposes; solely for testing or use as a model by a registered manufacturer; or solely for use as sample by a registered importer or registered dealer.[142] Importers may not import NFA firearms for tockpiling or warehousing them at their licensed premises for the purpose of filling future orders from ualifying agencies or dealers requesting sales samples. Imported NFA firearms may be stored in a Customs Bonded Warehouse (CBW) or a Foreign Trade Zone (FTZ). NFA firearms may only be vithdrawn from these facilities to fill specifically approved purchase requests.

.3.1 Importation of machineguns. Section 922(o), Title 18, U.S.C., makes it unlawful to possess or ransfer a machinegun, except for transfers to or possession by Federal and State agencies or the transfer or possession of a machinegun lawfully possessed before the effective date of the statute, May 19, 1986.

---

[8] 27 CFR 479.31(b)
[9] 18 U.S.C. 921(a)(9)
[0] 26 U.S.C. 5845(i)
[1] 27 CFR 479.11
[2] 26 U.S.C. 5844

So, machineguns "lawfully possessed" before the effective date are those manufactured before May 19, 1986 and registered in the NFRTR. Notwithstanding the prohibition in Section 922(o), licensed importers may import machineguns on or after May 19, 1986 for sale to Federal and State agencies or to FFLs/SOTs as "sales samples" for demonstration to prospective governmental customers. The weapon may be imported and stockpiled in CBWs or in FTZs in contemplation of future sales, but their registration and subsequent transfer are conditioned upon and restricted to sales only to government agencies or as "sales samples." See Chapter 9 for a discussion of the required documentation supporting these sales. Due to the restrictions in Section 922(o), machineguns may not be imported for other purposes, such as transfer to government defense contractors, security purposes, testing, research or use as a model by registered manufacturers.

8.3.2 Importation for use of the United States, qualifying political subdivisions, and law enforcement agencies. NFA firearms may be imported for sale to Federal, State, or local government agencies. To import NFA firearms for this purpose, an ATF Form 6, Application and Permit for Importation of Firearms, Ammunition and Implements of War, must be submitted to ATF in triplicate. Appendix C contains a copy of Form 6. The Form 6 must include the agency's letter or purchase order reflecting the purchase of the firearms.[143] Appendix E contains a sample letter for use by an agency ordering imported firearms.

8.3.3 Importation for use by Nuclear Regulatory Commission (NRC) licensees and authorized contractors. Congress recently amended the Atomic Energy Act of 1954 with passage of the Energy Policy Act of 2005.[144] Although the new provision does not amend the GCA or the NFA, it does allow for the possession of machineguns by NRC licensees and authorized contractors that provide security to these licensees at nuclear facilities. This new provision, 42 U.S.C. 2201a, is a departure from legal restrictions in Federal firearms laws that do not allow for machineguns to be imported for, transferred to or possessed by non-government entities. The law states that before the new statute may take effect, the NRC must establish guidelines for implementation that are approved by the Attorney General. As of January 30, 2006, the guidelines have not been prepared or approved. Therefore, ATF does not yet have the authority to process import or transfer applications filed pursuant to 42 U.S.C. 2201a.

8.3.4 Importation for authorized scientific or research purposes. NFA firearms, except machineguns, may be imported for authorized scientific or research purposes. To import NFA firearms for this purpose, a Form 6 must be submitted in triplicate. The Form 6 must include a statement describing the specific scientific or research purpose for which the firearm is needed.[145] The statement must also include a detailed explanation of why the importation of the firearm is needed for scientific or research purposes

8.3.5 Importation for use as a model by a registered manufacturer. NFA firearms, except machineguns, may be imported for use as a model by a registered manufacturer. To import NFA firearms for this purpose, a Form 6 must be submitted in triplicate. The Form 6 must include a detailed explanation of why the importation of the firearm is needed for use as a model.[146]

---

[143] 27 CFR 479.111
[144] Public Law 109-58
[145] 27 CFR 478.112(b)(vii)(A)
[146] 27 CFR 479.111

8.3.6 Importation for use as a "sales sample." NFA firearms may be imported for use as sales samples by qualified NFA importers and dealers to demonstrate the firearms to government agencies and generate possible future sales to such agencies. To import NFA firearms for this purpose, a Form 6 must be submitted to ATF, in triplicate. Information must be submitted with the Form 6 establishing that the firearm is suitable or potentially suitable for an agency's use; the expected governmental customers requiring a demonstration of the firearm; information as to the availability of the firearm to fill subsequent orders; and letters from agencies expressing a need for a particular model or interest in seeing a demonstration of a particular firearm.[147] Appendix D contains sample letters for use in acquiring imported sales samples.

8.3.7 Importing multiple quantities of the same model firearm as sales samples. An application to import two or more NFA firearms of a particular model for use as a sales sample by the importer or a qualified dealer will be approved if documentation shows the necessity for demonstration to government agencies.[148] A Form 6 must be submitted to ATF, in triplicate, along with documentation that the firearm is needed by the importer or dealer to demonstrate the weapon to all of the officers of a police department, SWAT team, or special operations unit. ATF may authorize the importation of more than 2 of the same model if an importer provides specific documentation to justify the need for additional weapons.

8.3.8 Importation of NFA weapons classified as curios or relics. An NFA firearm may not be imported as a sales sample for demonstration to a law enforcement agency if it is a curio or relic unless it is established that it is particularly suitable for use as a law enforcement weapon.[149] Suitability for law enforcement use may be established by providing detailed information as to why the weapon is particularly suitable for such use, the expected customers who would require a demonstration of the weapon, and information as to the availability of such firearms to fill subsequent orders. To import NFA firearms classified as curios or relics, a Form 6 must be submitted to ATF, in triplicate, including a statement showing that the weapons meet the above discussed criteria.

8.3.9 Conditional importation. ATF may impose conditions upon any application to import firearms.[150] In most cases, a conditional import permit is issued for the purpose of examining and testing a firearm to determine whether the weapon in question will be authorized for importation. The conditions to the permit will normally authorize only one sample of a particular firearm and require that the firearm be shipped directly from Customs to FTB for testing and examination. The importer must agree to either export or destroy the weapon if a final determination is made that the weapon may not be imported. If the weapon is found to qualify for importation for one of the authorized purposes, the firearm will be returned to Customs for release to the importer.

8.3.10 Temporary importation of NFA firearms. ATF has adopted an alternate procedure for importers to use when temporarily importing NFA weapons for inspection, testing, calibration, repair, or incorporation into another defense article.[151] This procedure requires that importers:

---

[147] 27 CFR 479.105(d)
[148] ATF Ruling 2002-5
[149] ATF Ruling 85-2
[150] 27 CFR 479.113
[151] ATF Rul. 2004-2

1. Must be qualified under the GCA and NFA to import the type of firearms at issue.

2. Obtain a temporary import license, DSP-61, from the Department of State in accordance with the ITAR (22 CFR 122.3) or qualify for a temporary import license exemption pursuant to 22 CFR 123.4.

3. Within 15 days of the release of the firearms from Customs custody, file an ATF Form 2. The DSP-61 must be attached to the Form 2. If the importation is subject to a licensing exemption under 22 CFR 123.4, the importer must submit with the Form 2 a statement, under penalties of perjury, attesting to the exemption and stating that the weapon will be exported within four years of its importation.

4. Maintain the firearm(s) in a secure place and manner to ensure that the firearm(s) is not diverted to criminal or terrorist use.

5. Export the firearm(s) within four years of importation.

NFA weapons temporarily imported may be temporarily provided to a contractor in the United States for the purposes stated above by obtaining advance approval of an ATF Form 5, Application for Tax Exempt Transfer and Registration of Firearm. As an alternative, the transfer may be accomplished with a letter from the importer to the contractor, who must be qualified to engage in an NFA firearms business, stating the following:

1. The weapon is being temporarily conveyed for inspection, testing, calibration, repair, or incorporation into another defense article; and

2. The approximate time period the weapon will be in the contractor's possession.

The above alternate procedure became effective on April 7, 2004.

8.3.11 Re-importation of NFA firearms temporarily exported from the United States by nonlicensees. A properly registered NFA firearm that has been temporarily exported from the United States may be re-imported by the same person who temporarily exported the firearm. The firearm may be returned to the United States without submission of a Form 6 or Form 2, provided that ATF Form 5320.20, Application to Transport Interstate or to Temporarily Export Certain National Firearms Act Weapons, has been submitted to and approved by ATF prior to the temporary exportation of the firearm. The firearm must be registered to the person who is re-importing the weapon.[152]

8.3.12 What is the difference between a Customs Bonded Warehouse (CBW) and a Foreign Trade Zone (FTZ)?

8.3.12.1 Customs Bonded Warehouse. A Customs Bonded Warehouse is a building or other secured area in which dutiable goods may be stored without payment of duty. Bringing items into a CBW is an importation. To bring NFA firearms into a CBW for storage requires an

[152] 27 CFR 479.111(c)

approved ATF Form 6, Application and Permit for Importation of Firearms, Ammunition and Implements of War. NFA weapons stored in a CBW may be manipulated, destroyed or otherwise altered in the CBW. To withdraw stored NFA firearms stored from a CBW, an additional Form 6 is required. NFA weapons may only be withdrawn from a CBW for the purposes discussed in Section 8.3.

8.3.12.2 Foreign Trade Zone. A Foreign Trade Zone is a specially designated area, in or adjacent to a U.S. Customs Port of Entry, which is considered to be outside the Customs Territory of the United States. For the purposes of the GCA and NFA, bringing items into an FTZ is not treated as an import because the FTZ is considered foreign soil. However, under the import provisions of the AECA, an FTZ is considered to be part of the United States. As a result, a Form 6 is required by AECA regulations to bring NFA firearms into an FTZ. Firearms stored in an FTZ may be manipulated, repackaged, destroyed, otherwise altered or exported. Moving an NFA firearm from an FTZ to a CBW for storage or for delivery to a qualifying customer is an "importation" requiring an additional approved Form 6.

.3.13 Preparation of ATF Forms 6. The form and its contents are prescribed by ATF regulations.[153] he importer must also establish to the satisfaction of ATF that the importation is for one of the urposes authorized by the statute. A detailed explanation of why the importation falls within the tandards must be attached to the application. Acceptable documentation may include purchase orders rom government agencies, a letter from a government agency requesting a demonstration of the firearm, r a letter from a dealer requesting an NFA weapon as a sales sample. Examples of the above letter equests can be found in Appendix D. The Form 6 must be submitted in triplicate.

.3.14 "eForm6." eForm6 is the system used to electronically file the Form 6 and track the status of ermit applications submitted for approval. In order to use the eFrom6 system a user ID and password ust be obtained from ATF. To register for eForm6, a registration form, eForm6 Access Request (ATF -form 5013.3),[154] must be submitted to the following address:

Firearms & Explosives Imports Branch
Bureau of Alcohol, Tobacco, Firearms and Explosives
244 Needy Road
Martinsburg, West Virginia 25405

dditional information concerning eForm6 may also be found on the ATF website, www.atf.gov .

.3.15 Preparation of ATF Forms 6A. This form and its contents are also prescribed by ATF egulations.[155] Upon release of the firearm(s) from Customs custody, the importer will prepare ATF orm 6A, in duplicate, and furnish the original to the Customs officer releasing the firearm(s). The ustoms officer will, after certification, forward the Form 6A to the address specified on the form. Vithin 15 days of the date of release from Customs, the importer must forward to the address specified

[3] 27 CFR 478.112(b)
[4] See Appendix C
[5] 27 CFR 478.112(c)

on the form a copy of the Form 6A on which must be reported any error or discrepancy appearing on th Form 6A certified by Customs and serial numbers if not previously provided on Form 6A.

Section 8.4 Registering the importation of NFA firearms. The importer must report and register each firearm imported by filing with the NFA Branch an accurate notice of the importation on ATF Form 2, Notice of Firearms Manufactured or Imported, executed under the penalties of perjury.[156] Appendix C contains a copy of Form 2. ATF's receipt of the form effectuates the registration of the firearms listed on the form.[157]

8.4.1 Preparation of ATF Forms 2. The form and its contents are prescribed by ATF regulation.[158] The regulation states that the form must set forth the name and address of the importer, identify the importer's special tax stamp and license, import permit number, the date of release of the firearm from Customs custody, the type, model, length of barrel, overall length, caliber, gauge or size, serial number and other marks of identification, and place where the firearms are kept. The Form 2 should list only firearms imported on one specifically approved Form 6.

8.4.2 Filing ATF Forms 2. A Form 2 must be filed by the importer no later than 15 days from the dat the firearm was released from Customs custody. If the importation involves more than one import permit, a separate Form 2 must be filed to report those firearms imported under each permit. The importer must prepare the form in duplicate, file the original with the NFA Branch, and keep the copy with the records required to be kept.

Section 8.5 The identification of firearms.

8.5.1 Serial numbers. Each importer of a firearm must legibly identify it by engraving, stamping (impressing), or otherwise conspicuously placing on the firearm's frame or receiver an individual serial number not duplicating any serial number placed by the importer on any other firearm.[159] An importer may adopt the serial number placed on the firearm by the foreign manufacturer provided that such seria number does not duplicate a number previously adopted or assigned by the importer to any other firearm. The requirement that the marking be "conspicuously" placed on the firearm means that the marking must be wholly unobstructed from plain view. For firearms imported on or after January 30, 2002, the serial number must be marked to a minimum depth of .003 inch and in a print size no smaller than 1/16 inch.

8.5.1.1 What is an acceptable serial number? Alpha characters (letters), for example a name are not acceptable as a serial number. A proper serial number may contain such characters or letters, but it must have at least one numeric character (number). ATF takes the view that marking "legibly" means using exclusively Roman letters (A, B, C, and so forth) and Arabic numerals (1, 2, 3, and so forth).[160] Deviations from this requirement have been found to seriously impair ATF's ability to trace firearms involved in crime.

---

[156] 27 CFR 479.112(a)
[157] 27 CFR 112(a)
[158] ibid
[159] 27 CFR 479.102
[160] ATF Ruling 2002-6

.5.2 Additional information. Certain additional information must also be conspicuously placed on ne frame, receiver, or barrel of the firearm by engraving, casting, stamping (impressing), that is, they nust be placed in such a manner that they are wholly unobstructed from plain view. For firearms nported on or after January 30, 2002, this information must be marked to a minimum depth of .003 nch. The additional information includes:

(1) The model, if such designation has been made;

(2) The caliber or gauge;

(3) The name of the importer;

(4) The name of the foreign manufacturer;

(5) The country in which the firearm was manufactured;

(6) The city and State (or recognized abbreviation) where the importer maintains . its place of business.[161]

.5.3 Measuring the depth of markings. The depth of all markings is measured from the flat surface f the metal and not the peaks or ridges. The height of serial numbers is measured as the distance etween the latitudinal ends of the character impression bottoms (bases).

.5.4 Obtaining variances to the marking requirements. Requests for variances from the marking equirements should be submitted by letter to ATF's Firearms Technology Branch (FTB) in duplicate. he requests should state the proposed variance and the reasons for the variance.

8.5.4.1 Variances in the name and location of the importer. As stated above, the regulations require firearms to be marked with the importer's name, city, and State or a recognized abbreviation of the information. If other than complete names or recognized abbreviations are proposed to be used, a variance must be obtained from ATF.

8.5.4.2 Marking frames or receivers that are not complete weapons at the time of disposition. Firearms frames or receivers that are not components of complete firearms at the time of disposition must be identified with all the required markings, that is, serial numbers and all the additional markings discussed in Sections 8.5.1 and 8.5.2.[162]

8.5.4.3 Marking destructive devices. In the case of a destructive device, ATF may authorize alternate means of identification upon receipt of an importer's letter application, in duplicate. The application should show that engraving, casting, or stamping the weapons would be dangerous or impractical.[163]

---

161 27 CFR 479.102
162 ibid
163 ibid

8.5.4.4 Marking parts, other than frames or receivers, defined as NFA firearms. ATF may authorize alternate means of identifying such parts, for example, parts defined as mufflers or silencers, upon receipt of the importer's letter application, in duplicate, showing that such other identification is reasonable and will not hinder the effective administration of the regulations.[164]

8.5.5 When must markings be applied to imported NFA firearms? All of the markings discussed in this section must be applied to imported NFA firearms no later than 15 days from the date the firearms were released from Customs custody. If the required markings cannot be applied within the 15 day period due to the quantity of weapons being imported or other factors, ATF may grant a variance extending the time for application of the markings. A request for an extension of the time should be submitted in the same manner as discussed in Section 8.5.4.

[164] 27 CFE 479.102

CHAPTER 9. TRANSFERS OF NFA FIREARMS

Section 9.1 Definition of "transfer." The term "transfer" is broadly defined by the NFA to include "selling, assigning, pledging, leasing, loaning, giving away, or otherwise disposing of" an NFA firearm.[165] The lawful transfer of an NFA firearm generally requires the filing of an appropriate transfer form with ATF, payment of any transfer tax imposed, approval of the form by ATF, and registration of the firearm to the transferee in the NFRTR. Approval must be obtained before a transfer may be made. See Section 9.5 for a discussion of certain NFA transactions not considered by ATF to be "transfers."

Section 9.2 Only previously registered firearms may be lawfully transferred. ATF will not approve the transfer of an NFA firearm unless it has been registered to the transferor in the NFRTR. NFA firearms may only be registered upon their lawful making, manufacture, or importation, or upon the transfer of firearms already registered. Generally, unregistered firearms may not be lawfully received, possessed, or transferred. They are contraband subject to seizure and forfeiture. Violators are also subject to criminal prosecution. However, see Sections 2.4 and 3.3 on removing NFA firearms from the scope of the NFA because of their status as collectors' items, modification, or elimination of certain component parts.

Section 9.3 Interstate transfers of NFA firearms. ATF will not approve the transfer of an NFA firearm to a non-FFL/SOT residing in a State other than the State in which the transferor's licensed business is located or the transferor resides. Such interstate transfers would violate the GCA. However, See section 9.5.4 regarding the custody of firearms by employees of FFLs/SOTs.

Section 9.4 ATF forms for use in transferring NFA firearms

9.4.1 ATF Form 3. Transfers by FFLs/SOTs to other FFLs/SOTs require the filing of ATF Forms 3, Application for Tax Exempt Transfer and Registration of a Firearm, to register firearms to the transferees.[166] See also Section 3.2.6.4. Appendix C contains a copy of the form. In these transactions, transferors have no liability for the transfer tax. As previously stated, Forms 3 must be approved by ATF before transfers may be made.

9.4.2 ATF Form 4. Forms 4, Application for Tax Paid Transfer and Registration of a Firearm, are for use in transferring serviceable NFA firearms in the following instances: transfers by non-FFLs/SOTs to other such persons; transfers by non-FFLs/SOTs to FFLs/SOTs; and transfers by FFLs/SOTs to non-FFLs/SOTs.[167] Appendix C contains a copy of the form. See also Sections 3.2.6.1 through 3.2.6.3. These transfers are subject to the NFA transfer tax, so the forms must be accompanied by the appropriate tax payment. Forms 4 transferring firearms to individuals other than FFLs/SOTs must also be accompanied by transferees' fingerprints and photographs on FBI Forms FD-258. If the individual's receipt or possession of the firearm would violate Federal, State, or local law, the form would be disapproved. In addition, an individual transferee must have an appropriate law enforcement official execute the certification on the form.[168] See Section 9.8 for more information on law enforcement

---

[165] 26 U.S.C. 5845(j)
[166] 27 CFR 479.88(b)
[167] 27 CFR 479.84
[168] 27 CFR 479.85

certifications. Forms 4 must be approved by ATF before the transfers may be made. The completed Form 4, in duplicate, with fingerprint cards, photographs of the transferee, and payment of the applicable transfer tax should be mailed to:

National Firearms Act Branch
Bureau of Alcohol, Tobacco, Firearms and Explosives
P.O. Box 530298
Atlanta, GA 30353-0298

Payment of the transfer tax is to be in the form of a check or money order payable to the Bureau of Alcohol, Tobacco, Firearms and Explosives.

9.4.2.1 Copies of transferees' State or local licenses or permits. If State or local law requires the transferee to have a State or local license or permit to possess the firearm and the requirement is imposed upon the person prior to receipt, the Form 4 application should also be accompanied by a copy of the license or permit.

9.4.2.2 Transfers of NFA firearms to persons other than an individual or an FFL and special (occupational) taxpayer. Section 479.85 of the Code of Federal Regulations requires the ATF Form 4 or Form 5 application to properly identify the transferee. Although transfers to natural persons (individuals) must include a recent photograph, duplicate fingerprint cards, and a certification from law enforcement, the NFA also defines a person to include a partnership, company, association, trust, estate, or corporation. The requirements for fingerprints, photographs, and the law enforcement certificate specified in § 479.85 are not applicable for transferee who is not an individual.

When an ATF Form 4 or Form 5 application is submitted to transfer a firearm to a partnership, company, association, trust, estate, or corporation (collectively, an entity), the transferee entity must be identified on the Form 4 using the complete, formal name of the entity, along with the entity's street address, city, and state. The Form 4 or Form 5 must not include an individual's proper name, unless the proper name is a part of the entity's name (e.g., The Irrevocable Trust of John Doe, John Smith, Inc., etc.). ATF requires that the Form 4 or Form 5 include documentation evidencing the existence of the entity. This documentation would include, without limitation, partnership agreements, articles of incorporation, corporate registration, a complete copy of the declaration of trust, schedules or attachments referenced in the trust, etc. If the firearm being transferred is a machinegun, short barreled rifle, short barreled shotgun, or destructive device and the transfer is from an FFL, a person authorized to act on behalf of the entity must complete item 15 of the Form 4 and Form 5.

Please see section 9.12 for information regarding the NICS background check.

9.4.3 ATF Form 5. Transferors of NFA firearms to government entities, Federal, State, or local, must file ATF Forms 5, Application for Tax Exempt Transfer and Registration of a Firearm, to transfer the firearms to such entities.[169] (*Note: The applicant may wish to include details regarding the receiving*

[169] 27 CFR 479.90

gency if the agency is obscure. Note also that there are no transfers to task forces.) Although Forms 5 are generally required to be filed and approved for transfer of firearms to U.S. agencies, firearms owned or possessed by Federal agencies are not required to be registered. Appendix C contains a copy of the form. In these transactions, the transferor has no liability for the transfer tax. As previously stated, the form must be approved by ATF before the transfer may be made. As discussed in more detail below, Forms 5 are also used to transfer unserviceable firearms tax free, transfer firearms to FFLs for repair and for their return, and for distribution of estate firearms to lawful heirs.

Section 9.5 Conveyances of NFA firearms not treated as "transfers" under the NFA

9.5.1 Repair of firearms. ATF does not consider the temporary conveyance of an NFA firearm to an FFL for repair to be a "transfer" under the NFA. Thus, a transfer application is not required to convey the firearm for repair or to return the repaired firearm to its owner/possessor. Nevertheless, in order to avoid any appearance that a "transfer" has taken place, ATF recommends that a Form 5 application be submitted for approval prior to conveying the firearm for repair. It is also recommended that the FFL making repairs obtain an approved Form 5 to return a repaired firearm. If Forms 5 are not used to convey a firearm for repair or return the repaired firearm to the owner, the parties should maintain documentation showing that the conveyance was for purposes of repair, identifying the firearm, and showing the anticipated time for repair. Approved Forms 5, or the recommended documentation, will show that an unlawful "transfer" did not take place and that the FFL making the repairs is not in unlawful possession of the firearm. A non-FFL who proposes to transport a destructive device, machinegun, or short-barrel shotgun or rifle interstate to an FFL for repair should first obtain an approved ATF Form 5320.20 before transporting the firearm.[170]

9.5.1.1 Repair of firearm silencers. ATF published FAQs on April 17, 2008, regarding the repair and replacement of silencers and silencer components. These FAQs are published on the ATF website and are included in Appendix B.

9.5.2 Possession of firearms by employees of FFLs/SOTs for employers' business purposes. No "transfer" under the NFA occurs when an FFL/SOT permits a bona fide employee to take custody of its registered NFA firearms for purposes within the employee's scope of employment and for the business purposes of the FFL/SOT. Therefore, no approved ATF transfer form is required when employees take custody of firearms under these circumstances. In addition, the interstate delivery of a firearm to the employee and the employee's receipt of the firearm would not violate the GCA.

9.5.3 Distribution of estate firearms. A decedent's registered NFA firearms may be conveyed tax-exempt to lawful heirs. These distributions are not treated as voluntary "transfers" under the NFA. Rather, they are considered to be involuntary "transfers by operation of law." Under this concept, ATF will honor State court decisions relative to the ownership and right to possess NFA firearms. So, when State courts authorize the distribution of estate firearms to decedents' lawful heirs, ATF will approve the distribution and registration to the heirs if the transactions are otherwise lawful. A lawful heir is anyone named in the decedent's will or, in the absence of a will, anyone entitled to inherit under the laws of the State in which the decedent last resided.

170 18 U.S.C. 922(a)(4); 27 CFR 478.28

9.5.3.1 Distributions to heirs. Although these distributions are not treated as "transfers" for purposes of the NFA, Form 5 must be filed by an executor or administrator to register a firearm to a lawful heir and the form must be approved by ATF prior to distribution to the heir. The form should be filed as soon as possible. However, ATF will allow a reasonable time to arrange for the transfer. This generally should be done before probate is closed. When a firearm is being transferred to an *individual* heir, his or her fingerprints on FBI Forms FD-258 must accompany the transfer application. The application will be denied if the heir's receipt or possession of the firearm would violate Federal, State, or local law. The law enforcement certification on the form need not be completed. The form should also be accompanied by documentation showing the executor's or administrator's authority to distribute the firearm as well as the heir's entitlement to the firearm. Distributions to heirs should not be made until Forms 5 are approved. Executors and administrators are not required to have estate firearms registered to them prior to distribution to lawful heirs.

9.5.3.2 Distributions to persons outside the estate. Distributions of NFA firearms by executors or administrators to persons outside the estate (not beneficiaries) are "transfers" under the NFA and require an ATF-approved transfer form. Transfers of serviceable firearms to other entities or persons require an approved Form 4. Form 4 applications must be accompanied by the applicable transfer tax, and, if the transferee is an individual, the transferee's fingerprints on FBI Forms FD-258. Applications will be denied if transferees' receipt or possession of the firearms would violate Federal, State, or local law. Also, Form 4 applications to transfer firearms to individuals must contain the law enforcement certification of an appropriate law enforcement official. See Section 9.8 for further information on these certifications. Form 4 applications to transfer firearms to non-FFLs residing outside the State in which the estate is being administered will be denied. Form 4 transfers should not be made until the transfers are approved.

9.5.3.3 Uncertainty about the registration status of decedents' firearms In some cases, an executor or administrator of an estate may be uncertain whether the decedent's firearms are registered to the decedent in the NFRTR. Perhaps the executor or administrator is unable to locate the decedent's registration documents. As discussed in Section 9.2, if the decedent's firearms are not registered to him/her in the NFRTR, the firearms are contraband and may not be lawfully possessed or transferred. If the executor or administrator cannot locate the decedent's registration documents, he/she should contact the NFA Branch in writing and inquire about the firearms' registration status. This inquiry should be accompanied by documents showing the executor's or administrator's authority under State law to represent the decedent and dispose of the decedent's firearms. Although ATF is generally prohibited from disclosing tax information, including the identity of persons to whom NFA firearms are registered, ATF may disclose such information to persons lawfully representing registrants of NFA firearms.

9.5.3.4 Unregistered estate firearms. Should an estate contain NFA firearms not registered to the decedent, these firearms are contraband that may not be lawfully possessed or transferred. Where these are found within an estate, the executor or administrator should contact his/her local ATF office and arrange for their disposal.

9.5.3.5 Distribution of decedents' "sales samples." If NFA firearms in a decedent's estate are "sales samples," that is, they were imported and distributed to the decedent as sales samples or were domestically manufactured machineguns distributed to the decedent as sales samples, the sale sample restriction continues in effect and lawful possession of the firearms requires that the firearms be held as "sales samples" for demonstration to government agencies.[171] Therefore, these firearms within an estate must be transferred to government agencies or FFLs/SOTs as sales samples for demonstration to such agencies.

ection 9.6 Manufacturers' use of contractors to perform work on firearms. As part of the ıanufacturing process to produce firearms for subsequent sale, some manufacturers contract with other ersons to perform steps in the manufacturing process. As discussed in Chapter 7, these contractors are lso manufacturers subject to licensing as firearms manufacturers and payment of NFA special tax. In ddition, a manufacturer's transfer of an NFA firearm to such a contractor and the return of the firearm ɔ the manufacturer are transfers required to be approved by ATF. These transfers require approved 'orms 3.

ection 9.7 Transfers of unserviceable NFA firearms. "Unserviceable firearms" are firearms incapable of discharging a shot by means of an explosive and incapable of being readily restored to a iring condition."[172] They are still "firearms" for purposes of the NFA and must be registered in the JFRTR to be lawfully possessed and transferred. However, their transfer is not subject to transfer ax.[173] To lawfully transfer unserviceable firearms, Form 5 transfer applications must be filed with ATF nd approved. Appendix C contains a copy of the form. Form 5 applications to transfer the firearms to ndividuals must be accompanied by transferees' fingerprints and photographs on FBI Forms FD-258. pplications will be disapproved if receipt or possession of the firearms would place transferees in iolation of Federal, State, or local law. In the case of transfers to individuals, the transferees must have n appropriate law enforcement official sign the law enforcement certification on the form. See Section .9 for further information on the law enforcement certification. A Form 5 transfer application will not e approved if the transferee is an individual residing outside the State in which the transferor resides; owever, as previously discussed, there is an exception for FFLs' over-the-counter transfers of rifles and hotguns to non-residents if the laws of the transferors' and transferees' States are complied with. 'ransfers pursuant to Forms 5 may not be made until approved.

ection 9.8 U.S. Government-owned firearms. Conveyances of U.S. Government-owned NFA irearms to FFLs/SOTs for repair, modification, or performing other work such as incorporating the irearms into a weapons system require no approved ATF transfer or registration. The same is true for ıe return of the firearms to U.S. Government entities.

ection 9.9 Law enforcement certifications. These certifications on Forms 1 and 4 must be signed by n appropriate law enforcement official when the forms seek the transfer or making of an NFA firearm ɔ or by an individual. However, as stated in Section 9.5.4.1, the certifications are not required when state firearms are distributed to lawful heirs. As provided by regulations, the certificate must state that ıe certifying official is satisfied that the individual's fingerprints and photographs accompanying the

[171] 26 U.S.C. 5844(3)
[172] 26 U.S.C. 5845(h)
[173] 27 CFR 479.91

application are those of the applicant and that the official has no information indicating that the receipt or possession of the firearm would place the transferee in violation of State or local law or that the transferee will use the firearm for other than lawful purposes. Acceptable certifying officials include chiefs of police, county sheriffs, heads of State police, State or local district attorneys, or "such other persons whose certificates may in a particular case be acceptable to the Director."[174] Examples of other officials whose certifications have been found acceptable include State attorneys general and judges of State courts having authority to conduct jury trials in felony cases.

If another official is proposed as an acceptable certifying official, the transferor may, in advance of filing the transfer form, submit a written request to the NFA Branch whether the official is an acceptable certifying official. Alternatively, the transfer form may be filed with the official's certificate. If the certification is unacceptable, ATF would disapprove the form and return it to the proposed transferor.

9.9.1 Is a law enforcement officer required to sign the certification? In some jurisdictions, officers whose certifications on a transfer form would be acceptable will not sign the certifications for reasons of their own. These officials cannot be compelled to sign the certifications.

9.9.2 Is a law enforcement certification acceptable if signed by an official outside the jurisdiction where the transferee resides? No. The certification must be signed by an official having jurisdiction where the transferee resides.

Section 9.10 Transfers of imported NFA firearms

9.10.1 Firearms imported for government agencies. As discussed in Chapter 8, NFA firearms may be imported for sale to Federal, State, or local government agencies. For approval of these imports, the importer's Form 6 permit application must be accompanied by the agency's letter or purchase order reflecting the purchase of the firearms. Appendix D contains a sample letter for use by an agency ordering imported firearms. Transfer of the firearms to the purchasing agency requires an approved Form 5. If a qualified NFA dealer received the agency's order and placed it with the importer, the importer may transfer the firearms to the dealer on an approved Form 3 and the dealer, in turn, would use an approved Form 5 to transfer the firearms to the agency.

9.10.2 "Sales samples." As discussed in Chapter 8, NFA firearms may be imported for use as sales samples by qualified NFA importers and dealers to demonstrate the firearms to government agencies and generate possible future sales to such agencies. As provided by the regulations, a Form 6 application to import such sample will be approved if it is established by specific information attached to the application that the firearm is suitable or potentially suitable for an agency's use; the expected governmental customers requiring a demonstration of the firearm; information as to the availability of the firearm to fill subsequent orders; and letters from agencies expressing a need for a particular model or interest in seeing a demonstration of a particular firearm.[175] Appendix D contains sample letters for use in acquiring imported sales samples, including a qualified NFA dealer's letter ordering a sales sample from an importer and an agency's letter requesting a demonstration of a sales sample. An importer's transfer of a sales sample to a dealer requires an approved Form 3.

[174] 27 CFR 479.85
[175] 27 CFR 479.112(c) and (d)

9.10.3 Transferring multiple quantities of the same firearm model as "sales samples." As provided by the regulations, applications to import or transfer more than one firearm of a particular model for use as a sales sample by an importer or dealer must establish the importer's or dealer's need for the quantity of samples sought to be imported.[176] In the case of machineguns imported on or after May 19, 1986 (as well as machineguns domestically manufactured after that date), ATF Ruling 2002-5 holds that if an FFL needs to demonstrate a particular model of machinegun to an entire police department or SWAT team, ATF will approve the transfer of two machineguns of that model to the dealer as sales samples. Additional quantities will be allowed if an FFL can document the need for more than two machineguns of a particular model.

Section 9.11 Transfers of machineguns imported or manufactured on or after May 19, 1986

9.11.1 Machinegun prohibition in 18 U.S.C. 922(o). This statute makes it unlawful to transfer or possess a machinegun, except for transfers to or by, or possession by or under the authority of, the United States or a State, or machineguns lawfully possessed before May 19, 1986 (that is, machineguns in the U.S. and registered in the NFRTR). Regulations implementing the statute allow domestic manufacturers to lawfully manufacture and stockpile machineguns for future sale to Federal and State agencies, for distribution to FFLs/SOTs as sales samples for demonstration to such agencies, or for exportation.[177] The procedures discussed in Section 9.8 for transferring imported firearms to government agencies or to FFLs for use as sales samples apply as well to domestically manufactured machineguns.

9.11.2 May machineguns subject to section 922(o) be transferred to government contractors? The statute provides no exception for the lawful possession of these machineguns by government contractors for use in testing, research, design, or other work in fulfilling government contracts. One specific exception to this general rule appears in the Atomic Energy Act of 1954, 42 U.S.C. 2201a. This recently enacted provision allows for machinegun possession by security personnel engaged in the protection of Nuclear Regulatory Commission facilities or radioactive materials. Note also that although the NFA provides for the importation of NFA firearms for scientific or research purposes or for testing or use as a model by a registered manufacturer, the prohibition in Section 922(o) contains no exception that would permit the lawful possession of machineguns in the U.S. for those purposes; thus, applications to import machineguns for those purposes would be denied.

Section 9.12 Are FFLs/SOTs required to initiate a background check of the transferee under the Brady Law in connection with the transfer of an NFA firearm? No. Although 18 U.S.C. § 922(t) requires an FFL to complete a National Instant Criminal Background Checks System (NICS) check of the firearm recipient prior to completing the transfer, subsection 922(t)(3)(B) removes ATF-approved transfers of NFA firearms from the NICS requirement for individuals.

9.12.1 NFA Transfers to other than individuals. Subsequent to the approval of an application requesting to transfer an NFA firearm to, or on behalf of, a partnership, company, association, trust, estate, or corporation, the authorized person picking up the firearm on behalf of, a

---

176 Ibid. See also 27 CFR 479.105(d)

177 27 CFR 479.105(c)

partnership, company, association, trust, estate, or corporation from the FFL must complete the Form 4473 with his/her personal information and undergo a NICS check. See also, question P60 in the ATF FAQs.

9.12.2 NFA transfer with other GCA firearm. An application to transfer an NFA firearm that includes a firearm regulated by the GCA, (e.g., silencer with a pistol, wallet-holster with pistol, etc.), requires the completion of a Form 4473 and NICS check prior to transfer of the GCA firearm. This includes the transfer of a firearm where the suppressor is permanently attached to the firearm. The GCA firearm can only be transferred after the required NICS check is completed. The serial number of the GCA firearm and the permanently attached NFA firearm must be included on the ATF form 4473. FFL/SOT payers are also responsible for and adhering to all applicable State and local requirements for NFA transfers. If the FFL/SOT is awaiting ATF approval of the transfer application, the unattached GCA firearm can be transfer prior to the completion of the NFA transfer, however, an additional NICS check may be necessary if the transfer is to a non-individual.

Section 9.13 May an FFL/SOT transfer a personally owned destructive device without qualifying to do business in destructive devices? Persons engaged in the business of dealing in firearms must have a GCA license authorizing them to deal in the type of firearms in which they deal. If they engage in the business of dealing in destructive devices, they must have a license to do so as required by 18 U.S.C. 923(a)(3)(A). However, ATF recognizes that persons licensed to deal in firearms other than destructive devices may lawfully maintain a personal collection of destructive devices and *occasionally* dispose of them as personal firearms on Forms 4 without having a license to deal in such devices. But if the dealer's receipt and disposition of these devices become *repetitive,* ATF may infer that the dealer is engaged in the business of dealing in the devices and require him/her to be licensed as a dealer in such devices. There is no precise number of transactions that would trigger the license requirement.

Section 9.14 Transferable Status and the Form 10. Unregistered firearms obtained by State or local government agencies through abandonment or forfeiture are registered on an ATF Form 10. See 27 CFR § 479.104. Upon registration the Form 10 is marked "official use only," and subsequent transfer and registration is limited to the official use of other State or local government entities. The firearms may not enter ordinary commercial channels. NFA firearms which were registered on Form 10 but were transferred to an FFL or individual prior to the effective date of ATF Ruling 74-8 may continue to be possessed in commercial or private channels. If the firearm was still registered to the State or local entity on Form 10 at the time of the effective date of ATF Ruling 74-8, the firearm may only be transferred to another government entity for "official use only."

CHAPTER 10. COLLECTORS OF NFA FIREARMS

ection 10.1 Curios or relics

0.1.1 Definition of curio or relic. Curios or relics include firearms which are of special interest to ollectors by reason of some quality other than is associated with firearms intended for sporting use or s offensive or defensive weapons.[178] To be recognized as curios or relics firearms must fall within one f the following categories:

- Firearms which were manufactured at least 50 years prior to the current date, but not including replicas thereof.

- Firearms which are certified by the curator of a municipal, State, or Federal museum which exhibits firearms to be curios or relics of museum interest.

- Any other firearms which derive a substantial part of their monetary value from the fact that they are novel, rare, bizarre, or because of their association with some historical figure, period or event. Proof of qualification of a particular firearm under this category may be established by evidence of present value and evidence that like firearms are not available except as collector's items, or that the value of like firearms available in ordinary commercial channels is substantially less.

0.1.2 Curio or relic classification. A formal ATF classification of a firearm as a curio or relic is not equired by the law or regulations, however, such official determination is recommended. By obtaining curio or relic classification, a firearm is officially recognized as a curio or relic and the weapon is dded to the ATF Firearms Curios or Relics List.[179] Firearms classified as curios or relics are still irearms as defined. A curio or relic classification may be obtained by submitting a written request, in uplicate, to ATF's Firearms Technology Branch (FTB).[180] The request must be executed under the enalties of perjury, contain a complete and accurate description of the firearm and documentation howing that the firearm fits within one of the qualifying categories listed in Section 10.1.1. FTB may equire submission of the firearm for examination and evaluation. If submission of the firearm is npractical, the person requesting the determination must advise FTB and designate the place where the irearm will be available for examination and evaluation.

n the case of firearms that are more than 50 years old, ATF has determined that a collector need not equest a classification for such weapons.[181] Firearms 50 years or older automatically qualify as curios r relics.

request for curio or relic determination should be submitted to:

---

[8] 27 CFR 478.11
[9] ATF P 5300.11
[0] 27 CFR 478.26
[1] FFL Newsletter, Vol. 1, Issue 2

Firearms Technology Branch
Bureau of Alcohol, Tobacco, Firearms and Explosives
244 Needy Road
Martinsburg, West Virginia 25405

10.1.3 NFA firearms classified as curios or relics. NFA firearms may be classified as curios or relic under the same requirements discussed in Section 10.1.1. NFA firearms classified as curios or relics ar still subject to the provisions of the NFA.

10.1.4 The Firearms Curios or Relics List (ATF P 5300.11). This publication contains a listing of those firearms that have been officially classified as curios or relics. The list contains the following sections:

1. Ammunition classified as curios or relics prior to the 1986 amendments of the GCA.

2. Firearms classified as curios or relics subject to the provisions of the GCA.

3. Weapons removed from the NFA as collector's items which are determined to be curios or relics subject to the provisions of the GCA.

4. Weapons removed from the NFA as collector's items which are antiques not subject to the provisions of the GCA.

5. NFA weapons classified as curios or relics subject to the provisions of the GCA and NFA.

The most recent edition of the Curios or Relics List is available on the ATF website www.ATF.gov. Periodic updates to the list are also available on the website.

10.1.5 DEWATS. Deactivated War Trophy (DEWAT) firearms are still firearms under the NFA, but have been rendered unserviceable (i.e., incapable of discharging a shot by means of an explosive and incapable of being readily restored to a firing condition). The deactivation may have been accomplishe by various means such as (but not limited to) welding of the chamber, cutting the barrel/chamber/breec plugging the barrel, welding the bolt to the chamber, or some combination of these actions which rendered the firearm incapable of firing a shot.

Regardless of being unserviceable, the DEWAT firearm must be registered and approved for transfer a any other NFA firearm. The process to transfer a DEWAT is set forth in §479.90. Because the DEWAT is unserviceable, it is transferred tax-exempt as a curio or ornament on an ATF Form 5. The Form 5 must identify the transferee and transferor, FFL/SOT, any importer, type, model, caliber, overa length, gauge, size, serial number or other marks of identification. Additionally, the transferor must check the block in item 1 of the Form 5 indicating the unserviceable status of the firearm and answer item 4(i) confirming how the firearm has been rendered unserviceable.

10.1.5.1 Reactivation of DEWATS. Deactivated War Trophy (DEWAT) firearms may be returned to a serviceable condition. A DEWAT that has been returned to a serviceable condition (or reactivated) is often referred to as Reactivated War Trophy (REWAT). The act of returning a DEWAT to a serviceable condition is considered the "making" or "manufacturing" of a firearm by ATF. The procedures for reactivation by a non-FFL or FFL with other than a manufacturer's SOT status or with no SOT status on Form 1 are found in section 6.6 of the handbook. The procedures for reactivation by an FFL with SOT status as a manufacturer are found in section 7.3.3 of the handbook.

Curio or Relic status: When the DEWAT firearm being reactivated is a curio and relic (C&R), the C&R status is retained. Please see 27 CFR 478.11 for the definition of curios and relics.

ection 10.2 Licensed collector

0.2.1 Collector of curios or relics license. There is no requirement to have a Federal license for the urpose of collecting firearms. However, the GCA has a license category for collectors of curios or elics.[182] The collector of curios or relics license serves one purpose - it enables the holder to receive urio or relic firearms interstate under the GCA. A collector's license does not authorize the holder to ngage in the business of dealing in firearms. If a collector wants to deal in curio or relic firearms, a lealer's license must be obtained. In the case of dealing in NFA curio or relic firearms, the SOT must lso be paid. Further, with respect to firearms other than curios or relics, a licensed collector is in the ame position as a nonlicensee. That is, the licensed collector may not lawfully receive them in interstate ommerce.

0.2.2 Receipt of NFA curios or relics by a licensed collector. A licensed collector may receive an IFA curio or relic firearm directly from any person in any State. Such receipt requires an approved 'orm 4 (or in the case of an unserviceable firearm, an approved Form 5) with law enforcement ertification and fingerprint cards. If the NFA curio or relic firearm being transferred is not specifically isted on the Curios or Relics List, the applicant or transferee should submit documentation that the irearm in question has been classified as a curio or relic. Including this information can reduce delays n processing a transfer application.

0.2.3 Transfer of NFA curios or relics by a licensed collector. A licensed collector may transfer an IFA curio or relic firearm in interstate commerce only to a licensed collector, licensed dealer, licensed nporter or licensed manufacturer. Such transfer requires an approved Form 4 (or in the case of an nserviceable firearm, an approved Form 5). If the transferee is not an SOT, the law enforcement ertification and fingerprint cards are also required. A licensed collector may transfer an NFA curio or elic firearm directly to a nonlicensee who resides in the same State as the licensed collector. A transfer f this type requires an approved Form 4 or Form 5, a law enforcement certification, and fingerprint ards.

---

[2] 18 U.S.C. 923(b)

Section 10.3 Recordkeeping requirements

10.3.1 Acquisition and disposition records. Licensed collectors must maintain a record of the acquisition and disposition of curio or relic firearms.[183] The regulations specify the format and information required to be maintained. Only firearms classified as curios or relics should be listed in th required records. A licensed collector is not required to execute a Form 4473, Firearms Transaction Record, or initiate a NICS background check when disposing of a curio or relic firearm.

10.3.2 Termination of a collector's license. A licensed collector who decides not to renew a collector's license is not required to surrender the acquisition and disposition records for curio or relic firearms to ATF.

[183] 27 CFR 478.125(f)

CHAPTER 11. EXPORTATION OF NFA FIREARMS

Section 11.1 Arms Export Control Act (AECA), 22 U.S.C. 2778. The AECA regulates the exportation of "defense articles," including firearms and firearm parts. Authority to administer and enforce the exportation provisions of the AECA rests with the U.S. State Department. The articles subject to AECA control are listed under various categories in the State Department's U.S. Munitions List contained in the regulations in 22 CFR 121.1. Firearms and component parts of firearms are listed under Categories I and II.

11.1.1 State Department regulations implementing the AECA. State Department's regulations implementing the AECA are known as the International Traffic in Arms Regulations (ITAR) (22 CFR Parts 120-130). The regulations can be found on the internet by querying "U.S. State Department" and then "Directorate of Defense Trade Controls." A hardcopy of the regulations is published annually and includes all amendments made to the ITAR from April 1 through March 31 of each year. The hardcopy can be obtained from the Government Printing Office.

11.1.2 Firearms and firearm component parts subject to the AECA. As stated in Section 11.1, firearms and component parts of firearms are included in Categories I and II of the U.S. Munitions List.

11.1.2.1 Category I. Category I covers nonautomatic and semi-automatic firearms to caliber .50 inclusive (12.7 mm), as well as fully automatic firearms to .50 caliber inclusive. This category also covers silencers, mufflers, sound and flash suppressors, and rifle scopes manufactured to military specifications. Excluded from Category I are non-combat shotguns with barrel lengths of 18 inches or longer. In other words, sporting shotguns not subject to the NFA are not defense articles subject to State Department export controls. Category I also covers components, parts, accessories and attachments for firearms. However, it excludes rifle scopes and sighting devices not manufactured to military specifications, as well as accessories and attachments (for example, belts, slings, after market rubber grips, cleaning kits) for firearms that do not enhance the usefulness, effectiveness, or capabilities of the firearm, components, and parts The Department of Commerce regulates the export of firearms and parts not covered by Category I. Therefore, see the Commerce's regulations in 15 CFR Parts 730-799 for export controls over those items.

11.1.2.2 Category II. This category covers guns over caliber .50, as well as all other components, parts, accessories, attachments and associated equipment specifically designed or modified for such guns.

Section 11.2 Registration of exporters and manufacturers under the AECA. As required by regulations in 22 CFR 122.1(a), any person engaged in the United States in the business of manufacturing or exporting defense articles on the U.S. Munitions List is required to register with State Department's Directorate of Defense Trade Controls (DDTC). Persons engaged in the business of manufacturing firearms and other defense articles must register even if they do not export such items. Certain exceptions to the registration requirement are listed in Section 122.1(b).

11.2.1 Submission of registration form, DSP-9. As provided by 22 CFR 122.2, Department of State Form DSP-9 must be submitted by the registrant with a check or money order payable to the Department of State of one of the fees prescribed in section 122.3(a).

11.2.2 Transmittal letter must accompany Form DSP-9. Form DSP-9 must be accompanied by a transmittal letter signed by a senior officer, as well as documentation showing that the registrant is incorporated or otherwise authorized to do business in the United States. The letter must contain the statements required by section 122.2(b). In summary, the letter must state whether any senior officer or member of the board of directors:

(1) Has ever been indicted or convicted of violating any criminal statute enumerated in 22 CFR 120.27;

(2) Is ineligible to contract with, or to receive a license or other approval to import defense articles from, or receive an export license or other approval from, any U.S. agency; and

(3) Whether the registrant is owned or controlled by foreign persons (as defined in 22 CFR 120.16). If the registrant is owned or controlled by foreign persons, the letter must also state whether the registrant is incorporated or otherwise authorized to engage in business in the United States. See Section 122.2(c) for definitions of "ownership" and "control."

11.2.3 Notification of changes in information furnished by registrants. See 22 CFR 122.4 for the requirement to report changes to DDTC in the information provided in registering with the State Department on Form DSP-9.

Section 11.3 License requirement for permanent export of a defense article. Any person or entity intending to permanently export a defense article must obtain approval of DDTC prior to the export. Approval must be obtained by filing with DDTC Form DSP-5. For more detailed information, see State Department regulations in 22 CFR 123.1.

Section 11.4 License requirement for temporary export of a defense article. Any person or entity intending to temporarily export a defense article must obtain approval of DDTC prior to the export. Approval must be obtained by filing with Form DSP-73. No ATF approval is required for a temporary exportation. For more detailed information, see State Department regulations in 22 CFR 123.5.

Section 11.5 Prohibited exports and sales to certain countries. It is the policy of the United States to deny licenses to export defense articles destined for certain countries. Therefore, the State Department will deny applications for licenses to export such articles, including firearms, to those countries. A list of these countries can be found in State Department's regulations in 22 CFR 126.1, for example, Cuba, Iran, Libya, North Korea, and Vietnam. This policy also applies to countries with respect to which the United States maintains an arms embargo, for example, Burma, China, and Liberia.

Section 11.6 ATF requirements relative to the exportation of firearms. Any person or entity desiring to permanently export an NFA firearm without payment of the transfer tax must file with ATF an application for a permit to export the firearm on ATF Form 9. Shipment may not be made until the permit is received. ATF will not act on the application unless it contains the information required by the regulations in 27 CFR 479.114. See ATF's regulations in 27 CFR 479.114 – 479.121 for detailed information on ATF's exportation requirements.

1.6.1 Applying for an export permit on ATF Form 9. Any person desiring to export an NFA irearm without paying the transfer tax must file with ATF a permit application to export the firearm on orm 9. Appendix C contains a copy of Form 9. The application requires submission of the following nformation:

(1) Name and address of the foreign consignee;

(2) Number of firearms covered by the application

(3) The intended port of exportation;

(4) A complete description of each firearm to be exported (an attached list of serial numbers/descriptions may be used);

(5) The name, address, State Department export license number (or date of application if not issued), and identification of the special (occupational) tax stamp of the transferor; and

(6) The application must be supported by a certified copy of a written order or contract of sale or other evidence showing that the firearm is to be shipped to a foreign destination. However, a copy of a State Department export license on DSP-5 is acceptable in lieu of such certification.

1.6.2 Transfers to another person for export. Where is desired to make a transfer free of tax to nother person who in turn will export the firearm, the transferor must likewise file an application on orm 9 supported by evidence that the transfer will start the firearm in the course of exportation. Iowever, where both the transferor and exporter are FFLs/SOTs, the transferor is not required to file a orm 9 application.

1.6.3 Proof of exportation. Within a six-month period from the date of issuance of the permit on orm 9, the exporter must provide ATF with the evidence of exportation specified by the regulations in 7 CFR 479.118. Where such evidence is not furnished, the exporter will be assessed the transfer tax on ne firearm.

CHAPTER 12. RECORDKEEPING

Section 12.1 Maintaining proof of registration. The NFA requires that a person possessing a firearm registered in the National Firearms Registration and Transfer Record (NFRTR) retain proof of registration which must be made available to the Attorney General, specifically an ATF agent or investigator, upon request.[184] Proof of registration would be on a Form 1 registering a firearm to its maker, Form 2 registering a firearm to an importer or manufacturer, or a Form 3, 4, or 5 showing registration of a firearm to a transferee.

12.1.1 Manner in which registration documents must be kept. NFA regulations require FFLs/SOT maintain their registration documents in chronological order at their place of business.[185]

12.1.2 Recordkeeping where registered firearms are kept on premises other than the registered owner's place of business. If the registered firearms are kept at a location other than the place of business shown on the registrant's special tax stamp, the forms showing the firearm's registration must be annotated to reflect the place where the firearms are kept or stored. ATF has advised in the past that documents (such as a statement, file card, or office form indicating the current location) may be kept with the registration records to show such off-premises location. In addition, an FFL/SOT's notation in the A & D Records disposition line is also acceptable to show such location if a notice is kept with the registration document stating "See bound book for storage location." ATF will clarify this policy in the near future.

Section 12.2 Verifying the description of firearms on registration documents. It is important to verify that the information on a registration document accurately describes the registered firearm. Where firearms were registered prior to the amended NFA in 1968 or during the 1968 amnesty period, was not uncommon for these firearms to have been inaccurately described on the registration forms. Owners of registered firearms should carefully compare each field in the "Description of Firearm" section on the registration document with their firearm. Of particular importance are the fields recordin the type of firearm, model, and serial number. Concerning the serial number, it is important to ensure that digits in the number have not been transposed, or, if the serial number contains letters as a prefix o suffix, that those letters are properly shown as part of the number.

Section 12.3 Correcting an error in the description of a registered firearm. If errors are found in the description of a firearm in the owner's registration document, the owner should write a letter to the Chief, NFA Branch and request the registration record for the firearm be corrected. The letter should describe the problem and provide the correct information. If the error being reported involves an incorrect serial number, ATF will require that close-up photographs or a pencil rubbing of the actual serial number be provided. A copy of the registration document should also be included with the letter request. Persons requesting a correction to the description of their registered firearm should also retain copy of the letter request with their registration document until ATF responds with an acknowledgeme that the registration record has been corrected.

---

[184] 26 U.S.C. 5841(e)
[185] 27 CFR 479.131

ection 12.4 Custody of NFA firearms by employees of FFLs/SOTs. It is common for employees of 'FLs/SOTs to take firearms registered to the FFLs/SOTs off-premises for display, demonstration, or ther purposes on behalf of their registered owners. This does not result in the "transfer" of IFAfirearms requiring ATF approval and registration of the firearms to the employees as long as the irearms are possessed for the employer's business purposes and not for employee's personal use. imilarly, employees taking custody of firearms under these circumstances would not constitute a firearms disposition" that is required to be recorded in FFL/SOT's acquisition and disposition records A & D Records). Although not legally required, FFLs/SOTs and their employees should consider aking the following actions as a matter of good business practice:

(1) Keep a copy of the NFA registration document with each registered firearm.

(2) If the firearm is located in an area having State or local registration requirements, keep a copy of the State/local registration document with the firearm.

(3) Where the firearm is taken off-premises for display, demonstration, or other purposes on behalf of the registered owner, keep a copy of the NFA and State/local registration documents with the firearm.

(4) Provide an employee having custody of the registered owner's firearm away from the licensed premises with a letter authorizing the employee to possess the firearm on the employer's behalf.

'hese actions would help convince law enforcement authorities who encounter the firearm that the 'FL/SOT's employee is in lawful possession of the firearm and that the firearm is not unlawfully ossessed and subject to seizure.

2.4.1 Who is an "employee" of an FFL/SOT? For purposes of this section, an "employee" is a bona ide employee of an FFL/SOT on the FFL/SOT's payroll. An "employee" is not a mere agent of the 'FL/SOT appointed temporarily to take possession of the registered owner's firearm for a particular nission. If an agent of an FFL/SOT, rather than a bona fide employee, takes possession of the 'FL/SOT's registered firearm, a "transfer" occurs subject to registration, tax, and other provisions of the IFA.

ection 12.5 Firearms acquisition and disposition records. NFA regulations require each nanufacturer, importer, and dealer in NFA firearms to maintain the records of the acquisition and isposition of firearms required by GCA regulations in 27 CFR Part478, Subpart H.[186] The regulations lentify specific formats and required information fields for recording and maintaining these records. 'he records must be maintained on the FFL/SOT's licensed premises.[187]

2.5.1 Commercial records of firearms received. When a commercial record is held by a licensed ealer or collector showing the acquisition of a firearm or curio or relic, and the record contains all cquisition information required to be recorded in the licensee's A & D Records, the licensee may, for a eriod not exceeding 7 days following the date of receipt, delay making the required entry in the A & D

---

[186] 27 CFR 479.131

[187] 27 CFR 478.121(a)

Records if the commercial record is (1) maintained separate from other commercial documents, and (2) is readily available for inspection on the licensed premises. When disposition is made of the firearm not entered into the A & D Records, the licensee must enter all required information in the A & D Records at the time of the transfer.[188]

12.5.2 Alternate acquisition and disposition records for low volume dealers. This is a little used but available alternative for low-volume firearms dealers to maintain records of the acquisition and disposition of firearms. In lieu of the A&D records discussed in Section 12.5, a licensed dealer contemplating the disposition of not more than 50 firearms within a succeeding 12-month period, whether to other FFLs or to nonlicensees, may maintain a record of the acquisition and disposition of firearms on firearms transaction records, Forms 4473(LV).[189] A dealer maintaining records on these forms, whose firearms dispositions exceed 50 firearms within the 12-month period, must make and maintain the prescribed A&D Records for each firearm exceeding 50. See the regulations in 27 CFR 478.124a for details on maintaining records on Forms 4473(LV).

12.5.2 Variances in the recordkeeping requirements for the acquisition and disposition of firearms. ATF may authorize an FFL to maintain alternate records of the acquisition and disposition of firearms if such records would accurately and readily disclose the required information.[190] To use an alternate method of recordkeeping, an FFL must submit a letter application, in duplicate, to the appropriate Director of Industry Operations (DIO) and describe the proposed alternate records. Alternate records may not be employed by the FFL until approval is received from ATF. With the exception of computerized recordkeeping variance requests presented by FFLs located within one ATF Field Division, DIOs may be required to forward variance requests to the ATF Firearms Programs Division for review and approval.

12.5.2.1 Computerized records. In lieu of the hard-copy, A & D Records required by the regulations, the DIO may authorize FFLs to maintain computerized records of their firearms acquisitions and dispositions records. DIOs will approve the use of a computerized recordkeeping system if:

(1) The system meets the regulatory requirements for recording the information in 27 CFR 478.122, 27 CFR 478.123, and/or 27 CFR 478.125.

(2) The system allows queries by serial number, acquisition date, and name of the manufacturer. Commonly recognized trade names may be used when denoting the manufacturer or importer.

(3) The system has an acceptable daily backup for all databases.

(4) The system must provide periodic printouts of all records at least semiannually, upon request of an ATF officer, when the database is purged, and when the license is

---

[188] 27 CFR 478.125(g)
[189] 27 CFR 478.124a
[190] 27 CFR 478.125(h)

terminated. The printouts must be limited to the information required by 27 CFR 478.122, 27 CFR 478.123, and/or 27 CFR 478.125 only. The printouts must be retained until the next required printout is prepared

(5) The computer printouts contain firearms in inventory, as well as all firearms transferred during the period covered, sequentially by date of acquisition.

(6) The system must record both the manufacturer and the importer of foreign made firearms.

(7) (7) Printouts may include antique firearms, but cannot include other non-firearm merchandise. However, firearms not subject to GCA requirements must be identified as "ANT" in the "firearm type" column.

(8) The name and address (if nonlicensee) or the name and FFL number (if licensee) of both the supplier/consignor and the purchaser/transferee must be included in the computer data. The name and ATF Form transaction number may be used in lieu of the address for recording the transfer of a firearm to a nonlicensee if the Forms 4473 are filed numerically.

(9) The system cannot rely upon invoices or other paper/manual systems to provide any of the required information.

(10) Any changes in an approved system must be reported to the DIO in writing with a copy of the original variance approval attached, for evaluation and consideration prior to implementation.

(11) All acquisition and disposition records must remain at the licensed premises.

Section 12.6 Forms 4473. A licensed firearms importer, manufacturer, or dealer may not sell or dispose of a firearm to a nonlicensee unless the transaction is recorded on ATF Form 4473.[191] See Section 12.6.2 for firearms transfers occurring as a result of the return of a repaired or replacement firearm.

12.6.1 Firearms acquired from an FFL by an organization. An FFL who sells or otherwise disposes of a firearm to a nonlicensee who is other an individual, that is, a corporation, partnership, or association, must obtain a Form 4473 from the individual acquiring the firearm on behalf of the organization.[192] The individual must attach a written statement to the form, executed under the penalties of perjury, stating that the firearm is being acquired for the use of his/her organization and the name and address of the organization. The FFL must also initiate a background check of the individual under the Brady Law.[193]

---

[191] 27 CFR 478.124(a)
[192] 27 CFR 478.124(g)
[193] Q&A P60, ATF Pub. 5300.4, Sept. 2005, p. 195

12.6.2 Return of firearms delivered to an FFL for repair or customizing. A Form 4473 is not required to record the disposition of a firearm delivered to an FFL for the sole purpose of repair or customizing when the firearm or a replacement firearm is returned to the person from whom it was received.[194] See Section 9.5.1 suggesting the use of Forms 5 to document the delivery of a firearm to an FFL for repair (or customizing) and the return of the firearm. The use of an ATF-approved Form 5 for these transactions is beneficial to the registered owner of the firearm and to the FFL having the firearm for repair or customizing. If the registered owner is inspected by ATF while the gun is away for repair or customizing, an ATF-approved Form 5 will satisfy the inspector that the absence of the gun from inventory is lawful and not due to a loss or unlawful transfer. If the FFL having the gun for repair or customizing has an approved Form 5 for receipt of the gun, the FFL can establish to an ATF inspector's satisfaction that the gun is lawfully possessed.

12.6.3 Maintaining Forms 4473. FFLs must retain in alphabetical (by name of purchaser), chronological (by date of disposition), or numerical (by transaction serial number) order, each Form 4473 obtained in the course of transferring firearms.[195] Forms 4473 obtained whereby the sale, delivery or transfer of the firearm did not take place must be separately retained in similar alphabetical, numerical, or chronological fashion.[196]

Section 12.7 Record retention period. The regulations provided under the NFA, specifically 27 CFR 479.131, provide that the retention periods for required records shall be in conformity with the requirements specified under Part 478. As provided by Part 478:

(1) Records of firearms transactions maintained by licensed importers and licensed manufacturers must maintain permanent records of the importation, manufacture, or other acquisition of firearms.[197]

(2) Licensed importer's and manufacturer's records of sales or other disposition of firearms over the age of 20 years may be discarded.[198]

(3) The A & D Records prepared by licensed dealers and licensed collectors over 20 years of age may be discarded.

(4) All FFLs shall retain each Form 4473 and 4473(LV) for a period not less than 20 years from the date of sale or disposition of the firearms. Forms 4473 obtained by FFLs where the NICS check was initiated, but the sale, delivery, or transfer of the firearm was not completed must be retained for a period of not less than 5 years.[199]

---

194 27 CFR 478.124(a)
195 27 CFR 478.124(b)
196 27 CFR 478.129(b)
197 27 CFR 478.129(d)
198 27 CFR 478.129(a); 478.129(d)
199 27 CFR 478.129(b)

(5) Licensees are also required to retain the ATF Form 3310.4, Multiple Sales or Other Disposition of Pistols and Revolvers, as well as ATF Form 3310.11, Federal Firearms Licensee Theft/Loss Report, for a period of not less than 5 years.[200]

(6) Retention of the records relating to transactions in semi-automatic assault weapons must be retained for a period not less 5 years

200 27 CFR 478.129(c)

CHAPTER 13. REQUIRED REPORTS AND NOTIFICATIONS TO ATF

Section 13.1 Change of business address.

13.1.1 GCA requirements. Technically, a change in an FFL's/SOT's business address does not trigg an ATF "notification" or "reporting" requirement under the GCA. Rather, it requires the FFL to apply for and receive an amended license before business may be resumed at the new business location. GC regulations require that an application for an amended license, ATF Form 5300.38, be filed not less tha 30 days prior to the change with the Chief, Federal Firearms Licensing Center (FFLC).[201] The application must be accompanied by the FFL's original license. The FFLC may also require the applicant to submit an original license application on ATF Form 7, or may issue an amended license if the applicant is determined to be qualified to do business at the new location. The amended license is valid for the remainder of the term of the original license.

13.1.1.1 Application for amended license. Moving a licensed business to a new location requires careful planning. In completing the application for an amended license, the FFL will be asked if there are any restrictions, covenants, or zoning ordinances prohibiting the conduct of business at the new premises. The FFL must also attach to the application copies of any State license or permit required to conduct the business, as well as evidence of payment of any required State or local business tax. The FFL must also certify that notification has been given t the chief law enforcement officer (the sheriff, chief of police, or equivalent officer) at the new location that the FFL intends to file the application. If ATF determines that the firearms busines may not be lawfully conducted at the new location, the application will be denied.

13.1.2 NFA requirements. If the FFL is qualified to do business in NFA firearms, the FFL may not lawfully conduct business in NFA firearms at the new location without the approval of the Chief, NFA Branch.[202] The regulations require the FFL/SOT to file Form 5630.7 with ATF, bearing the notation "Removal Registry" and showing the new address to be used. The FFL/SOT must also send the curren special tax stamp to ATF, as well as a letter application requesting the amendment of the registration. Upon approval, the NFA Branch will return the special tax stamp, amended to show the new business location.

Section 13.2 Change in trade name

13.2.1 GCA requirements. GCA regulations provide that an FFL is not required to obtain a new license by reason of a mere change in trade name, provided the license is sent for endorsement of the change to the Chief, Federal Firearms Licensing Center within 30 days from the date the business starts using the new name.[203]

13.2.2 NFA requirements. An FFL/SOT may not continue business in these types of firearms under new trade name without the approval of the Chief, NFA Branch.[204] To obtain approval, the regulations

---

[201] 27 CFR 478.52
[202] 27 CFR 479.46
[203] 27 CFR 478.53
[204] 27 CFR 479.47

equire the FFL/SOT to file ATF Form 5630.7 with ATF, bearing the notation "Amended" and showing e new trade name intended to be used. The FFL/SOT must also send the current NFA special tax tamp to ATF, as well as a letter application requesting amendment of the registration as a special occupational) taxpayer. Upon approval, the NFA Branch will return the amended special tax stamp howing the new trade name.

ection 13.3 Change in control. A "change in control" is defined in GCA regulations as a change in e actual or legal control of a licensed corporation or association, directly or indirectly, whether by eason of a change in stock ownership, by operation of law, or in any other manner.[205] The regulations equire the FFL to give written notification of the change, executed under the penalties of perjury, to the Chief, Federal Firearms Licensing Center, within 30 days of the change. Upon expiration of the license, e regulations require the FFL to file an application for an original license on ATF Form 7. Here are ome examples of a "change in control:"

(1) In the case of a licensed corporation, an individual's acquisition of 51% of the stock of the corporation.

(2) Corporation A, an FFL, is acquired by Corporation B. That is, Corporation B purchased 100% of the stock of Corporation A. Corporation A continues to exist as a distinct legal entity after the acquisition and may continue to engage in its licensed firearms business.

(3) The entire board of directors or slate of officers of a corporate FFL changes.

3.3.1 Changes in an FFL's "responsible persons." For a business entity such as a corporation, ssociation, or partnership to qualify for a GCA license, it must not have any person responsible for the anagement or policies of the firearms business under Federal disabilities, for example, felons.[206] Thus, e entity must list its "responsible persons" on the license application and provide ATF with the ingerprints and photographs of such responsible persons. ATF will conduct a background check of ese persons prior to issuing a license. Although a change in responsible persons may not result in a change in control" for the licensed entity, it is advisable that the FFL notify the Chief, Federal Firearms Licensing Center (FFLC) in writing of such a change and provide the FFLC with same identifying nformation as required for other responsible persons on the ATF Form 7, as well as the new responsible erson's photograph and fingerprints.

3.3.2 A "change in control" is distinguishable from the sale or transfer of a firearms business to nother person or entity. Where an FFL sells or transfers the firearms business to a different person or ntity, the FFL's license is not transferable to the new person or entity.[207] For example, if Corporation A purchases the assets of an FFL, Corporation B, and Corporation A intends to operate the new firearms usiness, Corporation A may not lawfully operate the business on Corporation B's license. Rather, Corporation A must apply for and receive its own GCA license to lawfully carry on the business. The

---

[205] 27 CFR 478.54
[206] 18 U.S.C 923(d)(1)(B); 27 CFR 478.47
[207] 27 CFR 478.51

NFA would also require Corporation A to pay the NFA special tax by filing ATF Form 5630.7, Special Tax Registration and Return.[208]

Section 13.4 Reporting firearms transaction information. Each licensee must, when required by letter issued by the ATF, submit on ATF Form 4483, Report of Firearms Transactions, for the periods and times specified in the letter, all record information called for by the letter.[209] This includes the requirement for firearms manufacturers to annually report their firearms production on ATF Form 5300.11. Even if there has been no production, the annual report must be filed. See the instructions on the form for more specific information on the reporting requirement. Should a manufacturer's production fall within the following categories, the report would reflect "o" (zero) for firearms produced:

(1) Firearms produced solely for the official use of the Armed Forces of the United States.

(2) Firearms disposed of to another licensed firearms manufacturer for the purpose of final finishing and assembly

(3) Destructive devices as defined in 18 U.S.C. 921(a)(4) and 26 U.S.C. 5845(f).

(4) Antique firearms as defined in 18 U.S.C. 921(a)(16).

(5) Firearms incorporating frames or receivers of foreign manufacture.

(7) Firearms remanufactured or customized and previously in the possession of non-FFLs.

Section 13.5 Reporting thefts or losses of firearms

13.5.1 GCA requirements. Each FFL must report the theft or loss of a firearm from the FFL's inventory (including any firearm that has been transferred from the inventory to the FFL's personal collection and held as a personal firearm for at least 1 year), or from the collection of a licensed collector, within 48 hours after the theft or loss is discovered.[210] Reports must be made by telephoning ATF at 1-888-930-9275 (nationwide toll free number) and by preparing ATF Form 3310.11, Federal Firearms Licensee Theft/Loss Report, in accordance with the instructions on the form. The original of the report must be forwarded to the ATF National Tracing Center at the address specified on the form, and a copy must be retained by the FFL in the FFL's permanent records for a period not less than 5 years.[211] Theft or loss of any firearm must also be reported to appropriate local authorities.

13.5.2 NFA requirements. When any registered firearm is stolen or lost, the person losing possession must, immediately upon discovery of the theft or loss, make a written report to ATF, specifically the NFA Branch, showing the following: (1) name and address of the person in whose name the firearm is

---

[208] 27 CFR 479.31
[209] 18 U.S.C 923(g)(5)(A); 27 CFR 478.126
[210] 18 U.S.C. 923(g)(6); 27 CFR 478.39a
[211] 27 CFR 478.129(c)

registered; (2) kind of firearm; (3) serial number; (4) model; (5) caliber; (6) manufacturer; (7) date and place of theft or loss; and (8) a complete statement of the facts and circumstances surrounding the theft or loss.[212]

Section 13.6 Reporting theft or loss of NFA registration documents. When any Form 1, 2, 3, 4, 5, 6A, or 10 evidencing possession of a firearm is stolen, lost, or destroyed, the person losing possession must immediately upon discovery of the theft, loss, or destruction report the matter in writing to ATF, specifically the NFA Branch.[213] The report must show the circumstances of the theft, loss, or destruction and include all known facts which may serve to identify the document. Upon receipt of the report, ATF may conduct an investigation and issue a duplicate document.

Section 13.7 Reporting information in response to ATF trace requests. The Gun Control Act of 1968 (GCA) requires an FFL to respond immediately, and in no event later than 24 hours after the receipt of, to a request by an ATF officer at the National Tracing Center (NTC) for information contained in the FFL's required GCA records for determining the disposition of one or more firearms in the course of a bona fide criminal investigation. (See 18 U.S.C. 923(g)(7) and 27 CFR 478.25a). The requested information must be provided orally to the ATF officer within the 24-hour period. Verification of the identity and employment of NTC personnel requesting this information may be established by telephoning the toll-free number 1-800-788-7133 or using the toll-free FAX number 1-800-578-7223.

Section 13.8. Requesting permission to transport certain firearms in interstate or foreign commerce. A person, other than an FFL/SOT, may not lawfully transport in interstate or foreign commerce any destructive device, machinegun, short-barreled shotgun, or short-barreled rifle, without prior written approval of ATF, specifically the NFA Branch.[214] For definitions of these firearms and devices, refer to 27 CFR 478.11. Licensed collectors are not required to obtain such approval if the firearms and devices being transported are "curio or relic" firearms under the GCA. Approval for the transportation may be obtained by (1) a written request or (2) an approved application filed with ATF on Form 5320.20.

13.8.1 A written (letter) request. A written (letter) request must contain:

(1) A complete description and identification of the device or firearm to be transported;

(2) A statement whether the transportation involves a transfer of title;

(3) The need for such transportation;

(4) The approximate date the transportation is to take place;

(5) The present location of the device or firearm and the place to which it is being transported.

---

[212] 27 CFR 479.141
[213] 27 CFR 479.37, 479.142
[214] 18 U.S.C. 922(a)(4); 27 CFR 478.28

(6) The mode of transportation to be used (including, if by common or contract carrier, the name and address of the carrier); and

(7) Evidence that the transportation or possession of the device or firearm is not inconsistent with the laws at the place of destination.

13.8.2 Form 5320.20. As stated above, transportation of these firearms and devices may also be approved by completing and filing this form and obtaining ATF approval of the form.

13.8.3 One-year approval. If a person will be transporting his/her firearm(s) to the same location on a continual basis, ATF will approve a transportation request for up to 1 year. *Example*: the person lives in State A, has a farm in State B (State B allows possession of the particular firearm), and wants to take the firearm to the farm throughout the year. The person may request permission for a 1-year period to transport the firearm interstate to the farm. Any other interstate transportation would still require a separate request and approval. *Second example*: the person lives in State A and wants to transport his/her firearm to a site in State B where competitions and shoots occur several times a year (State B allows possession of the particular firearm). The person may request permission for a 1-year period to transport the firearm interstate to the site for competitions and shoots. Any other interstate transportation would still require a separate request and approval.

CHAPTER 14. GOING OUT OF BUSINESS

ection 14.1 Disposition of GCA records.

4.1.1 "Records" for purposes of this section. Reference to the term "records" in this chapter refers o the acquisition and disposition records required by GCA and NFA regulations, Forms 4473, Forms 310.4 (Report of Multiple Sale or Other Disposition of Pistols and Revolvers), ATF Forms 3310.11 Federal Firearms Licensee Theft/Loss Report), records of transactions in semiautomatic assault veapons, records of importation (ATF Forms 6 and 6A), and law enforcement certification letters. If the icensee was granted a variance to use a computerized recordkeeping system, this term also refers to the equired printout of the all A & D records, as required by the variance approval.[215] As required by the egulations, these records must be surrendered to the ATF Out-of-Business Records Center (OBRC) or ransferred to any successor of the firearms business within 30 days of the discontinuance of such usiness.[216] [Handwritten changes add cite "27 C.F.R. § 478.129" but may just be the above referenced orms]

4.1.2 Transfer of the firearms business to a new owner. In the event that the firearms business is ransferred to a new owner (i.e. successor), the FFL going out of business may dispose of the records in ne of two ways: (1) close all open A & D Record disposition entries by recording the date of transfer, s well as the name and FFL of the succeeding licensee (or record that the firearm was transferred to the iscontinued FFL's personal inventory), underline the final entry in each bound book, and deliver all ecords to the business successor; or (2) deliver the records to the ATF Out-of-Business Records Center, 44 Needy Road, Martinsburg, West Virginia 25405, or to any ATF office in the division in which the usiness was located. A successor licensee, who receives the records from the original licensee, may hoose to forward the records to the ATF Out-of-Business Records Center.[217]

4.1.3 Discontinuance of the business. If the firearms business is discontinued and there is no uccessor, within 30 days of business' discontinuance, the FFL must ship the records to the ATF Out-of-Business Records Center or to any office in the ATF division in which the business was located. If the FL was granted a variance to use a computerized recordkeeping system, the FFL must provide a omplete printout of the acquisition and disposition records as stipulated in the variance approval.[218]

ection 14.2 Disposition of NFA firearms (other than "post-'86 machineguns").

4.2.1 Sole proprietors. FFLs licensed as sole proprietors, who have been qualified to deal in NFA irearms and who go out of the NFA business, may lawfully retain their inventory of these firearms, ncluding imported NFA "sales samples", in their individual capacity. No NFA transfer occurs that vould require an ATF-approved transfer because the firearms are still possessed by the same person to vhom they were previously transferred and registered. However, any firearm registered to the ndividual as a "sales sample" would continue to bear the "sales sample" restriction on any subsequent ransfer, unless the firearm is being transferred to a government agency. Thus, the transfer of a "sales

---

5 ATF P 5300.4 (09/05), Question C-7, Page 180

6 18 U.S.C. 923(g)(4); 27 CFR 478.127

7 Ibid.

8 Ibid.

sample" to other than a government agency will only be approved if the transferee is an FFL/SOT qualified to receive such samples.[219] Since the registered owner of the firearm is no longer an FFL/SOT the owner's subsequent transfer of the firearm to other than a government agency would be subject to transfer tax and require approval of an ATF Form 4. *NOTE: If an NFA firearm registered to a sole proprietor as a "sale sample" becomes part of the person's estate upon his/her death, the firearm may not be registered to a beneficiary of the estate unless the beneficiary is qualified to deal in NFA sales samples.*

14.2.2 Corporations, partnerships, and associations. FFLs licensed as corporations, partnerships, or associations, who have been qualified to deal in NFA firearms and who go out of the NFA business, may lawfully retain their inventory of these firearms, including imported NFA "sales samples," as long as the entity does not dissolve but continues to exist under State law. No NFA transfer occurs that would require an ATF-approved transfer because the firearms are still possessed by the same entity to which they were previously transferred and registered. However, any firearm registered to the entity as a "sales sample" would continue to bear the "sales sample" restriction on any subsequent transfer, unless the firearm is being transferred to a government agency. Thus, the transfer of a "sales sample" to other than a government agency will only be approved if the transferee is an FFL/SOT qualified such samples.[220]

14.2.2.1 Effect of dissolution of a corporation, partnership, or association. If an FFL licensed as a corporation, partnership, or association goes out of business and dissolves (ceases to exist under State law), the firm's NFA firearms will be considered to have been "transferred" to whomever takes custody of the firearms and possesses them after dissolution. To be lawful, such transfers must be approved in advance by ATF on ATF Forms 4 and are subject to NFA transfer tax. *However, ATF will not approve the transfer of an NFA "sales sample" unless the transfer is to a government agency or to an FFL/SOT qualified to receive such samples.*[221]

14.2.3 Avoiding NFA transfer tax. An FFL/SOT going out of business and contemplating the transfer of an inventory of NFA firearms or NFA "sales samples" should be mindful of the tax consequences of such transfers. If the FFL/SOT wishes to transfer the firearms to another such person or entity, the transfers may be made on approved ATF Forms 3 transfer tax free if both the transferor and transferee are licensed under the GCA and hold NFA special tax stamps. *Thus, a person or entity going out of business who allows his/her FFL or special tax stamp to expire before the transfers would be subject to the transfer tax on each firearm transferred and the transfers would have to be made on ATF-approved Forms 4.*[222]

Section 14.3 Disposition of "post-86 machineguns." Section 479.105(f) requires the FFL/SOT going out of business to transfer machinegun(s) manufactured after May 19, 1986 to a Federal, State or local government entity, qualified manufacturer, qualified importer, or, subject to the provisions of 479.105(d), to a qualified dealer. The transfers must be completed prior to the FFL/SOT going out of business. The transfer to a government entity would be on Form 5 and the transfer to a qualified

[219] Item 10g, ATF Pub. 5300.4 (9/05), Federal Firearms Regulations Reference Guide, p. 164
[220] Ibid.
[221] Ibid.
[222] 26 U.S.C. 5852(d)

ıanufacturer, importer, or dealer would be on Form 3. The transfer of a machinegun manufactured fter May 19, 1986 to a qualified manufacturer or qualified importer can be done without the 'law nforcement letters of interest' when the possessing FFL/SOT is going out of business. That fact that ıe transferring FFL is discontinuing business must be documented with the submission of the transfer pplications. Transfer of a machinegun manufactured after May 19, 1986 from an FFL/SOT going out f business to a qualified dealer requires compliance with section 479.105(d).

'he FFL/SOT going out of business may also choose to destroy the machinegun(s), transfer them to tate or local government agencies, or abandon them to ATF or State or local government agencies. If ıe firearms will be destroyed, ATF's Firearms Technology Branch can be contacted for procedures to roperly destroy a machinegun. Once the destruction has been accomplished, notify the NFA Branch of ıe destruction in writing, including a description of the machinegun(s) and the means of destruction. 'or abandonment to ATF, contact the local ATF office. [223]

[3] 27 CFR 478.36, 479.105; Item 10g, ATF Pub. 5300.4 (Sept. 2005), p. 164

CHAPTER 15. PENALTIES AND SANCTIONS

Section 15.1 NFA.

15.1.1 Criminal. The acts prohibited by the NFA and prosecutable as Federal offenses are listed in 26 U.S.C. 5861(a) through (l). As provided by 26 U.S.C. 5871, any person who commits an offense shall, upon conviction, be sentenced to imprisonment for not more than 10 years or fined. Although the fine specified in the statute is an amount not exceeding $10,000, an amendment to Federal law provides for fine of not more than $250,000 in the case of an individual or $500,000 in the case of an organization.[224]

15.1.2 Forfeiture. Any firearm involved in any violation of the NFA is subject to seizure and forfeiture.[225]

15.1.3 Assessment of NFA tax. ATF may assess tax liabilities under the NFA, including penalties and interest, as provided by the Internal Revenue Code.[226]

Section 15.2 GCA.

15.2.1 Criminal. The criminal penalties for violations of the GCA are provided for in 18 U.S.C. § 924. The criminal penalties in the GCA include both felonies and misdemeanors. For misdemeanors, the fines would be not more than $100,000 for individuals or $200,000 in the case of organizations. The criminal provisions of the NFA are found in 26 U.S.C. § 5871. As in the case of NFA offenses, fines for violation of the felony provisions would be not more than $250,000 in the case of an individual or $500,000 in the case of an organization.

15.2.2 Forfeiture. The GCA also provides for the forfeiture of firearms and ammunition involved in certain violations of the GCA and other violations of the criminal laws of the United States in 18 U.S.C. 924(d).

15.2.3 License denial or revocation. ATF may issue a notice of denial of an application for a Federal firearms license where it determines that the applicant fails to meet the licensing requirements of 18 U.S.C. 923(d).[227] It may also issue a notice of revocation of a license when it determines that an FFL has willfully violated the GCA or its implementing regulations.[228] For the applicable procedures, see ATF's regulations in
27 CFR Part 478, Subpart E. Note that the courts have held that a person's conduct was "willful" when the evidence showed that the FFL knew of his legal obligation and disregarded or was plainly indifferent to that obligation.[229]

---

[224] 18 U.S.C. 3571(b) and (c)
[225] 26 U.S.C. 5872
[226] 27 CFR 479.191
[227] 27 CFR 478.47, 478.71
[228] 18 U.S.C. 923(e); 27 CFR 478.73
[229] Bryan V. United States, 524 U.S. 184 (1998) [note: this is criminal case, not civil license revocation.]

Section 15.3 AECA

15.3.1 Criminal. Any person who (1) imports articles on the U.S. Munitions Import List without a permit, (2) engages in the business of importing articles on the List without registering as required, or (3) otherwise violates any provision of the regulations in 27 CFR Part 447 shall, upon conviction, be fined not more than $1,000,000 or imprisoned for not more than 10 years, or both.[230] The same penalties apply to a person who willfully, in a registration or permit application, makes a false statement of a material fact or fails to state a material fact.

15.3.2 Forfeiture. Whoever knowingly imports into the United States contrary to law any article on the U.S. Munitions Import List, or receives, conceals, buys, sells, or in any manner facilitates its transportation, concealment, or sale after importation, knowing the same to have been imported contrary to law, shall be fined not more than $10,000 or imprisoned not more than 5 years, or both; and the merchandise so imported or the value thereof shall be forfeited to the United States.[231]

---

[230] 22 U.S.C. 2778(c)
[231] 27 CFR 447.63

APPENDIX A

NFA, 26 U.S.C. Chapter 53

THE NATIONAL FIREARMS ACT

TITLE 26, UNITED STATES CODE, CHAPTER 53 INTERNAL REVENUE CODE

Editor's Note:

The National Firearms Act (NFA) is part of the Internal Revenue Code of 1986. The Internal Revenue Code, with the exception of the NF A, is administered and enforced by the Secretary of the Treasury. When ATF transferred to the Department of Justice under the Homeland Security Act of 2002, all its authorities, including the authority to administer and enforce the NFA, were transferred to the Attorney General In order to keep all the references throughout the Internal Revenue Code consistent, references to the Secretary of the Treasury in the NFA were left unchanged by the Homeland Security Act. However, section 7801(a)(2), Title 26, V.S.C., provides that references to the term "Secretary" or "Secretary of the Treasury" in the NFA shall mean the Attorney General.

CHAPTER 53 - MACHINE GUNS, DESTRUCTIVE DEVICES, AND CERTAIN OTHER FIREARMS

Subchapter A - Taxes

Part I - Special (occupational) taxes.

§ 5801 Imposition of tax.

§ 5802 Registration of importers, manufacturers, and dealers.

Part II - Tax on transferring firearms.

§ 5811 Transfer tax.

§ 5812 Transfers.

Part III - Tax on making firearms.

§ 5821 Making tax.

§ 5822 Making.

Subchapter B - General Provisions and Exemptions

Part I - General Provisions.

5841 Registration of firearms.

5842 Identification of firearms.

5843 Records and returns.

5844 Importation.

5845 Definitions.

5846 Other laws applicable.

5847 Effect on other laws..

5848 Restrictive use of information.

5849 Citation of chapter.

'art II - Exemptions.

5851 Special (occupational) taxes.

5852 General transfer and making tax exemption.

5853 Transfer and making tax exemption available to certain governmental entities.

5854 Exportation of firearms exempt from transfer tax.

ubchapter C - Prohibited Acts

5861 Prohibited acts.

ubchapter D- Penalties and Forfeitures

5871 Penalties.

5872 Forfeitures.

UBCHAPTER A - TAXES

'art I - Special Occupational Taxes

5801 Imposition of tax.

(a) General rule. On first engaging in business and thereafter on or before July 1 of each year, every nporter, manufacturer, and dealer in firearms shall pay a special (occupational) tax for each place of business at the following rates:

(1) Importers and manufacturers: $1,000 a year or fraction thereof.

(2) Dealers: $500 a year or fraction thereof.

(b) Reduced rates of tax for small importers and manufacturers.

(1) In general. Paragraph (1) of subsection (a) shall be applied by substituting "$500" for "$1,000" with respect to any taxpayer the gross receipts of which (for the most recent taxable year ending before the 1st day of the taxable period to which the tax imposed by subsection (a) relates) are less than $500,000.

(2) Controlled group rules. All persons treated as 1 taxpayer under section 5061 (e)(3) shall be treated as 1 taxpayer for purposes of paragraph (1).

(3) Certain rules to apply. For purposes of paragraph (1), rules similar to the rules of subparagraphs (B) and (C) of section 448(c)(3) shall apply.

§ 5802 Registration of importers, manufacturers, and dealers.

On first engaging in business and thereafter on or before the first day of July of each year, each importer, manufacturer, and dealer in firearms shall register with the Secretary in each internal revenue district in which such business is to be carried on, his name, including any trade name, and the address of each location in the district where he will conduct such business. An individual required to register under this section shall include a photograph and fingerprints of the individual with the initial application. Where there is a change during the taxable year in the location of, or the trade name used in such business, the importer, manufacturer, or dealer shall file an application with the Secretary to amend his registration. Firearms operations of an importer, manufacturer, or dealer may not be commenced at the new location or under a new trade name prior to approval of the application by the Secretary.

Part II - Tax on Transferring Firearms.

§ 5811 Transfer tax.

(a) Rate. There shall be levied, collected, and paid on firearms transferred a tax at the rate of $200 for each firearm transferred, except, the transfer tax on any firearm classified as any other weapon under section 5845(e) shall be at the rate of $5 for each such firearm transferred.

(b) By whom paid. The tax imposed by subsection (a) of this section shall be paid by the transferor.

(c) Payment. The tax imposed by subsection (a) of this section shall be payable by the appropriate stamps prescribed for payment by the Secretary.

§ 5812 Transfers.

(a) Application. A firearm shall not be transferred unless (1) the transferor of the firearm has filed vith the Secretary a written application, in duplicate, for the transfer and registration of the firearm to he transferee on the application form prescribed by the Secretary; (2) any tax payable on the transfer is aid as evidenced by the proper stamp affixed to the original application form; (3) the transferee is dentified in the application form in such manner as the Secretary may by regulations prescribe, except hat, if such person is an individual, the identification must include his fingerprints and his photograph; 4) the transferor of the firearm is identified in the application form in such manner as the Secretary may y regulations prescribe; (5) the firearm is identified in the application form in such manner as the ecretary may by regulations prescribe; and (6) the application form shows that the Secretary has pproved the transfer and the registration of the firearm to the transferee. Applications shall be denied if he transfer, receipt, or possession of the firearm would place the transferee in violation of law.

(b) Transfer of possession. The transferee of a firearm shall not take possession of the firearm unless he Secretary has approved the transfer and registration of the firearm to the transferee as required by ubsection (a) of this section.

art 111- Tax on Making Firearms.

5821 Making tax.

(a) Rate. There shall be levied, collected, and paid upon the making of a firearm a tax at the rate of 200 for each firearm made.

(b) By whom paid. The tax imposed by subsection (a) of this section shall be paid by the person naking the firearm.

(c) Payment. The tax imposed by subsection (a) of this section shall be payable by the stamp rescribed for payment by the Secretary.

5822 Making.

No person shall make a firearm unless

e has (a) filed with the Secretary a written application, in duplicate, to make and register the firearm on he form prescribed by the Secretary; (b) paid any tax payable on the making and such payment is vidEnced by the proper stamp affixed to the original application form; (c) identified the firearm to be nade in the application form in such manner as the Secretary may by regulations prescribe; (d) dentified himself in the application form in such manner as the Secretary may by regulations prescribe, xcept that, if such person is an individual, the identification must include his fingerprints and his hotograph; and (e) obtained the approval of the Secretary to make and register the firearm and the pplication form shows such approval. Applications shall be denied if the making or possession of the irearm would place the person making the firearm in violation of law.

SUBCHAPTER B GENERAL PROVISIONS AND EXEMPTIONS

Part I - General Provisions.

§ 5841 Registration of firearms.

(a) Central registry. The Secretary shall maintain a central registry of all firearms in the United States which are not in the possession or under the control of the United States. This registry shall be known as the National Firearms Registration and Transfer Record. The registry shall include

(1) identification of the firearm;

(2) date of registration; and

(3) identification and address of person entitled to possession of the firearm.

(b) By whom registered. Each manufacturer, importer, and maker shall register each firearm he manufactures, imports, or makes. Each firearm transferred shall be registered to the transferee by the transferor.

(c) How registered. Each manufacturer shall notify the Secretary of the manufacture of a firearm in such manner as may by regulations be prescribed and such notification shall effect the registration of the firearm required by this section. Each importer, maker, and transferor of a firearm shall, prior to importing, making, or transferring a firearm, obtain authorization in such manner as required by this chapter or regulations issued thereunder to import, make, or transfer the firearm, and such authorization shall effect the registration of the firearm required by this section.

(d) Firearms registered on effective date of this act. A person shown as possessing a firearm by the records maintained by the Secretary pursuant to the National Firearms Act in force on the day immediately prior to the effective date of the National Firearms Act of 1968 shall be considered to have registered under this section the firearms in his possession which are disclosed by that record as being in his possession.

(e) Proof of registration. A person possessing a firearm registered as required by this section shall retain proof of registration which shall be made available to the Secretary upon request.

§ 5842 Identification of firearms.

(a) Identification of firearms other than destructive devices. Each manufacturer and importer and anyone making a firearm shall identify each firearm, other than a destructive device, manufactured, imported, or made by a serial number which may not be readily removed, obliterated, or altered, the name of the manufacturer, importer, or maker, and such other identification as the Secretary may by regulations prescribe.

(b) Firearms without serial number. Any person who possesses a firearm, other than a destructive device, which does not bear the serial number and other information required by subsection (a) of this section shall identify the firearm with a serial number assigned by the Secretary and any other information the Secretary may by regulations prescribe.

(c) Identification of destructive device. Any firearm classified as a destructive device shall be identified in such manner as the Secretary may by regulations prescribe.

§ 5843 Records and returns.

Importers, manufacturers, and dealers shall keep such records of, and render such returns in relation to, the importation, manufacture, making, receipt, and sale, or other disposition, of firearms as the Secretary may by regulations prescribe.

§ 5844 Importation.

No firearm shall be imported or brought into the United States or any territory under its control or jurisdiction unless the importer establishes, under regulations as may be prescribed by the Secretary, that the firearm to be imported or brought in is

(1) being imported or brought in for the use of the United States or any department, independent establishment, or agency thereof or any State or possession or any political subdivision thereof; or

(2) being imported or brought in for scientific or research purposes; or

(3) being imported or brought in solely for testing or use as a model by a registered manufacturer or solely for use as a sample by a registered importer or registered dealer;

except that, the Secretary may permit the conditional importation or bringing in of a firearm for examination and testing in connection with classifying the firearm.

§ 5845 Definitions.

For the purpose of this chapter-

(a) Firearm. The term 'firearm' means (1) a shotgun having a barrel or barrels of less than 18 inches in length; (2) a weapon made from a shotgun if such weapon as modified has an overall length of less than 26 inches or a barrel or barrels of less than 18 inches in length; (3) a rifle having a barrel or barrels of less than 16 inches in length; (4) a weapon made from a rifle if such weapon as modified has an overall length of less than 26 inches or a barrel or barrels of less than 16 inches in length; (5) any other weapon, as defined in subsection (e); (6) a machinegun; (7) any silencer (as defined in section 921 of title 18, United States Code); and (8) a destructive device. The term 'firearm' shall not include an antique firearm or any device (other than a machinegun or destructive device) which, although designed as a weapon, the Secretary finds by reason of the date of its manufacture, value, design, and other characteristics is primarily a collector's item and is not likely to be used as a weapon.

(b) Machinegun. The term 'machinegun' means any weapon which shoots, is designed to shoot, or can be readily restored to shoot, automatically more than one shot, without manual reloading, by a single function of the trigger. The term shall also include the frame or receiver of any such weapon, any part designed and intended solely and exclusively, or combination of parts designed and intended, for use in converting a weapon into a machinegun, and any combination of parts from which a machinegun can be assembled if such parts are in the possession or under the control of a person.

(c) Rifle. The term 'rifle' means a weapon designed or redesigned, made or remade, and intended to be fired from the shoulder and designed or redesigned and made or remade to use the energy of the explosive in a fixed cartridge to fire only a single projectile through a rifled bore for each single pull of the trigger, and shall include any such weapon which may be readily restored to fire a fixed cartridge.

(d) Shotgun. The term 'shotgun' means a weapon designed or redesigned, made or remade, and intended to be fired from the shoulder and designed or redesigned and made or remade to use the energy of the explosive in a fixed shotgun shell to fire through a smooth bore either a number of projectiles (ball shot) or a single projectile for each pull of the trigger, and shall include any such weapon which may be readily restored to fire a fixed shotgun shell.

(e) Any other weapon. The term 'any other weapon' means any weapon or device capable of being concealed on the person from which a shot can be discharged through the energy of an explosive, a pistol or revolver having a barrel with a smooth bore designed or redesigned to fire a fixed shotgun shell, weapons with combination shotgun and rifle barrels 12 inches or more, less than 18 inches in length, from which only a single discharge can be made from either barrel without manual reloading, and shall include any such weapon which may be readily restored to fire. Such term shall not include a pistol or a revolver having a rifled bore, or rifled bores, or weapons designed, made, or intended to be fired from the shoulder and not capable of firing fixed ammunition.

(f) Destructive device. The term 'destructive device' means (1) any explosive, incendiary, or poison gas (A) bomb, (8) grenade, (C) rocket having a propellant charge of more than four ounces, (0) missile having an explosive or incendiary charge of more than one-quarter ounce, (E) mine, or (F) similar device; (2) any type of weapon by whatever name known which will, or which may be readily converted to, expel a projectile by the action of an explosive or other propellant, the barrel or barrels of which have a bore of more than one-half inch in diameter, except a shotgun or shotgun shell which the Secretary finds is generally recognized as particularly suitable for sporting purposes; and (3) any combination of parts either designed or intended for use in converting any device into a destructive device as defined in subparagraphs (1) and (2) and from which a destructive device may be readily assembled. The term 'destructive device' shall not include any device which is neither designed nor redesigned for use as a weapon; any device, although originally designed for use as a weapon, which is redesigned for use as a signaling, pyrotechnic, line throwing, safety, or similar device; surplus ordnance sold, loaned, or given by the Secretary of the Army pursuant to the provisions of section 4684(2), 4685, or 4686 of title 10 of the United States Code; or any other device which the Secretary finds is not likely to be used as a weapon, or is an antique or is a rifle which the owner intends to use solely for sporting purposes.

(g) Antique firearm. The term 'antique firearm' means any firearm not designed or redesigned for using rim fire or conventional center fire ignition with fixed ammunition and manufactured in or before 1898 (including any matchlock, flintlock, percussion cap, or similar type of ignition system or replica thereof, whether actually manufactured before or after the year 1898) and also any firearm using fixed ammunition manufactured in or before 1898, for which ammunition is no longer manufactured in the

Jnited States and is not readily available in the ordinary channels of commercial trade.

(h) Unserviceable firearm. The term 'unserviceable firearm' means a firearm which is incapable of ischarging a shot by means of an explosive and incapable of being readily restored to a firing condition.

(i) Make. The term 'make', and the various derivatives of such word, shall include manufacturing other than by one qualified to engage in such business under this chapter), putting together, altering, ny combination of these, or otherwise producing a firearm.

(j) Transfer. The term. 'transfer' and the various derivatives of such word, shall include selling, ssigning, pledging, leasing, loaning, giving away, or otherwise disposing of.

(k) Dealer. The term 'dealer' means any person, not a manufacturer or importer, engaged in the usiness of selling, renting, leasing, or loaning firearms and shall include pawnbrokers who accept rearms as collateral for loans.

(l) Importer. The term 'importer' means any person who is engaged in the business of importing or bringing firearms into the United States.

(m) Manufacturer. The term 'manufacturer' means any person who is engaged in the business of ıanufacturing firearms.

5846 Other laws applicable.

All provisions of law relating to special taxes imposed by chapter 51 and to engraving, issuance, sale, ccountability, cancellation, and distribution of stamps for tax payment shall, insofar as not inconsistent vith the provisions of this chapter, be applicable with respect to the taxes imposed by sections 5801, 811, and 5821.

5847 Effect on other laws.

Nothing in this chapter shall be construed as modifying or affecting the requirements of section 414 of ıe Mutual Security Act of 1954, as amended, with respect to the manufacture, exportation, and nportation of arms, ammunition, and implements of war.

5848 Restrictive use of information.

(a) General Rule. No information or evidence obtained from an application, registration, or records equired to be submitted or retained by a natural person in order to comply with any provision of this hapter or regulations issued thereunder, shall, except as provided in subsection (b) of this section, be sed, directly or indirectly, as evidence against that person in a criminal proceeding with respect to a iolation of law occurring prior to or concurrently with the filing of the application or registration, or the ompiling of the records containing the information or evidence.

(b) Furnishing false information. Subsection (a) of this section shall not preclude the use of any such formation or evidence in a prosecution or other action under any applicable provision of law with espect to the furnishing of false information.

§ 5849 Citation of chapter.

This chapter may be cited as the 'National Firearms Act' and any reference in any other provision of law to the 'National Firearms Act' shall be held to refer to the provisions of this chapter.

Part II - Exemptions.

§ 5851 Special (occupational) tax exemption.

(a) Business with United States. Any person required to pay special (occupational) tax under section 5801 shall be relieved from payment of that tax if he establishes to the satisfaction of the Secretary that his business is conducted exclusively with, or on behalf of, the United States or any department, independent establishment, or agency thereof. The Secretary may relieve any person manufacturing firearms for, or on behalf of, the United States from compliance with any provision of this chapter in th conduct of such business.

(b) Application. The exemption provided for in subsection (a) of this section may be obtained by filing with the Secretary an application on such form and containing such information as may by regulations be prescribed. The exemptions must thereafter be renewed on or before July 1 of each year. Approval of the application by the Secretary shall entitle the applicant to the exemptions stated on the approved application.

§ 5852 General transfer and making tax exemption.

(a) Transfer. Any firearm may be transferred to the United States or any department, independent establishment, or agency thereof, without payment of the transfer tax imposed by section 5811.

(b) Making by a person other than a qualified manufacturer. Any firearm may be made by, or on behalf of, the United States, or any department, independent establishment, or agency thereof, without payment of the making tax imposed by section 5821.

(c) Making by a qualified manufacturer. A manufacturer qualified under this chapter to engage in su business may make the type of firearm which he is qualified to manufacture without payment of the making tax imposed by section 5821.

(d) Transfers between special (occupational) taxpayers. A firearm registered to a person qualified under this chapter to engage in business as an importer, manufacturer, or dealer may be transferred by that person without payment of the transfer tax imposed by section 5811 to any other person qualified under this chapter to manufacture, import, or deal in that type of firearm.

(e) Unserviceable firearm. An unserviceable firearm may be transferred as a curio or ornament without payment of the transfer tax imposed by section 5811, under such requirements as the Secretary may by regulations prescribe.

(f) Right to exemption. No firearm may be transferred or made exempt from tax under the provisions of this section unless the transfer or making is performed pursuant to an application in such form and manner as the Secretary may by regulations prescribe.

5853 Transfer and making tax exemption available to certain governmental entities.

(a) Transfer. A firearm may be transferred without the payment of the transfer tax imposed by section 5811 to any State, possession of the United States, any political subdivision thereof, or any official police organization of such a government entity engaged in criminal investigations.

(b) Making. A firearm may be made without payment of the making tax imposed by section 5821 by, or on behalf of, any State, or possession of the United States, any political subdivision thereof, or any official police organization of such a government entity engaged in criminal investigations.

(c) Right to exemption. No firearm may be transferred or made exempt from tax under this section unless the transfer or making is performed pursuant to an application in such form and manner as the Secretary may by regulations prescribe.

5854 Exportation of firearms exempt from transfer tax. A firearm may be exported without payment of the transfer tax imposed under section 5811 provided that proof of the exportation is furnished in such form and manner as the Secretary may by regulations prescribe.

SUBCHAPTER C - PROHIBITED ACTS

5861 Prohibited acts.

It shall be unlawful for any person

(a) to engage in business as a manufacturer or importer of, or dealer in, firearms without having paid the special (occupational) tax required by section 5801 for his business or having registered as required by section 5802; or

(b) to receive or possess a firearm transferred to him in violation of the provisions of this chapter; or

(c) to receive or possess a firearm made in violation of the provisions of this chapter; or

(d) to receive or possess a firearm which is not registered to him in the National Firearms Registration and Transfer Record; or

(e) to transfer a firearm in violation of the provisions of this chapter; or

(f) to make a firearm in violation of the provisions of this chapter; or

(g) to obliterate, remove, change, or alter the serial number or other identification of a firearm required by this chapter; or

(h) to receive or possess a firearm having the serial number or other identification required by this chapter obliterated, removed, changed, or altered; or obliterated, or

(i) to receive or possess a firearm which is not identified by a serial number as required by this chapter; or

(j) to transport, deliver, or receive any firearm in interstate commerce which has not been registered a required by this chapter; or

(k) to receive or possess a firearm which has been imported or brought into the United States in violation of section 5844; or

(I) to make, or cause the making of, a false entry on any application, return, or record required by this chapter, knowing such entry to be false.

SUBCHAPTER D - PENALTIES AND FORFEITURES

§ 5871 Penalties.

Any person who violates or fails to comply with any provision of this chapter shall, upon conviction, be fined not more than $10,000, or be imprisoned not more than ten years, or both.

§ 5872 Forfeitures.

(a) Laws applicable. Any firearm involved in any violation of the provisions of this chapter shall be subject to seizure and forfeiture, and (except as provided in subsection (b» all the provisions of internal revenue laws relating to searches, seizures, and forfeitures of unstamped articles are extended to and made to apply to the articles taxed under this chapter, and the persons to whom this chapter applies.

(b) Disposal. In the case of the forfeiture of any firearm by reason of a violation of this chapter, no notice of public sale shall be required; no such firearm shall be sold at public sale; if such firearm is forfeited for a violation of this chapter and there is no remission or mitigation of forfeiture thereof, it shall be delivered by the Secretary to the Administrator of General Services, General Services Administration, who may order such firearm destroyed or may sell it to any State, or possession, or political subdivision thereof, or at the request of the Secretary, may authorize its retention for official use of the Treasury Department, or may transfer it without charge to any executive department or independent establishment of the Government for use by it.

TITLE 27 CFR CHAPTER II

PART 479-MACHINE GUNS, DESTRUCTIVE DEVICES, AND CERTAIN OTHER FIREARMS

(This part was formerly designated as Part 179)

Editor's Note:

Effective January 24, 2003, the Homeland Security Act transferred the Bureau of Alcohol, Tobacco and Firearms from the Department of the Treasury to the Department of Justice. In addition, the agency's name was changed to the Bureau of Alcohol, Tobacco. Firearms and Explosives. The regulations, as printed in this publication, do not yet reflect this change. The regulations will be amended to change the references from the "Bureau of Alcohol, Tobacco and Firearms," the "Department of the Treasury," and the "Secretary of the Treasury" to the "Bureau of Alcohol, Tobacco, Firearms and Explosives," the "Department of Justice" and the "Attorney General," respectively.

Subpart A -Scope of Regulations

§ 479.1 General.

Subpart B - Definitions

§ 479.11 Meaning of terms.

Subpart C - Administrative and Miscellaneous Provisions

§ 479.21 Forms prescribed.

§ 479.22 Right of entry and examination.

§ 479.23 Restrictive use of required information.

§ 479.24 Destructive device determination.

§ 479.25 Collector's items.

§ 479.26 Alternate methods or procedures; emergency variations from requirements.

Subpart D - Special (Occupational) Taxes

§ 479.31 Liability for tax.

§ 479.32 Special (occupational) tax rates.

§ 479.32a Reduced rate of tax for small importers and manufacturers.

§ 479.33 Special exemption

§ 479.34 Special tax registration and return.

§ 479.35 Employer identification number.

§ 479.36 The special tax stamp, receipt for special (occupational) taxes.

§ 479.37 Certification in lieu of stamps lost or destroyed.

§ 479.38 Engaging in business at more than one location.

§ 479.39 Engaging in more than one business at the same location.

§ 479.40 Partnership liability.

§ 479.41 Single sale.
Change of Ownership

§ 479.42 Changes through death of *owner.*

§ 479.43 Changes through bankruptcy of owner.

§ 479.44 Change in partnership or unincorporated association.

§ 479.45 Changes in corporation.

Change of Business Location

§ 479.46 Notice by taxpayer.

Change of Trade Name

§ 479.47 Notice by taxpayer.

Penalties and Interest

§ 479.48 Failure to pay special (occupational) tax.

§ 479.49 Failure to register change or removal.

§ 479.50 Delinquency.

§ 479.51 Fraudulent return.

Application of State Laws

479.52 State regulations.

ubpart E-Tax on Making Firearms

479.61 Rate of tax.

pplication to Make a Firearm

479.62 Application to make.

479.63 Identification of applicant.

479.64 Procedure for approval of application.

479.65 Denial of application.

479.66 Subsequent transfer of firearms.

479.67 Cancellation of stamp.

xceptions to Tax on Making Firearms

479.68 Qualified manufacturer.

479.69 Making a firearm for the United States.

479.70 Certain government entities.

egistration

479.71 Proof of registration.

ubpart F- Transfer Tax

479.81 Scope of tax.

479.82 Rate of tax.

479.83 Transfer tax in addition to import duty.

pplication and Order for Transfer of Firearm

479.84 Application to transfer.

479.85 Identification of transferee.

479.86 Action on application.

479.87 Cancellation of stamp.

Exemptions Relating to Transfers of Firearms

§ 479.88 Special (occupational) taxpayers.

§ 479.89 Transfers to the United States.

§ 479.90 Certain government entities.

§ 479.91 Unserviceable firearms.

§ 479.92 Transportation of firearms to effect transfer.

Other Provisions

§ 479.93 Transfers of firearms to certain persons.

Subpart G-Registration and Identification of Firearms

§ 479.101 Registration of firearms

§ 479.102 How must firearms be registered

§ 479.103 Registration of firearms manufactured

§ 479.104 Registration of firearms by certain governmental entities

Machine Guns

§ 479.105 Transfer and possession of machine guns.

Subpart H-Importation and Exportation

Importation

§ 479.111 Procedure.

§ 479.112 Registration of imported firearms.

§ 479.113 Conditional importation.

Exportation

§ 479.114 Application and permit for exportation of firearms.

§ 479.115 Action by Director.

§ 479.116 Procedure by exporter.

§ 479.182 Forfeitures.

Subpart O-Other Laws Applicable

§ 479.191 Applicability of other provisions of internal revenue laws.

§ 479.192 Commerce in firearms and ammunition.

§ 479.193 Arms Export Control Act.

Subpart A-Scope of Regulations

§ 479.1 General.

This part contains the procedural and substantive requirements relative to the importation, manufacture, making, exportation, identification and registration of, and the dealing in, machine guns, destructive devices and certain other firearms under the provisions of the National Firearms Act (26 U.S.C. Chapter 53).

Subpart B-Definitions

§ 479.11 Meaning of terms.

When used in this part and in forms prescribed under this part, where not otherwise distinctly expressed or manifestly incompatible with the intent thereof, terms shall have the meanings ascribed in this section. Words in the plural form shall include the singular, and vice versa, and words importing th masculine gender shall include the feminine. The terms "includes" and "including" do not exclude othe things not enumerated which are in the same general class or are otherwise within the scope thereof.

Antique firearm. Any firearm not designed or redesigned for using rim fire or conventional center fire ignition with fixed ammunition and manufactured in or before 1898 (including any matchlock, flintlock, percussion cap, or similar type of ignition system or replica thereof, whether actually manufactured before or after the year 1898) and also any firearm using fixed ammunition manufacturec in or before 1898, for which ammunition is no longer manufactured in the United States and is not readily available in the ordinary channels of commercial trade.

Any other weapon. Any weapon or device capable of being concealed on the person from which a shot can be discharged through the energy of an explosive, a pistol or revolver having a barrel with a smooth bore designed or redesigned to fire a fixed shotgun shell, weapons with combination shotgun and rifle barrels 12 inches or more, less than 18 inches in length, from which only a single discharge ca be made from either barrel without manual reloading, and shall include any such weapon which may be readily restored to fire. Such term shall not include a pistol or a revolver having a rifled bore, or rifled bores, or weapons designed, made, or intended to be fired from the shoulder and not capable of firing fixed ammunition.

ATF officer. An officer or employee of the Bureau of Alcohol, Tobacco and Firearms (ATF) authorized to perform any function relating to the administration or enforcement of this part.

Customs officer. Any officer of the Customs Service or any commissioned, warrant, or petty officer of the Coast Guard, or any agent or other person authorized by law or designated by the Secretary of the Treasury to perform any duties of an officer of the Customs Service.

Dealer. Any person, not a manufacturer or importer, engaged in the business of selling, renting, leasing, or loaning firearms and shall include pawnbrokers who accept firearms as collateral for loans.

Destructive device. (a) Any explosive, incendiary, or poison gas (1) bomb, (2) grenade, (3) rocket having a propellant charge of more than 4 ounces, (4) missile having an explosive or incendiary charge of more than one-quarter ounce, (5) mine, or (6) similar device; (b) any type of weapon by whatever name known which will, or which may be readily converted to, expel a projectile by the action of an explosive or other propellant, the barrel or barrels of which have a bore of more than one-half inch in diameter, except a shotgun or shotgun shell which the Director finds is generally recognized as particularly suitable for sporting purposes; and (c) any combination of parts either designed or intended for use in converting any device into a destructive device as described in paragraphs (a) and (b) of this definition and from which a destructive device may be readily assembled. The term shall not include any device which is neither designed or redesigned for use as a weapon; any device, although originally designed for use as a weapon, which is redesigned for use as a signaling, pyrotechnic, line throwing, safety, or similar device; surplus ordnance sold, loaned, or given by the Secretary of the Army under 10 U.S.C. 4684(2), 4685, or 4686, or any device which the Director finds is not likely to be used as a weapon, or is an antique or is a rifle which the owner intends to use solely for sporting purposes.

Director. The Director, Bureau of Alcohol, Tobacco and Firearms, the Department of the Treasury, Washington, DC.

Director of the Service Center. A director of an Internal Revenue Service Center in an internal revenue region.

District director. A district director of the Internal Revenue Service in an internal revenue district.

Executed under penalties of perjury. Signed with the prescribed declaration under the penalties of perjury as provided on or with respect to the return, form, or other document or, where no form of declaration is prescribed, with the declaration:

"I declare under the penalties of perjury that this-(insert type of document, such as, statement, application, request, certificate), including the documents submitted in support thereof, has been examined by me and, to the best of my knowledge and belief, is true, correct, and complete."

Exportation. The severance of goods from the mass of things belonging to this country with the intention of uniting them to the mass of things belonging to some foreign country.

Exporter. Any person who exports firearms from the United States.

Firearm. (a) A shotgun having a barrel or barrels of less than 18 inches in length; (b) a weapon made from a shotgun if such weapon as modified has an overall length of less than 26 inches or a barrel or barrels of less than 18 inches in length; (c) a rifle having a barrel or barrels of less than 16 inches in length; (d) a weapon made from a rifle if such weapon as modified has an overall length of less than 26 inches or a barrel or barrels of less than 16 inches in length; (e) any other weapon, as defined in this subpart; (f) a machine gun; (g) a muffler or a silencer for any firearm whether or not such firearm is included within this definition; and (h) a destructive device. The term shall not include an antique firearm or any device (other than a machine gun or destructive device) which, although designed as a weapon, the Director finds by reason of the date of its manufacture, value, design, and other characteristics is primarily a collector's item and is not likely to be used as a weapon. For purposes of this definition, the length of the barrel having an integral chamber(s) on a shotgun or rifle shall be determined by measuring the distance between the muzzle and the face of the bolt, breech, or breech block when closed and when the shotgun or rifle is cocked. The overall length of a weapon made from a shotgun or rifle is the distance between the extreme ends of the weapon measured along a line parallel to the center line of the bore.

Fixed ammunition. That self-contained unit consisting of the case, primer, propellant charge, and projectile or projectiles.

Frame or receiver. That part of a firearm which provides housing for the hammer, bolt or breechblock and firing mechanism, and which is usually threaded at its forward portion to receive the barrel.

Importation. The bringing of a firearm within the limits of the United States or any territory under its control or jurisdiction, from a place outside thereof (whether such place be a foreign country or territory subject to the jurisdiction of the United States), with intent to unlade. Except that, bringing a firearm from a foreign country or a territory subject to the jurisdiction of the United States into a foreign trade zone for storage pending shipment to a foreign country or subsequent importation into this country, under Title 26 of the United States Code, and this part, shall not be deemed importation.

Importer. Any person who is engaged in the business of importing or bringing firearms into the United States.

Machine gun. Any weapon which shoots, is designed to shoot, or can be readily restored to shoot, automatically more than one shot, without manual reloading, by a single function of the trigger. The term shall also include the frame or receiver of any such weapon, any part designed and intended solely and exclusively, or combination of parts designed and intended, for use in converting a weapon into a machine gun, and any combination of parts from which a machine gun can be assembled if such parts are in the possession or under the control of a person.

Make. This term and the various derivatives thereof shall include manufacturing (other than by one qualified to engage in such business under this part), putting together, altering, any combination of these, or otherwise producing a firearm.

Manual reloading. The inserting of a cartridge or shell into the chamber of a firearm either with the hands or by means of a mechanical device controlled and energized by the hands.

Manufacturer. Any person who is engaged in the business of manufacturing firearms.

Muffler or silencer. Any device for silencing, muffling, or diminishing the report of a portable firearm, including any combination of parts, designed or redesigned, and intended for the use in assembling or fabricating a firearm silencer or firearm muffler, and any part intended only for use in such assembly or fabrication.

Person. A partnership, company, association, trust, estate, or corporation, as well as a natural person.

Pistol. A weapon originally designed, made, and intended to fire a projectile (bullet) from one or more barrels when held in one hand, and having (a) a chamber(s) as an integral part(s) of, or permanently aligned with, the bore(s); and (b) a short stock designed to be gripped by one hand and at an angle to and extending below the line of the bore(s).

Regional director (compliance). The principal ATF regional official responsible for administering regulations in this part.

Revolver. A projectile weapon, of the pistol type, having a breechloading chambered cylinder so arranged that the cocking of the hammer or movement of the trigger rotates it and brings the next cartridge in line with the barrel for firing.

Rifle. A weapon designed or redesigned, made or remade, and intended to be fired from the shoulder and designed or redesigned and made or remade to use the energy of the explosive in a fixed cartridge to fire only a single projectile through a rifled bore for each single pull of the trigger, and shall include any such weapon which may be readily restored to fire a fixed cartridge.

Shotgun. A weapon designed or redesigned, made or remade, and intended to be fired from the shoulder and designed or redesigned and made or remade to use the energy of the explosive in a fixed shotgun shell to fire through a smooth bore either a number of projectiles (ball shot) or a single projectile for each pull of the trigger, and shall include any such weapon which may be readily restored to fire a fixed shotgun shell.

Transfer. This term and the various derivatives thereof shall include selling, assigning, pledging, leasing, loaning, giving away, or otherwise disposing of

United States. The States and the District of Columbia.

U.S.C. The United States Code.

Unserviceable firearm. A firearm which is incapable of discharging a shot by means of an explosive and incapable of being readily restored to a firing condition.

Subpart C-Administrative and Miscellaneous Provisions

479.21 Forms prescribed.

(a) The Director is authorized to prescribe all forms required by this part. All of the information calle for in each form shall be furnished as indicated by the headings on the form and the instructions on or pertaining to the form. In addition, information called for in each form shall be furnished as required by this part. Each form requiring that it be executed under penalties of perjury shall be executed under penalties of perjury.

(b) Requests for forms should be mailed to the ATF Distribution Center, 7943 Angus Court, Springfiel Virginia 22153.

§ 479.22 Right of entry and examination.

Any ATF officer or employee of the Bureau of Alcohol, Tobacco and Firearms duly authorized to perform any function relating to the administration or enforcement of this part may enter during busine hours the premises (including places of storage) of any importer or manufacturer of or dealer in firearms, to examine any books, papers, or records required to be kept pursuant to this part, and any firearms kept by such importer, manufacturer or dealer on such premises, and may require the production of any books, papers, or records necessary to determine any liability for tax under 26 U.S.C Chapter 53, or the observance of 26 U.S.C. Chapter 53, and this part.

§ 479.23 Restrictive use of required information.

No information or evidence obtained from an application, registration, or record required to be submitted or retained by a natural person in order to comply with any provision of 26 U.S.C. Chapter 5 or this part or section 207 of the Gun Control Act of 1968 shall be used, directly or indirectly, as evidence against that person in a criminal proceeding with respect to a violation of law occurring prior to or concurrently with the filing of the application or registration, or the compiling of the record containing the information or evidence: Provided, however, that the provisions of this section shall not preclude the use of any such information or evidence in a prosecution or other action under any applicable provision of law with respect to the furnishing of false information.

§ 479.24 Destructive device determination.

The Director shall determine in accordance with 26 U.S.C. 5845(f) whether a device is excluded fron the definition of a destructive device. A person who desires to obtain a determination under that provision of law for any device which he believes is not likely to be used as a weapon shall submit a written request, in triplicate, for a ruling thereon to the Director. Each such request shall be executed under the penalties of perjury and contain a complete and accurate description of the device, the name and address of the manufacturer or importer thereof, the purpose of and use for which it is intended, an such photographs, diagrams, or drawings as may be necessary to enable the Director to make his determination. The Director may require the submission to him, of a sample of such device for examination and evaluation. If the submission of such device is impracticable, the person requesting th ruling shall so advise the Director and designate the place where the device will be available for examination and evaluation.

§ 479.25 Collector's items.

The Director shall determine in accordance with 26 U.S.C. 5845(a) whether a firearm or device, which lthough originally designed as a weapon, is by reason of the date of its manufacture, value, design, and ther characteristics primarily a collector's item and is not likely to be used as a weapon. A person who esires to obtain a determination under that provision of law shall follow the procedures prescribed in § 79.24 relating to destructive device determinations, and shall include information as to date of ıanufacture, value, design and other characteristics which would sustain a finding that the firearm or evice is primarily a collector's item and is not likely to be used as a weapon.

479.26 Alternate methods or procedures; emergency variations from requirements.

(a) Alternate methods or procedures. Any person subject to the provisions of this part, on specific pproval by the Director as provided in this paragraph, may use an alternate method or procedure in lieu f a method or procedure specifically prescribed in this part. The Director may approve an alternate ıethod or procedure, subject to stated conditions, when it is found that:

(1) Good cause is shown for the use of the alternate method or procedure;

(2) The alternate method or procedure is within the purpose of, and consistent with the effect intended y, the specifically prescribed method or procedure and that the alternate method or procedure is ubstantially equivalent to that specifically prescribed method or procedure; and

(3) The alternate method or procedure will not be contrary to any provision of law and will not result ı an increase in cost to the Government or hinder the effective administration of this part. Where such erson desires to employ an alternate method or procedure, a written application shall be submitted to ıe appropriate regional director (compliance), for transmittal to the Director. The application shall pecifically describe the proposed alternate method or procedure and shall set forth the reasons for it. lternate methods or procedures may not be employed until the application is approved by the Director. uch person shall, during the period of authorization of an alternate method or procedure, comply with ıe terms of the approved application. Authorization of any alternate method or procedure may be ithdrawn whenever, in the judgment of the Director, the effective administration of this part is indered by the continuation of the authorization.

(b) Emergency variations from requirements. The Director may approve a method of operation other ıan as specified in this part, where it is found that an emergency exists and the proposed variation from ıe specified requirements are necessary and the proposed variations (1) will not hinder the effective dministration of this part, and (2) will not be contrary to any provisions of law. Variations from equirements granted under this paragraph are conditioned on compliance with the procedures, onditions, and limitations set forth in the approval of the application. Failure to comply in good faith ith the procedures, conditions, and limitations shall automatically terminate the authority for the ariations, and the person granted the variance shall fully comply with the prescribed requirements of egulations from which the variations were authorized. Authority for any variation may be withdrawn henever, in the judgment of the Director, the effective administration of this part is hindered by the ontinuation of the variation. Where a person desires to employ an emergency variation, a written pplication shall be submitted to the appropriate regional director (compliance) for transmittal to the irector. The application shall describe the proposed variation and set forth the reasons for it. Variations ıay not be employed until the application is approved.

(c) Retention of approved variations. The person granted the variance shall retain and make available for examination by A TF officers any application approved by the Director under this section.

Subpart D-Special (Occupational) Taxes

§ 479.31 Liability for tax.

(a) General. Every person who engages in the business of importing, manufacturing, or dealing in (including pawnbrokers) firearms in the United States shall pay a special (occupational) tax at a rate specified by § 479.32. The tax shall be paid on or before the date of commencing the taxable business, and thereafter every year on or before July 1. Special (occupational) tax shall not be prorated. The tax shall be computed for the entire tax year (July 1 through June 30), regardless of the portion of the year during which the taxpayer engages in business. Persons commencing business at any time after July 1 i any year are liable for the special (occupational) tax for the entire tax year.

(b) Each place of business taxable. An importer, manufacturer, or dealer in firearms incurs special ta liability at each place of business where an occupation subject to special tax is conducted. A place of business means the entire office, plant or area of the business in any one location under the same proprietorship. Passageways, streets, highways, rail crossings, waterways, or partitions dividing the premises are not sufficient separation to require additional special tax, if the divisions of the premises are otherwise contiguous. See also §§ 479.38-479.39.

§ 479.32 Special (occupational) tax rates.

(a) Prior to January 1, 1988, the special (occupational) tax rates were as follows:

	Per year or fraction thereof
Class 1 Importer of firearms	$500.
Class 2 Manufacturer of firearms	$500.
Class 3 Dealer in firearms	$200
Class 4 Importer only of weapons classified as "any other weapon"	$25.
Class 5 Manufacturer only of weapons classified as "any other weapon"	$25.

Class 6
Dealer only in weapons classified as "any other weapon" $10.

(b) Except as provided in § 479.32a, the special (occupational) tax rates effective January 1, 1988, are as follows:

	Per year or fraction Thereof
Class 1 Importer of firearms (including an importer only of weapons classified as "any other weapon")	$1000.
Class 2 Manufacturer of firearms (including a manufacturer only of weapons classified as "any other weapon")	$1000.
Class 3 Dealer in firearms (including a dealer only of weapons classified as "any other weapon")	$500.

(c) A taxpayer who was engaged in a business on January 1, 1988, for which a special (occupational) tax was paid for a taxable period which began before January 1, 1988, and included that date, shall pay an increased special tax for the period January 1, 1988, through June 30, 1988. The increased tax shall not exceed one half the excess (if any) of (1) the rate of special tax in effect on January 1, 1988, over (2) the rate of such tax in effect on December 31, 1987. The increased special tax shall be paid on or before April 1, 1988.

§ 479.32a Reduced rate of tax for small importers and manufacturers.

(a) General. Effective January 1, 1988, 26 U.S.C. 5801(b) provides for a reduced rate of special tax with respect to any importer or manufacturer whose gross receipts (for the most recent taxable year ending before the first day of the taxable period to which the special tax imposed by § 479.32 relates) are less than $500,000. The rate of tax for such an importer or manufacturer is $500 per year or fraction thereof. The "taxable year" to be used for determining gross receipts is the taxpayer's income tax year. All gross receipts of the taxpayer shall be included, not just the gross receipts of the business subject to special tax. Proprietors of new businesses that have not yet begun a taxable year, as well as proprietors of existing businesses that have not yet ended a taxable year, who commence a new activity subject to special tax, quality for the reduced special (occupational) tax rate, unless the business is a member of a "controlled group;" in that case, the rules of paragraph (b) of this section shall apply.

(b) Controlled group. All persons treated as one taxpayer under 26 U.S.C. 5061 (e)(3) shall be treated as one taxpayer for the purpose of determining gross receipts under paragraph (a) of this section. "Controlled group" means a controlled group of corporations, as defined in 26 U.S.C. 1563 and implementing regulations in 26 CFR 1.1563-1 through 1.1563-4, except that the words "at least 80 percent" shall be replaced by the words "more than 50 percent" in each place they appear in subsection (a) of 26 U.S.C. 1563, as well as in the implementing regulations. Also, the rules for a "controlled group of corporations" apply in a similar fashion to groups which include partnerships and/or sole proprietorships. If one entity maintains more than 50% control over a group consisting of corporations and one, or more, partnerships and/or sole proprietorships, all of the members of the controlled group are one taxpayer for the purpose of this section.

(c) Short taxable year. Gross receipts for any taxable year of less than 12 months shall be annualized by multiplying the gross receipts for the short period by 12 and dividing the result by the number of months in the short period, as required by 26 U.S.C. 448(c)(3).

(d) Returns and allowances. Gross receipts for any taxable year shall be reduced by returns and allowances made during that year under 26 U.S.C. 448(c)(3).

§ 479.33 Special exemption.

(a) Any person required to pay special (occupational) tax under this part shall be relieved from payment of that tax if he establishes to the satisfaction of the Director that his business is conducted exclusively with, or on behalf of, the United States or any department, independent establishment, or agency thereof. The Director may relieve any person manufacturing firearms for or on behalf of the United States from compliance with any provision of this part in the conduct of the business with respect to such firearms.

(b) The exemption in this section may be obtained by filing with the Director an application, in letter form, setting out the manner in which the applicant conducts his business, the type of firearm to be manufactured, and proof satisfactory to the Director of the existence of the contract with the United States, department, independent establishment, or agency thereof, under which the applicant intends to operate.

§ 479.34 Special tax registration and return.

(a) General. Special tax shall be paid by return. The prescribed return is ATF Form 5630.7, Special Tax

Registration and Return. Special tax returns, with payment of tax, shall be filed with A TF in accordance with instructions on the form. Properly completing, signing, and timely filing of a return (Form 5630.7) constitutes compliance with 26 U.S.C. 5802.

(b) Preparation of ATF Form 5630.7. All of the information called for on Form 5630.7 shall be provided, including:

(1) The true name of the taxpayer.

(2) The trade name(s) (if any) of the business(es) subject to special tax.

(3) The employer identification number (see § 479.35).

(4) The exact location of the place of business, by name and number of building or street, or if these do not exist, by some description in addition to the post office address. In the case of one return for two or more locations, the address to be shown shall be the taxpayer's principal place of business (or principal office, in the case of a corporate taxpayer).

(5) The class(es) of special tax to which the taxpayer is subject.

(6) Ownership and control information: That is, the name, position, and residence address of every owner of the business and of every person having power to control its management and policies with respect to the activity subject to special tax. "Owner of the business" shall include every partner, if the taxpayer is a partnership, and every person owning 10% or more of its stock, if the taxpayer is a corporation. However, the ownership and control information required by this paragraph need not be stated if the same information has been previously provided to ATF in connection with a license application under Part 478 of this chapter, and if the information previously provided is still current.

(c) Multiple locations and/or classes of tax. A taxpayer subject to special tax for the same period at more than one location or for more than one class of tax shall

1) File one special tax return,

ATF Form 5630.7, with payment of tax, to cover all such locations and classes of tax; and

(2) Prepare, in duplicate, a list identified with the taxpayer's name, address (as shown on ATF Form 5630.7), employer identification number, and period covered by the return. The list shall show, by states, the name, address, and tax class of each location for which special tax is being paid. The original of the list shall be filed with A TF in accordance with instructions on the return, and the copy shall be retained at the taxpayer's principal place of business (or principal office, in the case of a corporate taxpayer) for not less than 3 years.

(d) Signing of ATF Forms 5630.7- (1) Ordinary returns. The return of an individual proprietor shall be signed by the individual. The return of a partnership shall be signed by a general partner. The return of a corporation shall be signed by any officer. In each case, the person signing the return shall designate his or her capacity as "individual owner," "member of firm," or, in the case of a corporation, the title of the officer.

(2) Fiduciaries. Receivers, trustees, assignees, executors, administrators, and other legal representatives who continue the business of a bankrupt, insolvent, deceased person, etc., shall indicate the fiduciary capacity in which they act.

(3) Agent or attorney in fact. If a return is signed by an agent or attorney in fact, the signature shall be preceded by the name of the principal and followed by the title of the agent or attorney in fact. A return signed by a person as agent will not be accepted unless there is filed, with the ATF office with which the return is required to be filed, a power of attorney authorizing the agent to perform the act.

(4) Perjury statement. ATF Forms 5630.7 shall contain or be verified by a written declaration that the

return has been executed under the penalties of perjury.

(e) Identification of taxpayer. If the taxpayer is an individual, with the initial return such person shall securely attach to Form 5630.7 a photograph of the individual 2 x 2 inches in size, clearly showing a fu front view of the features of the individual with head bare, with the distance from the top of the head to the point of the chin approximately 1 1/4 inches, and which shall have been taken within 6 months prio to the date of completion of the return. The individual shall also attach to the return a properly completed FBI Form FD-258 (Fingerprint Card). The fingerprints must be clear for accurate classification and should be taken by someone properly equipped to take them: Provided, That the provisions of this paragraph shall not apply to individuals who have filed with A TF a properly execute Application for License under 18 U.S.C. Chapter 44, Firearms, ATF Form 7 (5310.12) (12-93 edition), as specified in § 478.44(a).

§ 479.35 Employer identification number.

(a) Requirement. The employer identification number (defined in 26 CFR 301.7701-12) of the taxpayer who has been assigned such a number shall be shown on each special tax return, including amended returns, filed under this subpart. Failure of the taxpayer to include the employer identification number may result in the imposition of the penalty specified in § 70.113 of this chapter.

(b) Application for employer identification number. Each taxpayer who files a special tax return, wh has not already been assigned an employer identification number, shall file IRS Form SS-4 to apply for one. The taxpayer shall apply for and be assigned only one employer identification number, regardless of the number of places of business for which the taxpayer is required to file a special tax return. The employer identification number shall be applied for no later than 7 days after the filing of the taxpayer's first special tax return. IRS Form SS-4 may be obtained from the director of an IRS service center or from any IRS district director.

(c) Preparation and filing of IRS Form SS-4. The taxpayer shall prepare and file IRS Form SS-4, together with any supplementary statement, in accordance with the instructions on the form or issued in respect to it.

§ 479.36 The special tax stamp, receipt for special (occupational) taxes.

Upon filing a properly completed and executed return (Form 5630.7) accompanied by remittance of the full amount due, the taxpayer will be issued a special tax stamp as evidence of payment of the special (occupational) tax.

§ 479.37 Certificates in lieu of stamps lost or destroyed.

When a special tax stamp has been lost or destroyed, such fact should be reported immediately to the regional director (compliance) who issued the stamp. A certificate in lieu of the lost or destroyed stamp will be issued to the taxpayer upon the submission of an affidavit showing to the satisfaction of the regional director (compliance) that the stamp was lost or destroyed.

479.38 Engaging in business at more than one location.

A person shall pay the special (occupational) tax for each location where he engages in any business axable under 26 U.S.C. 5801. However, a person paying a special (occupational) tax covering his rincipal place of business may utilize other locations solely for storage of firearms without incurring pecial (occupational) tax liability at such locations. A manufacturer, upon the single payment of the ppropriate special (occupational) tax, may sell firearms, if such firearms are of his own manufacture, at he place of manufacture and at his principal office or place of business if no such firearms, except amples, are kept at such office or place of business. When a person changes the location of a business or which he has paid the special (occupational) tax, he will be liable for another such tax unless the hange is properly registered with the regional director (compliance) for the region in which the special ax stamp was issued, as provided in § 479.46.

479.39 Engaging in more than one business at the same location.

If more than one business taxable under 26 U.S.C. 5801 is carried on at the same location during a axable year, the special (occupational) tax imposed on each such business must be paid. This section oes not require a qualified manufacturer or importer to qualify as a dealer if such manufacturer or nporter also engages in business on his qualified premises as a dealer. However, a qualified nanufacturer who engages in business as an importer must also qualify as an importer. Further, a ualified dealer is not entitled to engage in business as a manufacturer or importer.

479.40 Partnership liability.

Any number of persons doing business in partnership at anyone location shall be required to pay but ne special (occupational) tax.

479.41 Single sale.

A single sale, unattended by circumstances showing the one making the sale to be engaged in business, oes not create special (occupational) tax liability.

hange of ownership

479.42 Changes through death of owner.

Whenever any person who has paid special (occupational) tax dies, the surviving spouse or child, or xecutors or administrators, or other legal representatives, may carry on this business for the remainder f the term for which tax has been paid and at the place (or places) for which the tax was paid, without ny additional payment, subject to the following conditions. If the surviving spouse or child, or executor r administrator, or other legal representative of the deceased taxpayer continues the business, such erson shall, within 30 days after the date on which the successor begins to carry on the business, file a ew return, Form 5630.7, with ATF in accordance with the instructions on the form. The return thus xecuted shall show the name of the original taxpayer, together with the basis of the succession. (As to ability in case of failure to register, see § 479.49.)

§ 479.43 Changes through bankruptcy of owner.

A receiver or referee in bankruptcy may continue the business under the stamp issued to the taxpayer at the place and for the period for which the tax was paid. An assignee for the benefit of creditors may continue business under his assignor's special tax stamp without incurring additional special (occupational) tax liability. In such cases, the change shall be registered with A TF in a manner similar to that required by § 479.42.

§ 479.44 Change in partnership or unincorporated association.

When one or more members withdraw from a partnership or an unincorporated association, the remaining member, or members, may, without incurring additional special (occupational) tax liability, carry on the same business at the same location for the balance of the taxable period for which special (occupational) tax was paid, provided any such change shall be registered in the same manner as required by § 479.42. Where new member(s) are taken into a partnership or an unincorporated association, the new firm so constituted may not carry on business under the special tax stamp of the ol firm. The new firm must file a return, pay the special (occupational) tax and register in the same manne as a person who first engages in business is required to do under § 479.34 even though the name of the new firm may be the same as that of the old. Where the members of a partnership or an unincorporated association, which has paid special (occupational) tax, form a corporation to continue the business, a new special tax stamp must be taken out in the name of the corporation.

§ 479.45 Changes in corporation.

Additional special (occupational) tax is not required by reason of a mere change of name or increase i the capital stock of a corporation if the laws of the State of incorporation provide for such change or increase without the formation of a new corporation. A stockholder in a corporation, who after its dissolution continues the business, incurs new special (occupational) tax liability.

Change of Business Location

§ 479.46 Notice by taxpayer.

Whenever during the taxable year a taxpayer intends to remove his business to a location other than specified in his last special (occupational) tax return (see § 479.34), he shall file with ATF (a) a return, Form 5630.7, bearing the notation "Removal Registry," and showing the new address intended to be used, (b) his current special tax stamp, and (c) a letter application requesting the amendment of his registration. The regional director (compliance), upon approval of the application, shall return the special tax stamp, amended to show the new business location. Firearms operations shall not be commenced at the new business location by the taxpayer prior to the required approval of his applicatic to so change his business location.

Change of Trade Name

§ 479.47 Notice by taxpayer.

Whenever during the taxable year a taxpayer intends to change the name of his business, he shall file with ATF (a) a return, Form 5630.7, bearing the notation "Amended," and showing the trade name intended to be used, (b) his current special tax stamp, and (c) a letter application requesting the amendment of his registration. The regional director (compliance), upon approval of the application, shall return the special tax stamp, amended to show the new trade name. Firearms operations shall not be commenced under the new trade name by the taxpayer prior to the required approval of his application to so change the trade name.

Penalties and Interest

§ 479.48 Failure to pay special (occupational) tax.

Any person who engages in a business taxable under 26 U.S.C. 5801, without timely payment of the tax imposed with respect to such business (see § 479.34) shall be liable for such tax, plus the interest and penalties thereon (see 26 U.S.C. 6601 and 6651). In addition, such person may be liable for criminal penalties under 26 U.S.C. 5871.

§ 479.49 Failure to register change or removal.

Any person succeeding to and carrying on a business for which special (occupational) tax has been paid without registering such change within 30 days thereafter, and any taxpayer removing his business with respect to which special (occupational) tax has been paid to a place other than that for which tax was paid without obtaining approval therefor (see § 479.46), will incur liability to an additional payment of the tax, addition to tax and interest, as provided in sections 5801, 6651, and 6601, respectively, LRC., for failure to make return (see § 479.50) or pay tax, as well as criminal penalties for carrying on business without payment of special (occupational) tax (see section 5871 LRC.)

§ 479.50 Delinquency.

Any person liable for special (occupational) tax under section 5801, LR.C., who fails to file a return (Form 5630.7), as prescribed, will be liable for a delinquency penalty computed on the amount of tax due unless a return (Form 5630.7) is later filed and failure to file the return timely is shown to the satisfaction of the regional director (compliance), to be due to reasonable cause. The delinquency penalty to be added to the tax is 5 percent if the failure is for not more than 1 month, with an additional 5 percent for each additional month or fraction thereof during which failure continues, not to exceed 25 percent in the aggregate (section 6651, LRC.). However, no delinquency penalty is assessed where the 50 percent addition to tax is assessed for fraud (see § 479.51).

§ 479.51 Fraudulent return.

If any part of any underpayment of tax required to be shown on a return is due to fraud, there shall be added to the tax an amount equal to 50 percent of the underpayment, but no delinquency penalty shall be assessed with respect to the same underpayment (section 6653, I.R.C.).

Application of State Laws

§ 479.52 State regulations.

Special tax stamps are merely receipts for the tax. Payment of tax under Federal law confers no privilege to act contrary to State law. One to whom a special tax stamp has been issued may still be punishable under a State law prohibiting or controlling the manufacture, possession or transfer of firearms. On the other hand, compliance with State law confers no immunity under Federal law. Persons who engage in the business of importing, manufacturing or dealing in firearms, in violation of the law of a State, are nevertheless required to pay special (occupational) tax as imposed under the internal revenue laws of the United States. For provisions relating to restrictive use of information furnished to comply with the provisions of this part see § 479.23.

Subpart E- Tax on Making Firearms

§ 479.61 Rate of tax.

Except as provided in this subpart, there shall be levied, collected, and paid upon the making of a firearm a tax at the rate of $200 for each firearm made. This tax shall be paid by the person making the firearm. Payment of the tax on the making of a firearm shall be represented by a $200 adhesive stamp bearing the words "National Firearms Act: The stamps are maintained by the Director.

Application to Make a *Firearm*

§ 479.62 Application to make.

No person shall make a firearm unless the person has filed with the Director a written application on Form 1 (Firearms), Application to Make and Register a Firearm, in duplicate, executed under the penalties of perjury, to make and register the firearm and has received the approval of the Director to make the firearm which approval shall effectuate registration of the weapon to the applicant. The application shall identify the firearm to be made by serial number, type, model, caliber or gauge, length of barrel, other marks of identification, and the name and address of original manufacturer (if the applicant is not the original manufacturer). The applicant must be identified on the Form 1 (Firearms) by name and address and, if other than a natural person, the name and address of the principal officer or authorized representative and the employer identification number and, if an individual, the identification must include the date and place of birth and the information prescribed in § 479.63. Each applicant shall identify the Federal firearms license and special (occupational) tax stamp issued to the applicant, if any. The applicant shall also show required information evidencing that making or possession of the firearm would not be in violation of law. If the making is taxable, a remittance in the amount of $200 shall be submitted with the application in accordance with the instructions on the form. If the making is taxable and the application is approved, the Director will affix a National Firearms Act stamp to the original application in the space provided therefor and properly cancel the stamp (see § 479.67). The approved application will be returned to the applicant. If the making of the firearm is tax exempt under this part, an explanation of the basis of the exemption shall be attached to the Form 1 (Firearms).

§ 479.63 Identification of applicant..

f the applicant is an individual, the applicant shall securely attach to each copy of the Form 1 Firearms), in the space provided on the form, a photograph of the applicant 2 x 2 inches in size, clearly howing a full front view of the features of the applicant with head bare, with the distance from the top of the head to the point of the chin approximately 1 1/4 inches, and which shall have been taken within 1 ear prior to the date of the application. The applicant shall attach two properly completed FBI Forms 'D-258 (Fingerprint Card) to the application. The fingerprints must be clear for accurate classification nd should be taken by someone properly equipped to take them. A certificate of the local chief of olice, sheriff of the county, head of the State police, State or local district attorney or prosecutor, or uch other person whose certificate may in a particular case be acceptable to the Director, shall be ompleted on each copy of the Form 1 (Firearms). The certificate shall state that the certifying official is atisfied that the fingerprints and photograph accompanying the application are those of the applicant nd that the certifying official has no information indicating that possession of the firearm by the maker vould be in violation of State or local law or that the maker will use the firearm for other than lawful urposes.

479.64 Procedure for approval of application.

The application to make a firearm, Form 1 (Firearms), must be forwarded directly, in duplicate, by the naker of the firearm to the Director in accordance with the instructions on the form. The Director will onsider the application for approval or disapproval. If the application is approved, the Director will eturn the original thereof to the maker of the firearm and retain the duplicate. Upon receipt of the pproved application, the maker is authorized to make the firearm described therein. The maker of the irearm shall not, under any circumstances, make the firearm until the application, satisfactorily xecuted, has been forwarded to the Director and has been approved and returned by the Director with ne National Firearms Act stamp affixed. If the application is disapproved, the original Form 1(Firearms) nd the remittance submitted by the applicant for the purchase of the stamp will be returned to the pplicant with the reason for disapproval stated on the form.

479.65 Denial of application.

An application to make a firearm shall not be approved by the Director if the making or possession of ne firearm would place the person making the firearm in violation of law.

479.66 Subsequent transfer of firearms.

Where a firearm which has been made in compliance with 26 U.S.C. 5821, and the regulations ontained in this part, is to be transferred subsequently, the transfer provisions of the firearms laws and egulations must be complied with. (See subpart F of this part).

479.67 Cancellation of stamp.

The person affixing to a Form 1(Firearms) a "National Firearms Act stamp shall cancel it by writing or tamping thereon, in ink, his initials, and the day, month and year, in such manner as to render it unfit or reuse. The cancellation shall not so deface the stamp as to prevent its denomination and genuineness om being readily determined.

Exceptions to Tax on Making Firearms

§ 479.68 Qualified manufacturer.

A manufacturer qualified under this part to engage in such business may make firearms without payment of the making tax. However, such manufacturer shall report and register each firearm made in the manner prescribed by this part.

§ 479.69 Making a firearm for the United States.

A firearm may be made by, or on behalf of, the United States or any department, independent establishment, or agency thereof without payment of the making tax. However, if a firearm is to be mac on behalf of the United States, the maker must file an application, in duplicate, on Form 1 (Firearms) and obtain the approval of the Director in the manner prescribed in § 479.62.

§ 479.70 Certain government entities.

A firearm may be made without payment of the making tax by, or on behalf of, any State, or possession of the United States, any political subdivision thereof, or any official police organization of such a government entity engaged in criminal investigations. Any person making a firearm under this exemption shall first file an application, in duplicate, on Form 1 (Firearms) and obtain the approval of the Director as prescribed in § 479.62.

Registration

§ 479.71 Proof of registration.

The approval by the Director of an application, Form 1 (Firearms), to make a firearm under this subpart shall effectuate registration of the firearm described in the Form 1 (Firearms) to the person making the firearm. The original Form 1 (Firearms) showing approval by the Director shall be retained by the maker to establish proof of his registration of the firearm described therein, and shall be made available to any A TF officer on request.

Subpart F-Transfer Tax § 479.81 Scope of tax.

Except as otherwise provided in this part, each transfer of a firearm in the United States is subject to a tax to be represented by an adhesive stamp of the proper denomination bearing the words "National Firearms Act" to be affixed to the Form 4 (Firearms), Application for Transfer and Registration of Firearm, as provided in this subpart.

§ 479.82 Rate of tax.

The transfer tax imposed with respect to firearms transferred within the United States is at the rate of $200 for each firearm transferred, except that the transfer tax on any firearm classified as "any other veapon" shall be at the rate of $5 for each such firearm transferred. The tax imposed on the transfer of ne firearm shall be paid by the transferor.

479.83 Transfer tax in addition to import duty.

'he transfer tax imposed by section 5811, !.R.C., is in addition to any import duty.

pplication and Order for Transfer of Firearm

479.84 Application to transfer.

Except as otherwise provided in this subpart, no firearm may be transferred in the United States unless n application, Form 4 (Firearms), Application for Transfer and Registration of Firearm, in duplicate, xecuted under the penalties of perjury to transfer the firearm and register it to the transferee has been led with and approved by the Director. The application, Form 4 (Firearms), shall be filed by the ansferor and shall identify the firearm to be transferred by type; serial number; name and address of the nanufacturer and importer, if known; model; caliber, gauge or size; in the case of a short-barreled hotgun or a short-barreled rifle, the length of the barrel: in the case of a weapon made from a rifle or hotgun, the overall length of the weapon and the length of the barrel; and any other identifying marks n the firearm. In the event the firearm does not bear a serial number, the applicant shall obtain a serial umber from the Regional director (compliance) and shall stamp (impress) or otherwise conspicuously lace such serial number on the firearm in a manner not susceptible of being readily obliterated, altered r removed. The application, Form 4 (Firearms), shall identify the transferor by name and address; shall lentify the transferor's Federal firearms license and special (occupational) Chapter tax stamp, if any; nd if the transferor is other than a natural person, shall show the title or status of the person executing ne application. The application also shall identify the transferee by name and address, and, if the ansferee is a natural person not qualified as a manufacturer, importer or dealer under this part, he shall e further identified in the manner prescribed in § 479.85. The application also shall identify the special occupational) tax stamp and Federal firearms license of the transferee, if any. Any tax payable on the ansfer must be represented by an adhesive stamp of proper denomination being affixed to the pplication, Form 4 (Firearms), properly cancelled.

479.85 Identification of transferee.

If the transferee is an individual, such person shall securely attach to each copy of the application, orm 4 (Firearms), in the space provided on the form, a photograph of the applicant 2)(2 inches in size, learly showing a full front view of the features of the applicant with head bare, with the distance from ne top of the head to the point of the chin approximately 1 1/4 inches, and which shall have been taken ithin 1 year prior to the date of the application. The transferee shall attach two properly completed FBI orms FD258 (Fingerprint Card) to the application. The fingerprints must be clear for accurate lassification and should be taken by someone properly equipped to take them. A certificate of the local hief of police, sheriff of the county, head of the State police, State or local district attorney or rosecutor, or such other person whose certificate may in a particular case be acceptable to the Director, nall be completed on each copy of the Form 4 (Firearms). The certificate shall state that the certifying fficial is satisfied that the fingerprints and photograph accompanying the application are those of the

applicant and that the certifying official has no information indicating that the receipt or possession of the firearm would place the transferee in violation of State or local law or that the transferee will use th firearm for other than lawful purposes.
§ 479.86 Action on application.

The Director will consider a completed and properly executed application, Form 4 (Firearms), to transfer a firearm. If the application is approved, the Director will affix the appropriate National Firearms Act stamp, cancel it, and return the original application showing approval to the transferor wh may then transfer the firearm to the transferee along with the approved application. The approval of an application, Form 4 (Firearms), by the Director will effectuate registration of the firearm to the transferee. The transferee shall not take possession of a firearm until the application, Form 4 (Firearms) for the transfer filed by the transferor has been approved by the Director and registration of the firearm is effectuated to the transferee. The transferee shall retain the approved application as proof that the firearm described therein is registered to the transferee, and shall make the approved Form 4 (Firearms) available to any A TF officer on request. If the application, Form 4 (Firearms), to transfer a firearm is disapproved by the Director, the original application and the remittance for purchase of the stamp will be returned to the transferor with reasons for the disapproval stated on the application. An application, Form 4 (Firearms), to transfer a firearm shall be denied if the transfer, receipt, or possession of a firear would place the transferee in violation of law. In addition to any other records checks that may be conducted to determine whether the transfer, receipt, or possession of a firearm would place the transferee in violation of law, the Director shall contact the National Instant Criminal Background Check System.

§ 479.87 Cancellation of stamp.
The method of cancellation of the stamp required by this subpart as prescribed in § 479.67 shall be use

Exemptions Relating to Transfers of Firearms

§ 479.88 Special (occupational) taxpayers.

(a) A firearm registered to a person qualified under this part to engage in business as an importer, manufacturer, or dealer may be transferred by that person without payment of the transfer tax to any other person qualified under this part to manufacture, import, or deal in firearms.

(b) The exemption provided in paragraph (a) of this section shall be obtained by the transferor of the firearm filing with the Director an application, Form 3 (Firearms), Application for Tax-exempt Transfe of Firearm and Registration to Special (Occupational) Taxpayer, in duplicate, executed under the penalties of perjury. The application, Form 3 (Firearms), shall (1) show the name and address of the transferor and of the transferee, (2) identify the Federal firearms license and special (occupational) tax stamp of the transferor and of the transferee, (3) show the name and address of the manufacturer and th importer of the firearm, if known, (4) show the type, model, overall length (if applicable), length of barrel, caliber, gauge or size, serial number, and other marks of identification of the firearm, and (5) contain a statement by the transferor that he is entitled to the exemption because the transferee is a person qualified under this part to manufacture, import, or deal in firearms. If the Director approves an application, Form 3 (Firearms), he shall return the original Form 3 (Firearms) to the transferor with the approval noted thereon. Approval of an application, Form 3 (Firearms), by the Director shall remove registration of the firearm reported thereon from the transferor and shall effectuate the registration of that firearm to the transferee. Upon receipt of the approved Form 3 (Firearms), the transferor shall deliver same with the firearm to the transferee. The transferor shall not transfer the firearm to the transferee until his application, Form 3 (Firearms), has been approved by the Director and the original thereof has been returned to the transferor. If the Director disapproves the application, Form 3 (Firearms), he shall return the original Form 3 (Firearms) to the transferor with the reasons for the disapproval stated thereon.

(c) The transferor shall be responsible for establishing the exempt status of the transferee before making a transfer under the provisions of this section. Therefore, before engaging in transfer negotiations with the transferee, the transferor should satisfy himself as to the claimed exempt status of the transferee and the bona fides of the transaction. If not fully satisfied, the transferor should communicate with the Director, report all circumstances regarding the proposed transfer, and await the Director's advice before making application for the transfer. An unapproved transfer or a transfer to an unauthorized person may subject the transferor to civil and criminal liabilities. (See 26 U.S.C. 5852, 5861, and 5871.)

§ 479.89 Transfers to the United States.

A firearm may be transferred to the United States or any department, independent establishment or agency thereof without payment of the transfer tax. However, the procedures for the transfer of a firearm as provided in § 479.90 shall be followed in a tax-exempt transfer of a firearm under this section, unless the transferor is relieved of such requirement under other provisions of this part.

§ 479.90 Certain government entities.

(a) A firearm may be transferred without payment of the transfer tax to or from any State, possession of the United States, any political subdivision thereof, or any official police organization of such a governmental entity engaged in criminal investigations.

(b) The exemption provided in paragraph (a) of this section shall be obtained by the transferor of the firearm filing with the Director an application, Form 5 (Firearms), Application for Tax-exempt Transfer and Registration of Firearm, in duplicate, executed under the penalties of perjury. The application shall (1) show the name and address of the transferor and of the transferee, (2) identify the Federal firearms license and special (occupational) tax stamp, if any, of the transferor and of the transferee, (3) show the name and address of the manufacturer and the importer of the firearm, if known, (4) show the type, model, overall length (if applicable), length of barrel, caliber, gauge or size, serial number, and other marks of identification of the firearm, and (5) contain a statement by the transferor that the transferor is entitled to the exemption because either the transferor or the transferee is a governmental entity coming within the purview of paragraph (a) of this section. In the case of a transfer of a firearm by a governmental entity to a transferee who is a natural person not qualified as a manufacturer, importer, or dealer under this part, the transferee shall be further identified in the manner prescribed in § 479.85. If the Director approves an application, Form 5 (Firearms), the original Form 5 (Firearms) shall be returned to the transferor with the approval noted thereon. Approval of an application, Form 5 (Firearms), by the Director shall effectuate the registration of that firearm to the transferee. Upon receipt of the approved Form 5 (Firearms), the transferor shall deliver same with the firearm to the transferee. The transferor shall not transfer the firearm to the transferee until the application, Form 5 (Firearms), has been approved by the Director and the original thereof has been returned to the transferor. If the Director disapproves the application, Form 5 (Firearms), the original Form 5 (Firearms) shall be returned to the transferor with the reasons for the disapproval stated thereon. An application by a governmental entity to transfer a firearm shall be denied if the transfer, receipt, or possession of a firearm would place the transferee in violation of law.

(c) The transferor shall be responsible for establishing the exempt status of the transferee before making a transfer under the provisions of this section. Therefore, before engaging in transfer negotiations with the transferee, the transferor should satisfy himself of the claimed exempt status of the transferee and the bona fides of the transaction. If not fully satisfied, the transferor should communicate with the Director, report all circumstances regarding the proposed transfer, and await the Director's advice before making application for transfer. An unapproved transfer or a transfer to an unauthorized person may subject the transferor to civil and criminal liabilities. (See 26 U.S.C. 5852, 5861, and 5871.)

§ 479.91 Unserviceable firearms.

An unserviceable firearm may be transferred as a curio or ornament without payment of the transfer tax. However, the procedures for the transfer of a firearm as provided in § 479.90 shall be followed in a tax exempt transfer of a firearm under this section, except a statement shall be entered on the transfer application, Form 5 (Firearms), by the transferor that he is entitled to the exemption because the firearm to be transferred is unserviceable and is being transferred as a curio or ornament. An unapproved transfer, the transfer of a firearm under the provisions of this section which is in fact not an unserviceable firearm, or the transfer of an unserviceable firearm as something other than a curio or ornament, may subject the transferor to civil and criminal liabilities. (See 26 U.S.C. 5811, 5852, 5861, and 5871.)

§ 479.92 Transportation of firearms to effect transfer.

Notwithstanding any provision of § 478.28 of this chapter, it shall not be required that authorization be obtained from the Director for the transportation in interstate or foreign commerce of a firearm in order to effect the transfer of a firearm authorized under the provisions of this subpart.

Other Provisions

§ 479.93 Transfers of firearms to certain persons.

Where the transfer of a destructive device, machine gun, shortbarreled shotgun, or short-barreled rifle is to be made by a person licensed under the provisions of Title I of the Gun Control Act of 1968 (82 Stat. 1213) to a person not so licensed, the sworn statement required by § 478.98 of this chapter shall be attached to and accompany the transfer application required by this subpart.

Subpart G-Registration and Identification of Firearms

§ 479.101 Registration of Firearms.

(a) The Director shall maintain a central registry of all firearms in the United States which are not in he possession of or under the control of the United States. This registry shall be known as the National 'irearms Registration and Transfer Record and shall include:

(1) Identification of the firearm as required by this part;

2) Date of registration; and

(3) Identification and address of person entitled to possession of the firearm as required by this part.

(b) Each manufacturer, importer, and maker shall register each firearm he manufactures, imports, or nakes in the manner prescribed by this part. Each firearm transferred shall be registered to the transferee y the transferor in the manner prescribed by this part. No firearm may be registered by a person nlawfully in possession of the firearm except during an amnesty period established under section 207 f the Gun Control Act of 1968 (82 Stat. 1235).

(c) A person shown as possessing firearms by the records maintained by the Director pursuant to the Jational Firearms Act (26 U.S.C. Chapter 53) in force on October 31, 1968, shall be considered to have egistered the firearms in his possession which are disclosed by that record as being in his possession on)ctober 31, 1968.

(d) The National Firearms Registration and Transfer Record shall include firearms registered to the ossessors thereof under the provisions of section 207 of the Gun Control Act of 1968.

(e) A person possessing a firearm registered to him shall retain proof of registration which shall be nade available to any ATF officer upon request.

(f) A firearm not identified as required by this part shall not be registered.

479.102 How must firearms be identified?

(a) You, as a manufacturer, importer, or maker of a firearm, must legibly identify the firearm as ollows:

(1) By engraving, casting, stamping (impressing), or otherwise conspicuously placing or causing to be ngraved, cast, stamped (impressed) or placed on the frame or receiver thereof an individual serial umber. The serial number must be placed in a manner not susceptible of being readily obliterated, ltered, or removed, and must not duplicate any serial number placed by you on any other firearm. For irearms manufactured, imported, or made on and after January 30, 2002, the engraving, casting, or tamping (impressing) of the serial number must be to a minimum depth of .003 inch and in a print size o smaller than 1/16 inch; and

(2) By engraving, casting, stamping (impressing), or otherwise conspicuously placing or causing to be ngraved, cast, stamped (impressed), or placed on the frame, receiver, or barrel thereof certain additional nformation. This information must be placed in a manner not susceptible of being readily obliterated, ltered or removed. For firearms manufactured, imported, or made on and after January 30, 2002, the ngraving, casting, or stamping (impressing) of this information must be to a minimum depth of .003

inch. The additional information includes:

(i) The model, if such designation has been made;

(U) The caliber or gauge;

(iii) Your name (or recognized abbreviation) and also, when applicable, the name of the foreign manufacturer or maker;

(iv) In the case of a domestically made firearm, the city and State (or recognized abbreviation thereof where you as the manufacturer maintain your place of business, or where you, as the maker, made the firearm; and

(v) In the case of an imported firearm, the name of the country in which it was manufactured and the city and State (or recognized abbreviation thereof) where you as the importer maintain your place of business. For additional requirements relating to imported firearms, see Customs regulations at 19 CFR part 134.

(b) The depth of all markings required by this section will be measured from the flat surface of the metal and not the peaks or ridges. The height of serial numbers required by paragraph (a)(1) of this section will be measured as the distance between the latitudinal ends of the character impression bottoms (bases).

(c) The Director may authorize other means of identification upon receipt of a letter application from you, submitted in duplicate, showing that such other identification is reasonable and will not hinder the effective administration of this part.

(d) In the case of a destructive device, the Director may authorize other means of identifying that weapon upon receipt of a letter application from you, submitted in duplicate, showing that engraving, casting, or stamping (impressing) such a weapon would be dangerous or impracticable.

(e) A firearm frame or receiver that is not a component part of a complete weapon at the time it is sold, shipped, or otherwise disposed of by you must be identified as required by this section.

(f)(1) Any part defined as a machine gun, muffler, or silencer for the purposes of this part that is not component part of a complete firearm at the time it is sold, shipped, or otherwise disposed of by you must be identified as required by this section.

(2) The Director may authorize other means of identification of parts defined as machine guns other than frames or receivers and parts defined as mufflers or silencers upon receipt of a letter application from you, submitted in duplicate, showing that such other identification is reasonable and will not hind the effective administration of this part.

(Approved by the Office of Management and Budget under control number 1512--D550)

§ 419.103 Registration of firearms manufactured.

Each manufacturer qualified under this part shall file with the Director an accurate notice on Form 2 Firearms), Notice of Firearms Manufactured or Imported, executed under the penalties of perjury, to how his manufacture of firearms. The notice shall set forth the name and address of the manufacturer, lentify his special (occupational) tax stamp and Federal firearms license, and show the date of nanufacture, the type, model, length of barrel, overall length, caliber, gauge or size, serial numbers, and ther marks of identification of the firearms he manufactures, and the place where the manufactured rearms will be kept. All firearms manufactured by him during a single day shall be included on one otice, Form 2 (Firearms), filed by the manufacturer no later than the close of the next business day. The nanufacturer shall prepare the notice, Form 2 (Firearms), in duplicate, file the original notice as rescribed herein and keep the copy with the records required by subpart I of this part at the premises overed by his special (occupational) tax stamp. Receipt of the notice, Form 2 (Firearms), by the Director shall effectuate the registration of the firearms listed on that notice. The requirements of this art relating to the transfer of a firearm are applicable to transfers by qualified manufacturers.

419.104 Registration of firearms by certain governmental entities.

Any State, any political subdivision thereof, or any official police organization of such a government ntity engaged in criminal investigations, which acquires for official use a firearm not registered to it, uch as by abandonment or by forfeiture, will register such firearm with the Director by filing Form 10 Firearms), Registration of Firearms Acquired by Certain Governmental Entities, and such registration hall become a part of the National Firearms Registration and Transfer Record. The application shall lentify the applicant, describe each firearm covered by the application, show the location where each rearm usually will be kept, and, if the firearm is unserviceable, the application shall show how the rearm was made unserviceable. This section shall not apply to a firearm merely being held for use as vidence in a criminal proceeding. The Form 10 (Firearms) shall be executed in duplicate in accordance vith the instructions thereon. Upon registering the firearm, the Director shall return the original Form 10 Firearms) to be returned to the applicant showing such approval and he will present the approved pplication, Form 6 (Firearms), to the Customs officer at the port of importation. The approval of an pplication to import a firearm shall be automatically terminated at the expiration of one year from the ate of approval unless, upon request, it is further extended by the Director. If the firearm described in ne approved application is not imported prior to the expiration of the approval, the Director shall be so otified. Customs officers will not permit release of a firearm from Customs custody, except for xportation, unless covered by an application which has been approved by the Director and which is urrently effective. The importation or bringing in of a firearm not covered by an approved application nay subject the person responsible to civil and criminal liabilities. (26 U.S.C. 5861, 5871, and 5872.)

(b) Part 478 of this chapter also contains requirements and procedures for the importation of firearms nto the United States. A firearm may not be imported into the United States under this part unless those equirements and procedures are also complied with by the person importing the firearm.

(c) The provisions of this subpart shall not be construed as prohibiting the return to the United States r any territory under its control or jurisdiction of a firearm by a person who can establish to the atisfaction of Customs that (1) the firearm was taken out of the United States or any territory under its control or jurisdiction by such person, (2) the firearm is registered to that person, and (3) if appropriate the authorization required by Part 478 of this chapter for the transportation of such a firearm in intersta or foreign commerce has been obtained by such person.

§ 479.112 Registration of imported firearms.

(a) Each importer shall file with the Director an accurate notice on Form 2 (Firearms), Notice of Firearms Manufactured or Imported, executed under the penalties of perjury, showing the importation a firearm. The notice shall set forth the name and address of the importer, identify the importer's specia (occupational) tax stamp and Federal firearms license, and show the import permit number, the date of release from Customs custody, the type, model, length of barrel, overall length, caliber, gauge or size, serial number, and other marks of identification of the firearm imported, and the place where the imported firearm will be kept. The Form 2 (Firearms) covering an imported firearm shall be filed by th importer no later than fifteen (15) days from the date the firearm was released from Customs custody. The importer shall prepare the notice, Form 2 (Firearms), in duplicate, file the original return as prescribed herein, and keep the copy with the records required by subpart I of this part at the premises covered by the special (occupational) tax stamp. The timely receipt by the Director of the notice, Form (Firearms), and the timely receipt by the Director of the copy of Form 6A (Firearms), Release and Receipt of Imported Firearms, Ammunition and Implements of War, required by § 478.112 of this chapter, covering the weapon reported on the Form 2 (Firearms) by the qualified importer, shall effectuate the registration of the firearm to the importer.

(b) The requirements of this part relating to the transfer of a firearm are applicable to the transfer of imported firearms by a qualified importer or any other person.

(c) Subject to compliance with the provisions of this part, an application, Form 6 (Firearms), to impc a firearm by an importer or dealer qualified under this part, for use as a sample in connection with sales of such firearms to Federal, State or local governmental entities, will be approved if it is established by specific information attached to the application that the firearm is suitable or potentially suitable for us by such entities. Such information must show why a sales sample of a particular firearm is suitable for such use and the expected governmental customers who would require a demonstration of the firearm. Information as to the availability of the firearm to fill subsequent orders and letters from governmental entities expressing a need for a particular model or interest in seeing a demonstration. of a particular firearm would establish suitability for governmental use. Applications to import more than one firearm of a particular model for use as a sample by an importer or dealer must also establish the importer's or dealer's need for the quantity of samples sought to be imported.

(d) Subject to compliance with the provisions of this part, an application, Form 6 (Firearms), to impc a firearm by an importer or dealer qualified under this part, for use as a sample in connection with sales of such firearms to Federal, State or local governmental entities, will be approved if it is established by specific information attached to the application that the firearm is particularly suitable for use by such entities. Such information must show why a sales sample of a particular firearm is suitable for such use and the expected governmental customers who would require a demonstration of the firearm. Information as to the availability of the firearm to fill subsequent orders and letters from governmental entities expressing a need for a particular model or interest in seeing a demonstration of a particular firearm would establish suitability for governmental use. Applications to import more than one firearm of a particular model for use as a sample by an importer or dealer must also establish the importer's or dealer's need for the quantity of samples sought to be imported.

§ 479.113 Conditional importation.

The Director shall permit the conditional importation or bringing into the United States of any firearm for the purpose of examining and testing the firearm in connection with making a determination as to whether the importation or bringing in of such firearm will be authorized under this subpart. An application under this section shall be filed on Form 6 (Firearms), in triplicate, with the Director. The Director may impose conditions upon any importation under this section including a requirement that the firearm be shipped directly from Customs custody to the Director and that the person importing or bringing in the firearm must agree to either export the weapon or destroy it if a final determination is made that it may not be imported or brought in under this subpart. A firearm so imported or brought into the United States may be released from Customs custody in the manner prescribed by the conditional authorization of the registrant with notification thereon that registration of the firearm has been made. The registration of any firearm under this section is for official use only and a subsequent transfer will be approved only to other governmental entities for official use.

Machine Guns

§ 479.105 Transfer and possession of machine guns.

(a) General. As provided by 26 U.S.C. 5812 and 26 U.S.C. 5822, an application to make or transfer a firearm shall be denied if the making, transfer, receipt, or possession of the firearm would place the maker or transferee in violation of law. Section 922(0), Title 18, U.S.C., makes it unlawful for any person to transfer or possess a machine gun, except a transfer to or by, or possession by or under the authority of, the United States or any department or agency thereof or a State, or a department, agency, or political subdivision thereof; or any lawful transfer or lawful possession of a machine gun that was lawfully possessed before May 19, 1986. Therefore, notwithstanding any other provision of this part, no application to make, transfer, or import a machine gun will be approved except as provided by this section.

(b) Machine guns lawfully possessed prior to May 19, 1986. A machine gun possessed in compliance with the provisions of this part prior to May 19, 1986, may continue to be lawfully possessed by the person to whom the machine gun is registered and may, upon compliance with the provisions of this part, be lawfully transferred to and possessed by the transferee.

(c) Importation and manufacture. Subject to compliance with the provisions of this part, importers and manufacturers qualified under this part may import and manufacture machine guns on or after May 19, 1986, for sale or distribution to any department or agency of the United States or any State or political subdivision thereof, or for use by dealers qualified under this part as sales samples as provided in paragraph (d) of this section. The registration of such machine guns under this part and their subsequent transfer shall be conditioned upon and restricted to the sale or distribution of such weapons for the official use of Federal, State or local governmental entities. Subject to compliance with the provisions of this part, manufacturers qualified under this part may manufacture machine guns on or after May 19, 1986, for exportation in compliance with the Arms Export Control Act (22 U.S.C. 2778) and regulations prescribed thereunder by the Department of State.

(d) Dealer sales samples. Subject to compliance with the provisions of this part, applications to transfer and register a machine gun manufactured or imported on or after May 19, 1986, to dealers qualified under this part will be approved if it is established by specific information the expected governmental customers who would require a demonstration of the weapon, information as to the availability of the machine gun to fill subsequent orders, and letters from governmental entities expressing a need for a particular model or interest in seeing a demonstration of a particular weapon. Applications to transfer more than one machine gun of a particular model to a dealer must also establish the dealer's need for the quantity of samples sought to be transferred.

(e) The making of machine guns on or after May 19, 1986. Subject to compliance with the provisions of this part, applications to make and register machine guns on or after May 19, 1986, for the benefit of a Federal, State or local governmental entity (e.g., an invention for possible Mure use of a governmental entity or the making of a weapon in connection with research and development on behalf of such an entity) will be approved if it is established by specific information that the machine gun is particularly suitable for use by Federal, State or local governmental entities and that the making of the weapon is at the request and on behalf of such an entity .

(f) Discontinuance of business. Since section 922(0), Title 18, U.S.C., makes it unlawful to transfer or possess a machine gun except as provided in the law, any qualified manufacturer, importer, or dealer intending to discontinue business shall, prior to going out of business, transfer in compliance with the provisions of this part any machine gun manufactured or imported after May 19, 1986, to a Federal, State or local governmental entity, qualified manufacturer, qualified importer, or, subject to the provisions of paragraph (d) of this section, dealer qualified to possess such, machine gun.

Subpart H-Importation and Exportation

Importation

§479.111 Procedure.

(a) No firearm shall be imported or brought into the United States or any territory under its control or jurisdiction unless the person importing or bringing in the firearm establishes to the satisfaction of the Director that the firearm to be imported or brought in is being imported or brought in for:

(1) The use of the United States or any department, independent establishment, or agency thereof or any State or possession or any political subdivision thereof; or

(2) Scientific or research purposes; or

(3) Testing or use as a model by a registered manufacturer or solely for use as a sample by a registered importer or registered dealer.

The burden of proof is affirmatively on any person importing or bringing the firearm into the United States or any territory under its control or jurisdiction to show that the firearm is being imported or rought in under one of the above paragraphs. Any person desiring to import or bring a firearm into the Jnited States under this paragraph shall file with the Director an application on Form 6 (Firearms), application and Permit for Importation of Firearms, Ammunition and Implements of War, in triplicate, xecuted under the penalties of perjury. The application shall show the information required by subpart i of Part 478 of this chapter. A detailed explanation of why the importation of the firearm falls within ne standards set out in this paragraph shall be attached to the application. The person seeking to import r bring in the firearm will be notified of the approval or disapproval of his application. If the pplication is approved, the original Form 6 (Firearms) will be returned to the applicant showing such pproval and he will present the approved application, Form 6 (Firearms), to the Customs officer at the ort of importation. The approval of an application to import a firearm shall be automatically terminated t the expiration of one year from the date of approval unless, upon request, it is further extended by the Director. If the firearm described in the approved application is not imported prior to the expiration of ne approval, the Director shall be so notified. Customs officers will not permit release of a firearm from Customs custody, except for exportation, unless covered by an application which has been approved by ne Director and which is currently effective. The importation or bringing in of a firearm not covered by n approved application may subject the person responsible to civil and riminal liabilities. (26 U.S.C. 5861, 5871, and 5872.)

b) Part 478 of this chapter also contains requirements and procedures for the importation of firearms nto the United States. A firearm may not be imported into the United States under this part unless those equirements and procedures are also complied with by the person importing the firearm.

c) The provisions of this subpart shall not be construed as prohibiting the return to the United States or ny territory under its control or jurisdiction of a firearm by a person who can establish to the atisfaction of Customs that **(1)** the firearm was taken out of the United States or any territory under its ontrol or jurisdiction by such person, **(2)** the firearm is registered to that person, and **(3)** if appropriate, ne authorization required by Part 478 of this chapter for the transportation of such a firearm in interstate r foreign commerce has been obtained by such person.

479.112 Registration of imported firearms.

a) Each importer shall file with the Director an accurate notice on Form 2 (Firearms), Notice of irearms Manufactured or Imported, executed under the penalties of perjury, showing the importation of firearm. The notice shall set forth the name and address of the importer, identify the importer's special occupational) tax stamp and Federal firearms license, and show the import permit number, the date of elease from Customs custody, the type, model, length of barrel, overall length, caliber, gauge or size, erial number, and other marks of identification of the firearm imported, and the place where the mported firearm will be kept. The Form 2 (Firearms) covering an imported firearm shall be filed by the mporter no later than fifteen (15) days from the date the firearm was released from Customs custody. he importer shall prepare the notice, Form 2 (Firearms), in duplicate, file the original return as rescribed herein, and keep the copy with the records require by subpart I of this part at the premises overed by the special (occupational) tax stamp. The timely receipt by the Director of the notice, Form 2 Firearms), and the timely receipt by the Director of the copy of Form 6A (Firearms), Release and eceipt of Imported Firearms, Ammunition and Implements of War, required by § 478.112 of this napter, covering the weapon reported on the Form 2 (Firearms) by the qualified importer, shall ffectuate the registration of the firearm to the importer.

(b) The requirements of this part relating to the transfer of a firearm are applicable to the transfer of imported firearms by a qualified importer or any other person.

(c) Subject to compliance with the provisions of this part, an application, Form 6 (Firearms), to import firearm by an importer or dealer qualified under this part, for use as a sample in connection with sales c such firearms to Federal, State or local governmental entities, will be approved if it is established by specific information attached to the application that the firearm is suitable or potentially suitable for us by such entities. Such information must show why a sales sample of a particular firearm is suitable for such use and the expected governmental customers who would require a demonstration of the firearm. Information as to the availability of the firearm to fill subsequent orders and letters from governmental entities expressing a need for a particular model or interest in seeing a demonstration of a particular firearm would establish suitability for governmental use. Applications to import more than one firearm of a particular model for use as a sample by an importer or dealer must also establish the importer's or dealer's need for the quantity of samples sought to be imported.

(d) Subject to compliance with the provisions of this part, an application, Form 6 (Firearms), to import firearm by an importer or dealer qualified under this part, for use as a sample in connection with sales c such firearms to Federal, State or local governmental entities, will be approved if it is established by specific information attached to the application that the firearm is particularly suitable for use by such entities. Such information must show why a sales sample of a particular firearm is suitable for such use and the expected governmental customers who would require a demonstration of the firearm. Information as to the availability of the firearm to fill subsequent orders and letters from governmental entities expressing a need for a particular model or interest in seeing a demonstration of a particular firearm would establish suitability for governmental use. Applications to import more than one firearm of a particular model for use as a sample by an importer or dealer must also establish the importer's or dealer's need for the quantity of samples sought to be imported.

§ 479.113 Conditional importation.

The Director shall permit the conditional importation or bringing into the United States of any firearm for the purpose of examining and testing the firearm in connection with making a determination as to whether the importation or bringing in of such firearm will be authorized under this subpart. An application under this section shall be filed on Form 6 (Firearms), in triplicate, with the Director. The Director may impose conditions upon any importation under this section including a requirement that the firearm be shipped directly from Customs custody to the Director and that the person importing or bringing in the firearm must agree to either export the weapon or destroy it if a final determination is made that it may not be imported or brought in under this subpart. A firearm so imported or brought in the United States may be released from Customs custody in the manner prescribed by the conditional authorization of the Director.

Exportation

§ 479.114 Application and permit for exportation of firearms.

Any person desiring to export a firearm without payment of the transfer tax must file with the Direct an application on Form 9 (Firearms), Application and Permit for Exportation of Firearms, in uadruplicate, for a permit providing for deferment of tax liability. Part 1 of the application shall show ne name and address of the foreign consignee, number of firearms covered by the application, the ntended port of exportation, a complete description of each firearm to be exported, the name, address, tate Department license number (or date of application if not issued), and identification of the special occupational) tax stamp of the transferor. Part 1 of the application shall be executed under the penalties f perjury by the transferor and shall be supported by a certified copy of a written order or contract of ale or other evidence showing that the firearm is to be shipped to a foreign designation. Where it is esired to make a transfer free of tax to another person who in turn will export the firearm, the transferor hall likewise file an application supported by evidence that the transfer will start the firearm in course f exportation, except, however, that where such transferor and exporter are registered special taxpayers ne transferor will not be required to file an application on Form 9 (Firearms).

479.115 Action by Director.

If the application is acceptable, the Director will execute the permit, Part 2 of Form 9 (Firearms), to xport the firearm described on the form and return three copies thereof to the applicant. Issuance of the ermit by the Director will suspend assertion of tax liability for a period of six (6) months from the date f issuance. If the application is disapproved, the Director will indicate thereon the reason for such ction and return the forms to the applicant.

479.116 Procedure by exporter.

Shipment may not be made until the permit, Form 9 (Firearms), is received from the Director. If xportation is to be made by means other than by parcel post, two copies of the form must be addressed the District Director of Customs at the port of exportation, and must precede or accompany the hipment in order to permit appropriate inspection prior to lading. If exportation is to be made by parcel ost, one copy of the form must be presented to the postmaster at the office receiving the parcel who vill execute Part 4 of such form and return the form to the exporter for transmittal to the Director. In the vent exportation is not effected, all copies of the form must be immediately returned to the Director for ancellation.

479.117 Action by Customs.

Upon receipt of a permit, Form 9 (Firearms), in duplicate, authorizing the exportation of firearms, the District Director of Customs may order such inspection as deemed necessary prior to lading of the nerchandise. If satisfied that the shipment is proper and the information contained in the permit to xport is in agreement with information shown in the shipper's export declaration, the District Director f Customs will, after the merchandise has been duly exported, execute the certificate of exportation Part 3 of Form 9 (Firearms». One copy of the form will be retained with the shipper's export declaration nd the remaining copy thereof win be transmitted to the Director.

479.118 Proof of exportation.

Within a six-month's period from date of issuance of the permit to export firearms, the exporter shall urnish or cause to be furnished to the Director (a) the certificate of exportation (Part 3 of Form 9

(Firearms» executed by the District Director of Customs as provided in § 479.117, or (b) the certificate of mailing by parcel post (Part 4 of Form 9 (Firearms» executed by the postmaster of the post office receiving the parcel containing the firearm, or (c) a certificate of lading executed by a Customs officer the foreign country to which the firearm is exported, or Cd) a sworn statement of the foreign consignee covering the receipt of the firearm, or (e) the return receipt, or a reproduced copy thereof, signed by the addressee or his agent, where the shipment of a firearm was made by insured or registered parcel post. Issuance of a permit to export a firearm and furnishing of evidence establishing such exportation under this section will relieve the actual exporter and the person selling to the exporter for exportation from transfer tax liability. Where satisfactory evidence of exportation of a firearm is not furnished within the stated period, the transfer tax will be assessed.

§ 479.119 Transportation of firearms to effect exportation.

Notwithstanding any provision of § 478.28 of this chapter, it shall not be required that authorization b obtained from the Director for the transportation in interstate or foreign commerce of a firearm in order to effect the exportation of a firearm authorized under the provisions of this subpart.

§479.120 Refunds.

Where, after payment of tax by the manufacturer, a firearm is exported, and satisfactory proof of exportation (see § 479.118) is furnished, a claim for refund may be submitted on Form 843 (see § 479.172). If the manufacturer waives all claim for the amount to be refunded, the refund shall be made to the exporter. A claim for refund by an exporter of tax paid by a manufacturer should be accompanie by waiver of the manufacturer and proof of tax payment by the latter.

§ 479.121 Insular possessions.

Transfers of firearms to persons in the insular possessions of the United States are exempt from transfer tax, provided title in cases involving change of title (and custody or control, in cases not involving change of title), does not pass to the transferee or his agent in the United States. However, such exempt transactions must be covered by approved permits and supporting documents corresponding to those required in the case of firearms exported to foreign countries (see §§ 479.114 ar 479.115), except that the Director may vary the requirements herein set forth in accordance with the requirements of the governing authority of the insular possession. Shipments to the insular possessions will not be authorized without compliance with the requirements of the governing authorities thereof. I the case of a nontaxable transfer to a person in such insular possession, the exemption extends only to such transfer and not to prior transfers.

Arms Export Control Act

§ 479.122 Requirements.

(a) Persons engaged in the business of importing firearms are required by the Arms Export Control Act (22 U.S.C. 2778) to register with the Director. (See Part 447 of this chapter.)

(b) Persons engaged in the business of exporting firearms caliber .22 or larger are subject to the requirements of a license issued by the Secretary of State. Application for such license should be made to the Office of Munitions Control, Department of State, Washington, DC 20502, prior to exporting firearms.

Subpart I-Records and Returns

§ 479.131 Records.

For the purposes of this part, each manufacturer, importer, and dealer in firearms shall keep and maintain such records regarding the manufacture, importation, acquisition (whether by making, transfer, or otherwise), receipt, and disposition of firearms as are prescribed, and in the manner and place required, by part 478 of this chapter. In addition, each manufacturer, importer, and dealer shall maintain, in chronological order, at his place of business a separate record consisting of the documents required by this part showing the registration of any firearm to him. If firearms owned or possessed by a manufacturer, importer, or dealer are stored or kept on premises other than the place of business shown on his special (occupational) tax stamp, the record establishing registration shall show where such firearms are stored or kept. The records required by this part shall be readily accessible for inspection at all reasonable times by ATF officers.

(Approved by the Office of Management and Budget under control number *1512-0387)*

Subpart J-Stolen or Lost Firearms or Documents

§ 479.141 Stolen or lost firearms.

Whenever any registered firearm is stolen or lost, the person losing possession thereof will, immediately upon discovery of such theft or loss, make a report to the Director showing the following: (a) Name and address of the person in whose name the firearm is registered, (b) kind of firearm, (c) serial number, (d) model, (e) caliber, (I) manufacturer of the firearm, (g) date and place of theft or loss, and (h) complete statement of facts and circumstances surrounding such theft or loss.

§ 479.142 Stolen or lost documents.

When any Form 1,2,3,4,5, 6A, or 10 (Firearms) evidencing possession of a firearm is stolen, lost, or destroyed, the person losing possession will immediately upon discovery of the theft, loss, or destruction report the matter to the Director. The report will show in detail the circumstances of the theft, loss, or destruction and will include all known facts which may serve to identify the document. Upon receipt of the report, the Director will make such investigation as appears appropriate and may issue a duplicate document upon such conditions as the circumstances warrant.

Subpart K-Examination of Books and Records

§ 479.151 Failure to make returns: Substitute returns.

If any person required by this part to make returns shall fail or refuse to make any such return within the time prescribed by this part or designated by the Director, then the return shall be made by an ATF officer upon inspection of the books, but the making of such return by an A TF officer shall not relieve the person from any default or penalty incurred by reason of failure to make such return.

§ 479.152 Penalties (records and returns).

Any person failing to keep records or make returns, or making, or causing the making of, a false entry on any application, return or record, knowing such entry to be false, is liable to fine and imprisonment as provided in section 5871, I.R.C.

Subpart L-Distribution and Sale of Stamps

§ 479.161 stamps.

National Firearms Act

"National Firearms Act" stamps evidencing payment of the transfer tax or tax on the making of a firearm are maintained by the Director. The remittance for purchase of the appropriate tax stamp shall be submitted with the application. Upon approval of the application, the Director will cause the appropriate tax to be paid by affixing the appropriate stamp to the application.

§ 479.162 Stamps authorized.

Adhesive stamps of the $5 and $200 denomination, bearing the words "National Firearms Act," have been prepared and only such stamps shall be used for the payment of the transfer tax and for the tax on the making of a firearm.

§ 479.163 Reuse of stamps prohibited.

A stamp once affixed to one document cannot lawfully be removed and affixed to another. Any person willfully reusing such a stamp shall be subject to the penalty prescribed by 26 U.S.C. 7208.

Subpart M-Redemption of or Allowance for Stamps or Refunds

§ 479.171 Redemption of or allowance for stamps.

Where a National Firearms Act stamp is destroyed, mutilated or rendered useless after purchase, and before liability has been incurred, such stamp may be redeemed by giving another stamp in lieu thereof. Claim for redemption of the stamp should be filed on ATF Form 2635 (5620.8) with the Director. Such claim shall be accompanied by the stamp or by a satisfactory explanation of the reasons why the stamp cannot be returned, and shall be filed within 3 years after the purchase of the stamp.

§ 479.172 Refunds.

As indicated in this part, the transfer tax or tax on the making of a firearm is ordinarily paid by the purchase and affixing of stamps, while special tax stamps are issued in payment of special ɔccupational) taxes. However, in exceptional cases, transfer tax, tax on the making of firearms, and/or pecial (occupational) tax may be paid pursuant to assessment. Claims for refunds of such taxes, paid ursuant to assessment, shall be filed on ATF Form 2635 (5620.8) within 3 years next after payment of ıe taxes. Such claims shall be filed with the regional director (compliance) serving the region in which ıe tax was paid. (For provisions relating to hand-carried documents and manner of filing, see 26 CFR 01.60911 (b) and 301.6402-2(a).) When an applicant to make or transfer a firearm wishes a refund of ıe tax paid on an approved application where the firearm was not made pursuant to an approved Form 1 Firearms) or transfer of the firearm did not take place pursuant to an approved Form 4 (Firearms), the pplicant shall file a claim for refund of the tax on ATF Form 2635 (5620.8) with the Director. The laim shall be accompanied by the approved application bearing the stamp and an explanation why the ax liability was not incurred. Such claim shall be filed within 3 years next after payment of the tax.

ubpart N-Penalties and Forfeitures

479.181 Penalties.

Any person who violates or fails to comply with the requirements of 26 U.S.C. Chapter 53 shall, upon onviction, be subject to the penalties imposed under 26 U.S.C. 5871.

479.182 Forfeitures.

Any firearm involved in any violation of the provisions of 26 U.S.C. Chapter 53, shall be subject to eizure, and forfeiture under the internal revenue laws: Provided, however, That the disposition of ɔrfeited firearms shall be in conformance with the requirements of 26 U.S.C. 5872. In addition, any essel, vehicle or aircraft used to transport, carry, convey or conceal or possess any firearm with respect which there has been committed any violation of any provision of 26 U.S.C. Chapter 53, or the egulations in this part issued pursuant thereto, shall be subject to seizure and forfeiture under the ustoms laws, as provided by the act of August 9, 1939 (49 U.S.C. App., Chapter 11).

ubpart O-Other Laws Applicable

479.191 Applicability of other provisions of internal revenue laws.

All of the provisions of the internal revenue laws not inconsistent with the provisions of 26 U.S.C. hapter 53 shall be applicable with respect to the taxes imposed by 26 U.S.C. 5801, 5811, and 5821 (see 6 U.S.C.5846).

479.192 Commerce in firearms and ammunition.

For provisions relating to commerce in firearms and ammunition, including the movement of estructive devices, machine guns, short-barreled shotguns, or short-barreled rifles, see 18 U.S.C. hapter 44, and Part 478 of this chapter issued pursuant thereto.

479.193 Arms Export Control

For provisions relating to the registration and licensing of persons engaged in the business of manufacturing, importing or exporting arms, ammunition, or implements of war, see the Arms Export Control Act (22 U.S.C. 2778), and the regulations issued pursuant thereto. (See also Part 447 of this chapter.)

APPENDIX B

RULINGS AND ATF ARTICLES

7 CFR 479.11[232]: MEANING OF TERMS

'he AR15 auto sear is a machine gun as defined by 26 U.S.C. 5845(b)

.TF Rul. 81-4

'he Bureau of Alcohol, Tobacco and Firearms has examined an auto sear known by the various trade ames including "AR15 Auto Sear," "Drop In Auto Sear," and "Auto Sear II," which consists of a sear ıounting body, sear, return spring, and pivot pin. The Bureau finds that the single addition of this auto ear to certain AR15 type semiautomatic rifles, manufactured with M16 internal components already ıstalled, will convert such rifles into machine guns.

'he National Firearms Act, 26 U.S.C. 5845(b) defines "machine gun" to include any combination of arts designed and intended for use in converting a weapon to shoot automatically more than one shot, ithout manual reloading, by a single function of the trigger.

Ield: The auto sear known by various trade names including "AR15 Auto Sear," "Drop In Auto Sear," nd "Auto Sear II," is a combination of parts designed and intended for use in converting a weapon to hoot automatically more than one shot, without manual reloading, by a single function of the trigger. 'onsequently, the auto sear is a machine gun as defined by 26 U.S.C. 5845(b).

Vith respect to the machine gun classification of the auto sear under the National Firearms Act, ursuant to 26 U.S.C. 7805(b), this ruling will not be applied to auto sears manufactured before Iovember 1, 1981. Accordingly, auto sears manufactured on or after November 1, 1981, will be subject) all of the provisions of the National Firearms Act and 27 C.F.R. Part 479.

.ditor's Note: Regardless of the date of manufacture of a drop in auto sear, possession of such a ear and certain M16 fire control parts is possession of a machine gun as defined by the NFA. pecifically, these parts are a combination of parts designed and intended for use in converting a 'eapon into a machine gun as defined in the NFA. (See "Important Information Concerning .R15-Type rifles" which follows)

MPORTANT INFORMATION CONCERNING AR-15 TYPE RIFLES

.TF has encountered various AR-15 type rifles such as those manufactured by Colt, E.A. Company, GW, Sendra and others which have been assembled with fire control components designed for use in 116 machine guns.

---

[2] The citations to 27 CFR part 179 have been rewritten as 27 CFR part 479 in conformance with the reorganization of title 7 CFR which became effective on January 24, 2003.

The vast majority of these rifles which have been assembled with an M16 bolt carrier, hammer, trigger disconnector and selector will fire automatically merely by manipulation of the selector or removal of the disconnector. Many of these rifles using less than the 5 M16 parts listed above will also shoot automatically by manipulation of the selector or removal of the disconnector.

Any weapon which shoots automatically, more than 1 shot, without manual reloading, by a single function of the trigger, is a machine gun as defined in 26 U.S.C. 5845(b), the National Firearms Act (NFA). In addition, the definition of machine gun also includes any combination of parts from which a machine gun may be assembled, if such parts are in possession or under the control of a person. An AR 15 type assault rifle which fires more than one shot by a single function o the trigger is a machine gun under the NFA. Any machine gun is subject to the NFA and the possession of an unregistered machine gun could the possessor to criminal prosecution.

Additionally, these rifles could pose a safety hazard in that they may fire automatically without the user being aware that the weapon will fire more than 1 shot with a single pull of the trigger.

In order to avoid violations of the NFA, M16, hammers, triggers, disconnectors, selectors and bolt carriers must not be used in assembly of AR-15 type semiautomatic rifles, unless the M16 parts have been modified to AR-15 Model SP1 configuration. Any AR-15 type rifles which have been assembled with M16 internal components should have those parts removed and replaced with AR-15 Model SP1 type parts which are available commercially. The M16 components also may be modified to AR-15 Model SP1 configuration.

It is important to note that any modification of the M16 parts should be attempted by fully qualified personnel only.

Should you have any questions concerning AR-15 type rifles with M16 parts, please contact your nearest ATF office. Our telephone numbers are listed in the **"United States Government"** section of your telephone directory under the **"United States Department of Justice."**

27 CFR 479.11: MEANING OF TERMS

The KG-9 pistol is a machinegun as defined in the National Firearms Act.

ATF Rul. 82-2

The Bureau of Alcohol, Tobacco and Firearms has examined a firearm identified as the KG-9 pistol. The KG-9 is a 9 millimeter caliber, semiautomatic firearm which is blowback operated and which fires from the open bolt position with the bolt incorporating a fixed firing pin. In addition, a component part of the weapon is a disconnector which prevents more than one shot being fired with a single function of the trigger.

The disconnector is designed in the KG-9 pistol in such a way that a simple modification to it, such as cutting, filing, or grinding, allows the pistol to operate automatically. Thus, this simple modification to the disconnector together with the configuration of the above design features (blowback operation, firing from the open bolt position, and fixed firing pin) in the KG-9 permits the firearm to shoot automatically more than one shot, without manual reloading, by a single function of the trigger. The above combination of design features as employed in the KG-9 is normally not found in the typical sporting firearm.

The National Firearms Act, 26 U.S.C. 5845(b), defines a machine gun to include any weapon which shoots, is designed to shoot, or can be readily restored to shoot, automatically more than one shot, without manual reloading, by a single function of the trigger.

The "shoots automatically" definition covers weapons that will function automatically. The "readily restorable" definition defines weapons which previously could shoot automatically but will not in their present condition. The "designed" definition includes those weapons which have not previously functioned as machine guns but possess design features which facilitate full automatic fire by simple modification or elimination of existing component parts.

Held: The KG-9 pistol is designed to shoot automatically more than one shot, without manual reloading, by a single function of the trigger. Consequently, the KG-9 pistol is a machine gun as defined in section 5845(b) of the Act.

With respect to the machine gun classification of the KG-9 pistol under the National Firearms Act, pursuant to 26 U.S.C. § 7805(b), this ruling will not be applied to KG-9 pistols manufactured before January 19, 1982. Accordingly, KG-9 pistols manufactured on or after January 19, 1982, will be subject to all the provisions of the National Firearms Act and 27 C.F.R., Part 479.

27 CFR 479.11: MEANING OF TERMS

The SM10 and SM11A1 pistols and SAC carbines are machineguns as defined in the National Firearms Act.

ATF Rul. 82-8

The Bureau of Alcohol, Tobacco and Firearms has reexamined firearms identified as SM10 pistols, SM11A1 pistols, and SAC carbines .The SM10 is a 9 millimeter or .45ACP caliber, semiautomatic firearm; the SM11A1 is a .380ACP caliber, semiautomatic firearm. And the SAC carbine is a 9 millimeter or .45ACP caliber, semiautomatic firearm. The weapons are blowback operated, fire from the open bolt position with the bolt incorporating a fixed firing pin, and the barrels of the pistols are threaded to accept a silencer. In addition, component parts of the weapons are a disconnector and a trip which prevent more than one shot being fired with a single function of the trigger.

The disconnector and trip are designed in the SM10 and SM11A1 pistols and in the SAC carbine (firearms) in such a way that a simple modification to them, such as cutting, filing, or grinding, allows the firearms to operate automatically. Thus, this simple modification to the disconnector or trip together with the configuration of the above design features (blowback operation, firing from the open bolt position, and fixed firing pin) in the SM10 and SM11A1pistols and in the SAC carbine, permits the firearms to shoot automatically, more than one shot, without manual reloading, by a single function of the trigger. The above combination of design features as employed in the SM10 and SM11A1 pistols and the SAC carbine are normally not found in the typical sporting firearm.

The National Firearms Act, 26 U.S.C. § 5845(b), defines a machine gun to include any weapon which shoots, is designed to shoot, or can be readily restored to shoot, automatically more than one shot, without manual reloading, by a single function of the trigger.

The "shoots automatically" definition covers weapons that will function automatically. The "readily restorable" definition defines weapons which previously could shoot automatically but will not in their present condition. The "designed" definition includes those weapons which have not previously functioned as machine guns but possess design features which facilitate full automatic fire by simple modification or elimination of existing component parts.

Held: The SM10 and SM11A1 pistols and the SAC carbine are designed to shoot automatically more than one shot, without manual reloading, by a single function of the trigger. Consequently, the SM10 and SM11A1 pistols and SAC carbines are machine guns as defined in Section 5845(b) of the Act.

With respect to the machine gun classification of the SM10 and SM11A1 pistols and SAC carbines, under the National Firearms Act, pursuant to 26 U.S.C. 7805(b), this ruling will not be applied to SM10 and SM11A1 pistols and SAC carbines manufactured or assembled before June, 21, 1982. Accordingly, SM10 and SM11A1 pistols and SAC carbines, manufactured or assembled on or after June 21, 1982, will be subject to all the provisions of the National Firearms Act and 27 C.F.R., Part 479.

27 CFR 479.11: MEANING OF TERMS

The YAC STEN MK II carbine is a machine gun as defined in the National Firearms Act.

ATF Rul. 83-5

The Bureau of Alcohol, Tobacco and Firearms has examined a firearm identified as YAC STEN MK II carbine. The YAC STEN MKII carbine is a 9 millimeter caliber, semiautomatic firearm which has identical design characteristics to the original selective fire STEN submachine gun designed by Reginald Vernon Shepherd and Harold John Turpin. The weapon is blowback operated and fires from the open bolt position with the bolt incorporating a fixed firing pin. In addition, a component part of the weapon is a trip lever (disconnector) which has been modified to prevent more than one shot being fired with a single function of the trigger.

The trip lever (disconnector) is designed in such a way that a simple modification to it, such as bending, breaking or cutting, allows the weapon to operate automatically. Thus, this simple modification to the trip lever (disconnector) or trip together with STEN submachine gun design features and components in the YAC STEN MK II carbine, permits the firearm to shoot automatically, more than one shot, without manual reloading, by a single function of the trigger. The above combination of machine gun design features as employed in the YAC STEN MK II carbine are normally not found in the typical sporting firearm.

The National Firearms Act, 26 U.S.C. 5845(b), defines a machine gun to include any weapon which shoots, is designed to shoot, or can be readily restored to shoot, automatically more than one shot, without manual reloading, by a single function of the trigger.

'he "shoots automatically" definition covers weapons that will function automatically. The "readily :storable" definition defines weapons which previously could shoot automatically but will not in their resent condition. The "designed" definition includes those weapons which have not previously inctioned as machine guns but possess specific machine gun design features which facilitate full utomatic fire by simple modification or elimination of existing component parts.

Ield: The YAC STEN MK II carbine is designed to shoot automatically more than one shot, without ianual reloading, by a single function of the trigger. Consequently, the YAC STEN MK II :miautomatic carbine is a machine gun as defined in Section 5845(b) of the Act.

7 CFR 479.111: IMPORTATION PROCEDURE

National Firearms Act (NFA) firearm may not be imported for use as a sample for sales to law nforcement agencies if the firearm is a curio or relic unless it is established that the firearm is articularly suitable for use as a law enforcement weapon.

TF Rul. 85-2

'he Bureau of Alcohol, Tobacco and Firearms has approved a number of applications to import lational Firearms Act (NFA) firearms for the use of registered importers to generate orders for such rearms from law enforcement agencies.

review of the characteristics of the NFA firearms approved for importation as sales samples indicates iat some of the firearms are not being imported for the purposes contemplated by the statute. Some of ie NFA firearms imported are, in fact, curios or relics and are more suitable for use as collector's items ian law enforcement weapons.

nportations of NFA firearms are permitted by 26 U.S.C. 5844, which provides in pertinent part:

No firearms shall be imported or brought into the United States or any territory under its control or irisdiction unless the importer establishes, under regulations as may be prescribed by the Secretary, that ie firearm to be imported or brought in is:

(1) being imported or brought in for the use of the United States or any department., independent establishment, or agency thereof or any State or possession or any political subdivision thereof; or
(2) ***
(3) being imported or brought in solely for ... use as a sample by a registered importer or registered dealer;

xcept that, the Secretary may permit the conditional importation or bringing in of a firearm for xamination and testing in connection with classifying the firearm."

'he sole purpose of the statute permitting the importation of NFA firearms as sales samples is to permit :gistered importers to generate orders for firearms from government agencies, on the basis of the ample.

The implementing regulation, 27 CFR Section 479.111, provides that the person importing or bringing firearm into the United States or any territory under its control or jurisdiction has the burden of proof t affirmatively establish that the firearm is being imported or brought in for one of the authorized purposes. In addition, a detailed explanation of why the importation falls within one of the authorized purposes must be attached to the application to import. The mere statement that an NFA firearm is bein imported as a sales sample for demonstration to law enforcement agencies does not meet the required burden of proof and is not a detailed explanation of why the importation falls within the import standards.

Held, an application to import a National Firearms Act firearm as a sample in connection with sales of such firearms to law enforcement agencies will not be approved if the firearm is determined to be a cur or relic unless it is established by specific information that the firearm is particularly suitable for use as law enforcement weapon. For example, the importer must provide detailed information as to why a sal sample of a particular weapon is suitable for law enforcement purposes and the expected customers wh would require a demonstration of the weapon. Information as to the availability of firearms to fill subsequent orders would help meet the burden of establishing use as a sales sample. Also, letters from law enforcement agencies expressing a need for a particular model or interest in seeing a demonstratio of a particular firearm would be relevant.

Editor's Note:

The importation of machine guns for use as sales samples must also meet the requirements of 27 CFR 479.105(d).

26 U.S.C. § 5845(f)(2): DESTRUCTIVE DEVICE (Nonsporting shotgun having a bore of more than one-half inch in diameter)

The USAS-12 shotgun has a bore of more than one-half inch in diameter and is not generally recognized as particularly suitable for sporting purposes. Therefore, it is classified as a destructi device for purposes of the National Firearms Act, 26 U.S.C. Chapter 53.

ATF Rul. 94-1

The Bureau of Alcohol, Tobacco and Firearms (ATF) has examined a firearm identified as the USAS-1 shotgun to determine whether it is a destructive device as that term is used in the National Firearms Ac (NFA), 26 U.S.C. Chapter 53.

The USAS-12 is a 12 gauge, gas-operated, autoloading semiautomatic shotgun which is chambered for 12 gauge 2 ¾-inch ammunition. It has an 18 ¼-inch barrel, is approximately 38 inches long, and weig 12.4 pounds unloaded and approximately 15 pounds with a loaded magazine, depending on the capacit of the magazine. The USAS-12 is equipped with a 12 round detachable box magazine, but a 28 round detachable drum magazine is also available. The shotgun is approximately 11 inches deep with a box magazine. There is an integral carrying handle on top of the receiver which houses a rifle –type apertur rear and adjustable post-type front sight. The USAS-12 has a separate combat-style pistol grip located on the bottom of the receiver, forward of the buttstock. An optional telescopic sight may be attached t the carrying handle. The barrel is located below the operating mechanism in such fashion that the barre is in a straight line with the center of the buttstock.

ection 5845(f), Title 26 U.S.C., classifies certain weapons as "destructive devices" which are subject to he registration and tax provisions of the NFA. Section 5845(f)(2) provides the following:

(f) **Destructive Device** – The term "destructive device' means* * *

2) any type of weapon by whatever name known which will, or which may be readily converted to, xpel a projectile by the action of an explosive or other propellant, the barrel or barrels of which have a ore of more than one-half inch in diameter, except a shotgun or shotgun shell the Secretary or his elegate finds is generally recognized as particularly suitable for sporting purposes.

A "sporting purpose" test which is almost identical to that in section 5845(f)(2) appears in 18 U.S.C. § 25(d)(3). This provision of the Gun Control Act of 21968 (GCA) provides that the Secretary shall uthorize a firearm to be imported into the United States if the firearm is "generally recognized as articularly suitable for or readily adaptable to sporting purposes." With the exception of the "readily daptable" language, this provision is identical to the sporting shotgun exception to the destructive evice definition. The definition of "destructive device" in the GCA (18 U.S.C. § 921(a)(4)) is identical o that in the NFA.

n determining whether shotguns with a bore diameter of more than one-half inch in diameter are generally recognized as particularly suitable for sporting purposes" and thus are not destructive devices nder the NFA, we believe it is appropriate to use the same criteria used for evaluating shotguns under e "sporting purposes" test of section 925(d)(3). Congress used virtually identical language in escribing the weapons subject to the two statutory schemes, and the language was added to the GCA nd the NFA at the same time.

n connection with the determination of importability, ATF determined that the USAS-12 shotgun was ot eligible for importation under the sporting purposes test in section 925(d)(3). In reaching this etermination, ATF evaluated the weight, size, bulk, designed magazine capacity, configuration, and ther characteristics of the USAS-12. It was determined that the weight of the USAS-12, 12.4 pounds, nade it much heavier than traditional 12 gauge sporting shotguns, which made it awkward to carry for xtended periods, as in hunting, and cumbersome to fire at multiple small moving targets, as in skeet and ap shooting. The width of the USAS-12 with drum magazine, approximately 6 inches, and the depth ith box magazine, in excess of 11 inches, far exceed that of traditional sporting shotguns, which do not xceed 3 inches in width and 4 inches in depth. The large size and bulk of the USAS-12 made it xtremely difficult to maneuver quickly enough to engage moving targets as is necessary in hunting, keet, and trap shooting. The detachable box magazine with 12 cartridge capacity and the detachable rum magazine with 28 cartridge capacity were or a larger capacity than traditional repeating sporting hotguns, which generally contain tubular magazines with a capacity of 3 – 5 cartridges. Additionally, etachable magazines permit more rapid reloading than do tubular magazines. Finally, the combat-style istol grip, the barrel-to-buttstock configuration, the bayonet lug, and the overall appearance and general hape of the weapon were radically different from traditional sporting shotguns and strikingly similar to hotguns designed specifically for or modified for combat and law enforcement use.

ection 7805(b), title 26 U.S.C., provides that the Secretary may prescribe the extent, if any, to which ny ruling relating to the internal revenue laws shall be applied without retroactive effect. Accordingly, all rulings issued under the Internal Revenue Code are applied retroactively unless they specifically provide otherwise. Pursuant to section 7805(b), the Director, as the delegate of the Secretary, may prescribe the extent to which any ruling will apply without retroactive effect.

Held: TheUSAS-12 is a shotgun with a bore of more than one-half inch in diameter which is not particularly suitable for sporting purposes. The weight, size, bulk, designed magazine capacity, configuration, and other factors indicate that the USAS-12 is a semiautomatic version of a military type assault shotgun. Accordingly, the USAS-12 is a destructive device as that term is used in 26 U.S.C. § 5845(f)(2). Pursuant to section 7805(b), this ruling is applied prospectively effective March 1, 1994, with respect to the making, transfer, and special (occupational) taxes imposed by the NFA. All other provisions of the NFA apply retroactively effective March 1, 1994.

26 U.S.C. § 5845(f)(2): DESTRUCTIVE DEVICE (Non-sporting shotgun having a bore of more than one-half inch in diameter)

The Striker-12/Streetsweeper shotgun has a bore of more than one-half inch in diameter and is not generally recognized as particularly suitable for sporting purposes. Therefore, it is classified as a destructive device for purposes of the National Firearms Act, 26 U.S.C. Chapter 53.

ATF Rul. 94-2

The Bureau of Alcohol, Tobacco and Firearms (ATF) has examined a firearm identified as the Striker-12/Streetsweeper shotgun to determine whether it is a destructive device as that term is used in the National Firearms Act (NFA), 26 U.S.C. Chapter 53.

The Striker-12 and Streetsweeper shotguns are virtually identical 12 - gauge shotguns with a spring-driven revolving magazine. The magazine has a 12 round capacity, The shotgun has a fixed stock or folding stock and may be fired with the folding stock collapsed. The shotgun with 18-inch barrel is 37 inches in length with the stock extended, and 26 .5 inches in length with the stock folded. The shotgun 5.7 inches in width and weighs 9.24 pounds unloaded. The Striker/Streetsweeper has two pistol grips, one in the center of the firearm below the buttstock, and one on the forearm. The Striker/Streetsweeper was designed and developed in South Africa as a military, security, and anti-terrorist weapon. Various types of 12-gauge cartridges can be fired from the shotgun, and a rapid indexing procedure allows various types of ammunition to be loaded into the cylinder and selected for firing. All 12 rounds can be fired from the shotgun in 3 seconds or less.

Section 5845(f), Title 26 U.S.C., classifies certain weapons as "destructive devices" which are subject the registration and tax provisions of the NFA. Section 5845(f)(2) provides as follows:

(f) **Destructive Device** – The term "destructive device" means* * *

(2) any type of weapon by whatever name known which will, or which may be readily converted to, expel a projectile by the action of an explosive or other propellant, the barrel or barrels of which have a bore of more than one-half inch in diameter, except a shotgun or shotgun shell the Secretary or his delegate finds is generally recognized as particularly suitable for sporting purposes.

A "sporting purpose" test which is almost identical to that in section 5845(f)(2) appears in 18 U.S.C. § 925(d)(3). This provision of the Gun Control Act of 21968 (GCA) provides that the Secretary shall authorize a firearm to be imported into the United States if the firearm is "generally recognized as particularly suitable for or readily adaptable to sporting purposes." With the exception of the "readily adaptable" language, this provision is identical to the sporting shotgun exception to the destructive device definition. The definition of "destructive device" in the GCA (18 U.S.C. § 921(a)(4)) is identical to that in the NFA.

In determining whether shotguns with a bore diameter of more than one-half inch in diameter are "generally recognized as particularly suitable for sporting purposes" and thus are not destructive devices under the NFA, we believe it is appropriate to use the same criteria used for evaluating shotguns under the "sporting purposes" test of section 925(d)(3). Congress used virtually identical language in describing the weapons subject to the two statutory schemes, and the language was added to the GCA and the NFA at the same time.

In 1984, ATF ruled that the Striker – 12 was not eligible for importation under section 925(d)(3) since it is not particularly suitable for or readily adaptable to sporting purposes. In making this determination, the 1984 letter-ruling notes that the Striker was being used in a number of "combat" shooting events. In a letter dated June 30, 1986, ATF again denied importation of the Striker – 12, on the basis that it did not meet the "sporting purposes" test of section 925(d)(3). This letter states that, "We believe the weapon to have been specifically designed for military and law enforcement uses."

In evaluating the physical characteristics of the Striker – 12 /Streetsweeper, ATF concludes that the weight, bulk, designed magazine capacity, configuration, and other features indicate that it was designed primarily for military and law enforcement use and is not particularly suitable for sporting purposes.

The weight of the Striker – 12/Streetsweeper, 9.24 pounds unloaded, is on the high end for traditional 12-gauge sporting shotguns, which generally weigh between 7 and 10 pounds. Thus, the weight of the Striker – 12/Streetsweepe, makes it awkward to carry for extended periods, as in hunting, and cumbersome to fire at multiple small moving targets, as in skeet and trap shooting. The width of the Striker – 12/Streetsweeper, 5.7 inches, far exceeds that of traditional sporting shotguns, which do not exceed three inches in width or four inches in depth. The large size and bulk of the Striker – 12/Streetsweeper make it extremely difficult to maneuver quickly enough to engage moving targets as is necessary in hunting, skeet and trap shooting. The spring driven revolving magazine with 12 –cartridge capacity is a much larger capacity than traditional repeating shotguns, which generally contain tubular magazines with a capacity of 3 – 5 cartridges. The folding shoulder stock and the two pistol grips are not typical of sporting-type shotguns. Finally, the overall appearance and general shape of the weapon are radically different from traditional sporting shotguns and strikingly similar to shotguns designed specifically for or modified for combat and law enforcement use.

Section 7805(b), title 26 U.S.C., provides that the Secretary may prescribe the extent, if any, to which any ruling relating to the internal revenue laws shall be applied without retroactive effect. Accordingly, all rulings issued under the Internal Revenue Code are applied retroactively unless they specifically provide otherwise. Pursuant to section 7805(b), the Director, as the delegate of the Secretary, may prescribe the extent to which any ruling will apply without retroactive effect.

Held: The Striker – 12/Streetsweeper is a shotgun with a bore of more than one-half inch in diameter which is not particularly suitable for sporting purposes. The weight, size, bulk, designed magazine capacity, configuration, and other factors indicate that the Striker – 12/Streetsweeper is a military-type shotgun, as opposed to a shotgun particularly suitable for sporting purposes. Accordingly, the Striker – 12/Streetsweeper is a destructive device as that term is used in 26 U.S.C. § 5845(f)(2). Pursuant to section 7805(b), this ruling is applied prospectively effective March 1, 1994, with respect to the making, transfer, and special (occupational) taxes imposed by the NFA. All other provisions of the NFA apply retroactively effective March 1, 1994.

18 U.S.C. § 921(a)(4)
DESTRUCTIVE DEVICE

26 U.S.C. § 5845(f)(2):
DESTRUCTIVE DEVICE

(Firearm having a bore of more than one-half inch in diameter)

37/38 mm gas/flare guns possessed with cartridges containing wood pellets, rubber pellets or balls, or bean bags are classified as destructive devices for purposes of the Gun Control Act, 18 U.S.C. Chapter 44, and the National Firearms Act, 26 U.S.C. Chapter 53.

ATF Rul. 95-3

The Bureau of Alcohol, Tobacco and Firearms (ATF) has examined various 37/38 mm gas/flare guns in combination with certain types of ammunition to determine whether these are destructive devices as defined in the Gun Control Act (GCA), 18 U.S.C. Chapter 44, and the National Firearms Act (NFA), 26 U.S.C. Chapter 53.

Section 5845(f), Title 26 U.S.C., classifies certain weapons as "destructive devices" which are subject to the registration and tax provisions of the NFA. Section 5845(f)(2) provides the following:

(f) **Destructive Device** – The term "destructive device" means* * *

(2) any type of weapon by whatever name known which will, or which may be readily converted to, expel a projectile by the action of an explosive or other propellant, the barrel or barrels of which have a bore of more than one-half inch in diameter, except a shotgun or shotgun shell the Secretary or his delegate finds is generally recognized as particularly suitable for sporting purposes: . ."

Section 5845(f)(3) excludes from the term "destructive device" any device which is neither designed or redesigned for use as a weapon and any device, although originally designed for use as a weapon, which is redesigned for use as a signaling, pyrotechnic,Line throwing, safety, or similar device.

The definition of "destructive device" in the GCA (18 U.S.C. § 921(a)(4)) is identical to that in the NFA.

ATF has previously held that devices designed for expelling tear gas or pyrotechnic signals are not weapons and are exempt from the destructive device definition. However, ammunition designed to be used against individuals is available for these 37/38 mm devices. This "anti-personnel" ammunition consists of cartridges containing wood pellets, rubber pellets or balls, and bean bags.

When a gas/flare gun is possessed with "anti-personnel" type ammunition, it clearly becomes an instrument of offensive or defensive combat and is capable of use as a weapon. Since these gas/flare guns have a bore diameter greater than one-half inch, fire a projectile by means or an explosive, and, when possessed with "anti-personnel" ammunition, are capable of use as weapons, the combination of the gas/flare gun and "anti-personnel" ammunition is a destructive as defined in the GCA and the NFA. As a result, registration as a destructive device is required. Any person possessing a gas/flare gun with which "anti-personnel" ammunition will be used must register the making of a destructive device prior to the acquisition of any "anti-personnel" ammunition. In addition, the gas/flare guns are classified as firearms as defined by the GCA when possessed with "anti-personnel" type ammunition.

Each gas/flare gun possessed with ant–personnel ammunition will be required to be identified as required by law and regulations (27 C.F.R. §§ 478.92and 479.102), including a serial number. Any person manufacturing the gas/flare gun and "anti-personnel" ammunition must, if selling them in combination, have the appropriate Federal firearms license as a manufacturer of destructive devices and must have paid the special (occupational) tax as a manufacturer of National Firearms Act firearms. Any person importing the gas/flare gun and the "anti-personnel" type ammunition must, if importing them in combination, have the appropriate Federal firearms license as an importer of destructive devices and must have paid the special (occupational) tax as an importer of National Firearms Act firearms.

Further, the "anti-personnel" ammunition to be used in the gas/flare launchers is ammunition for destructive devices for purposes of the GCA. Any person manufacturing the "anti-personnel" ammunition ,must have the appropriate Federal firearms license as a manufacturer of ammunition for destructive devices. Any person importing the "anti-personnel" ammunition must have the appropriate Federal firearms license as an importer of ammunition for destructive devices.

HELD: 37/38 mm gas/flare guns possessed with "anti-personnel" ammunition, consisting of cartridges containing wood pellets, rubber pellets or balls, or bean bags, are destructive devices as that term is used in 18 U.S.C. § 921(a)(4) and 26 U.S.C. 5845(f)(2).

26 USC 5844, 18 USC 922(o), 22 USC 2778: IMPORTATION OF BROWNING M1919 TYPE RECEIVERS FOR UNRESTRICTED COMMERCIAL SALE.

An ATF – approved method of destruction for the Browning M 1919 type machinegun will result in the severed portions of the receiver being importable for unrestricted commercial sale.

ATF Rul. 2003-1

The Bureau of Alcohol, Tobacco and Firearms (ATF) has received inquiries about modifications necessary to the receiver of a Browning M1919 type machinegun to make it importable under 26 U.S.C. 5844 and 18 U.S.C. 922(o) for unrestricted commercial sale.

The Browning M1919 is a machinegun as defined in 26 U.S.C. 5845(b). The receiver of a Browning M1919 is also a machinegun as defined. Various manufacturers made Browning M1919 style machineguns in caliber .30-06 and 7.62x51mm (.308). The M1919 is a recoil operated, belt-fed machinegun designed to be fired from a mount.

Section 5844 of title 26, United States Code, makes it unlawful to import any firearm into the United States, unless the firearm to be imported or brought in is: (1) being imported for use by the United State or any department, independent establishment, or agency thereof or any State or possession or any political subdivision thereof; or (2) the firearm is being imported for scientific or research purposes; or (3) it is being imported solely for testing or use as a model by a registered or dealer. Additionally, the Secretary may permit the conditional importation of a firearm for examination and testing in connection with classifying the firearm.

Section 922(o) of Title 18, United States Code, makes it unlawful for any person to transfer or possess machinegun, except a transfer to or by the United States or any department or agency thereof or a State or department, agency or political subdivision thereof; or any lawful transfer or lawful possession of a machinegun lawfully possessed before May 19, 1986.

A review of the statutes above indicates that machinegun and machinegun receivers cannot be lawfully imported for unrestricted commercial sale. Accordingly. Machinegun receivers may be imported for commercial sale only if they are destroyed in a manner that will prevent their function and future use as a firearm. The resulting severed receiver portions would not be subject to the provisions of 26 U.S.C. 5844 or 18 U.S.C. 922(o); however, these articles would be subject the provisions of the Arms Export Control Act, 18 U.S.C. 925, 22 U.S.C. 2778, and implementing regulation at 27 CFR Part 47. It is important to note that these machinegun receivers must be destroyed and cannot be imported whether they are serviceable or unserviceable.

An ATF-approved method of destruction for a Browning M1919 type machinegun receiver requires three diagonal torch cuts that sever or pass through the following areas: (1) the trunion or barrel mounting block (corner to corner), (2) the center area of the bolt handle slot, and (3) the cover catch an back plate spline. All cutting must be done with a cutting torch having a tip of sufficient size to displac at least ¼ inch of material at each location. Each cut must completely sever the receiver in the designated areas and must be done with a diagonal torch cut. Using a bandsaw or a cut-off wheel to destroy the receiver does not ensure destruction of the weapon.

his method of destruction is illustrated in the diagram below.

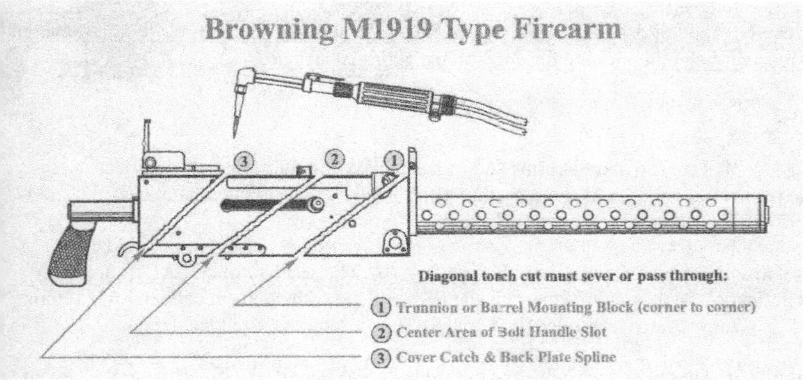

lternative methods of destruction may also be acceptable. These alternative methods must be quivalent in degree to the approved method of destruction. Receivers that are not sufficiently modified annot be approved for importation. To ensure compliance with the law, it is recommended that the nporter submit in writing the alternative method of destruction to the ATF Firearms Technology ranch (FTB) for review and approval prior to importation.

eld, an ATF-approved method of destruction for a Browning M1919 type machinegun receiver will esult in the severed portions of the receiver being importable for unrestricted commercial sale. The evered articles would not be subject to the provisions of 26 U.S.C. 5844 or 18 U.S.C. 922(o), but would ontinue to be subject to the provisions of the Arms Export Control Act, 22 U.S.C. 2778. Alternative ethods of destruction may also be acceptable. It is recommended that such methods be reviewed and pproved by the ATF Firearms Technology Branch prior to the weapon's importation.

ate signed: January 24, 2003

26 USC 5844, 18 USC 922(o), 22 USC 2778: IMPORTATION OF FN FAL TYPE RECEIVERS FOR UNRESTRICTED COMMERCIAL SALE.

An ATF – approved method of destruction for the FN FAL type machinegun will result in the severed portions of the receiver being importable for unrestricted commercial sale.

ATF Rul. 2003-2

The Bureau of Alcohol, Tobacco and Firearms (ATF) has received inquiries about modifications necessary to the receiver of an FN FAL type machinegun to make it importable under 26 U.S.C. 5844 and 18 U.S.C. 922(o) for unrestricted commercial sale.

The FN FAL is a machinegun as defined in 26 U.S.C. 5845(b). The receiver of an FAL is also a machinegun as defined. Various manufacturers made FAL style machineguns in caliber 7.62x51mm (.308). The FAL is a gas operated, shoulder fired, magazine fed, selective fire machinegun.

Section 5844 of title 26, United States Code, makes it unlawful to import any firearm into the United States, unless the firearm to be imported or brought in is: (1) being imported for use by the United State or any department, independent establishment, or agency thereof or any State or possession or any political subdivision thereof; or (2) the firearm is being imported for scientific or research purposes; or (3) it is being imported solely for testing or use as a model by a registered or dealer. Additionally, the Secretary may permit the conditional importation of a firearm for examination and testing in connectio with classifying the firearm.

Section 922(o) of Title 18, United States Code, makes it unlawful for any person to transfer or possess machinegun, except a transfer to or by the United States or any department or agency thereof or a State or department, agency or political subdivision thereof; or any lawful transfer or lawful possession of a machinegun lawfully possessed before May 19, 1986.

A review of the statutes above indicates that machinegun and machinegun receivers cannot be lawfully imported for unrestricted commercial sale. Accordingly. Machinegun receivers may be imported for commercial sale only if they are destroyed in a manner that will prevent their function and future use a a firearm. The resulting severed receiver portions would not be subject to the provisions of 26 U.S.C. 5844 or 18 U.S.C. 922(o); however, these articles would be subject the provisions of the Arms Export Control Act, 18 U.S.C. 925, 22 U.S.C. 2778, and implementing regulation at 27 CFR Part 47. It is important to note that these machinegun receivers must be destroyed and cannot be imported whether they are serviceable or unserviceable.

An ATF-approved method of destruction for an FN FAL type machinegun receiver requires three diagonal torch cuts that sever or pass through the following areas: (1) the threaded portion of the receiver ring and magazine well opening at bottom, (2) the hinge pin, ejector block and bolt guide rails and (3) the body locking lug and bolt guide rails. All cutting must be done with a cutting torch having tip of sufficient size to displace at least ¼ inch of material at each location. Each cut must completely sever the receiver in the designated areas and must be done with a diagonal torch cut. Using a bandsaw or a cut-off wheel to destroy the receiver does not ensure destruction of the weapon.

This method of destruction is illustrated in the diagram below.

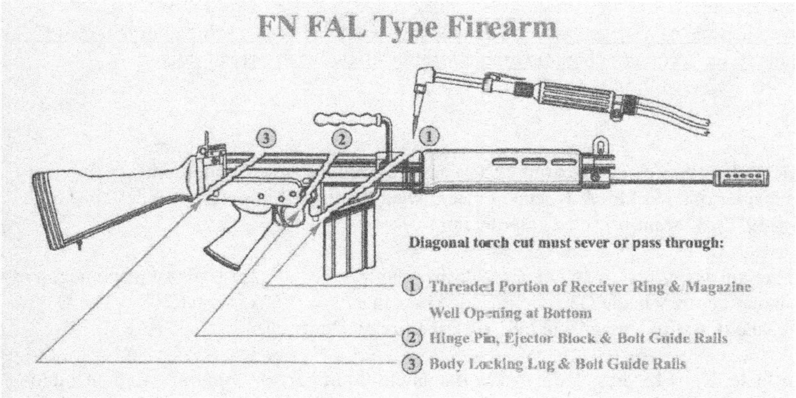

Alternative methods of destruction may also be acceptable. These alternative methods must be equivalent in degree to the approved method of destruction. Receivers that are not sufficiently modified cannot be approved for importation. To ensure compliance with the law, it is recommended that the importer submit in writing the alternative method of destruction to the ATF Firearms Technology Branch (FTB) for review and approval prior to importation.

Held, an ATF-approved method of destruction for an FN FAL type machinegun receiver will result in the severed portions of the receiver being importable for unrestricted commercial sale. The severed articles would not be subject to the provisions of 26 U.S.C. 5844 or 18 U.S.C. 922(o), but would continue to be subject to the provisions of the Arms Export Control Act, 22 U.S.C. 2778. Alternative methods of destruction may also be acceptable. It is recommended that such methods be reviewed and approved by the ATF Firearms Technology Branch prior to the weapon's importation.

Date signed: January 24, 2003

26 USC 5844, 18 USC 922(o), 22 USC 2778: IMPORTATION OF HECKLER & KOCH G3 TYPE RECEIVERS FOR UNRESTRICTED COMMERCIAL SALE.

An ATF – approved method of destruction for the Heckler & Koch G3 type machinegun will result in the severed portions of the receiver being importable for unrestricted commercial sale.

ATF Rul. 2003-3

The Bureau of Alcohol, Tobacco and Firearms (ATF) has received inquiries about modifications necessary to the receiver of a Heckler & Koch G3 machinegun to make it importable under 26 U.S.C. 5844 and 18 U.S.C. 922(o) for unrestricted commercial sale.

The G3 is a machinegun as defined in 26 U.S.C. 5845(b). The receiver of a G3 is also a machinegun as defined. Various manufacturers made G3 style machineguns in caliber 7.62x51mm (.308). The G3 is a delayed blowback, shoulder fired, magazine fed, selective fire machinegun.

Section 5844 of title 26, United States Code, makes it unlawful to import any firearm into the United States, unless the firearm to be imported or brought in is: (1) being imported for use by the United States or any department, independent establishment, or agency thereof or any State or possession or any political subdivision thereof; or (2) the firearm is being imported for scientific or research purposes; or (3) it is being imported solely for testing or use as a model by a registered or dealer. Additionally, the Secretary may permit the conditional importation of a firearm for examination and testing in connection with classifying the firearm.

Section 922(o) of Title 18, United States Code, makes it unlawful for any person to transfer or possess a machinegun, except a transfer to or by the United States or any department or agency thereof or a State or department, agency or political subdivision thereof; or any lawful transfer or lawful possession of a machinegun lawfully possessed before May 19, 1986.

A review of the statutes above indicates that machinegun and machinegun receivers cannot be lawfully imported for unrestricted commercial sale. Accordingly. Machinegun receivers may be imported for commercial sale only if they are destroyed in a manner that will prevent their function and future use as a firearm. The resulting severed receiver portions would not be subject to the provisions of 26 U.S.C. 5844 or 18 U.S.C. 922(o); however, these articles would be subject the provisions of the Arms Export Control Act, 18 U.S.C. 925, 22 U.S.C. 2778, and implementing regulation at 27 CFR Part 47. It is important to note that these machinegun receivers must be destroyed and cannot be imported whether they are serviceable or unserviceable.

An ATF-approved method of destruction for a Heckler & Koch G3 type machinegun receiver requires four diagonal torch cuts that sever or pass through the following areas: (1) the chamber area, (2) the grip assembly locking pin hole, (3) the ejection port, and (4) the buttstock locking pin hole. All cutting must be done with a cutting torch having a tip of sufficient size to displace at least ¼ inch of material at each location. Each cut must completely sever the receiver in the designated areas and must be done with a diagonal torch cut. Using a bandsaw or a cut-off wheel to destroy the receiver does not ensure destruction of the weapon.

'his method of destruction is illustrated in the diagram below.

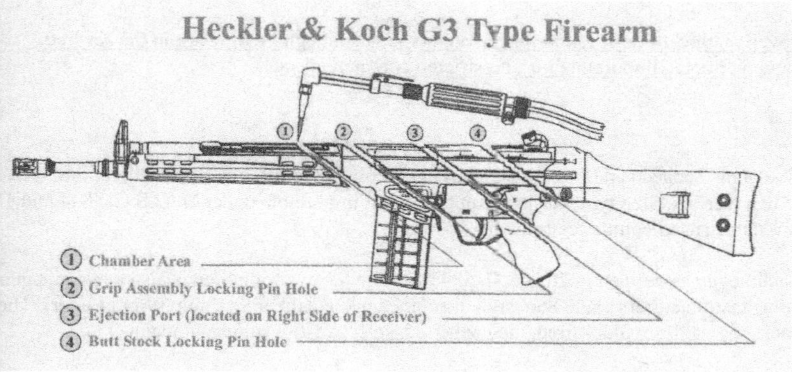

.lternative methods of destruction may also be acceptable. These alternative methods must be quivalent in degree to the approved method of destruction. Receivers that are not sufficiently modified annot be approved for importation. To ensure compliance with the law, it is recommended that the nporter submit in writing the alternative method of destruction to the ATF Firearms Technology ranch (FTB) for review and approval prior to importation.

eld, an ATF-approved method of destruction for a Heckler & Koch G3 type machinegun receiver will sult in the severed portions of the receiver being importable for unrestricted commercial sale. The vered articles would not be subject to the provisions of 26 U.S.C. 5844 or 18 U.S.C. 922(o), but ould continue to be subject to the provisions of the Arms Export Control Act, 22 U.S.C. 2778. lternative methods of destruction may also be acceptable. It is recommended that such methods be viewed and approved by the ATF Firearms Technology Branch prior to the weapon's importation.

ate signed: January 24, 2003

26 USC 5844, 18 USC 922(o), 22 USC 2778: IMPORTATION OF STEN TYPE RECEIVERS FOR UNRESTRICTED COMMERCIAL SALE.

An ATF – approved method of destruction for the Sten type machinegun will result in the severed portions of the receiver being importable for unrestricted commercial sale.

ATF Rul. 2003-4

The Bureau of Alcohol, Tobacco and Firearms (ATF) has received inquiries about modifications necessary to the receiver of a Sten type machinegun to make it importable under 26 U.S.C. 5844 and 18 U.S.C. 922(o) for unrestricted commercial sale.

The Sten is a machinegun as defined in 26 U.S.C. 5845(b). The receiver of a Sten is also a machinegun as defined. Various manufacturers made Sten style machineguns in caliber 9x19mm (9mm Luger). The Sten is a blowback-operated, shoulder fired, magazine fed, selective-fire submachinegun.

Section 5844 of title 26, United States Code, makes it unlawful to import any firearm into the United States, unless the firearm to be imported or brought in is: (1) being imported for use by the United States or any department, independent establishment, or agency thereof or any State or possession or any political subdivision thereof; or (2) the firearm is being imported for scientific or research purposes; or (3) it is being imported solely for testing or use as a model by a registered or dealer. Additionally, the Secretary may permit the conditional importation of a firearm for examination and testing in connection with classifying the firearm.

Section 922(o) of Title 18, United States Code, makes it unlawful for any person to transfer or possess machinegun, except a transfer to or by the United States or any department or agency thereof or a State or department, agency or political subdivision thereof; or any lawful transfer or lawful possession of a machinegun lawfully possessed before May 19, 1986.

A review of the statutes above indicates that machinegun and machinegun receivers cannot be lawfully imported for unrestricted commercial sale. Accordingly. Machinegun receivers may be imported for commercial sale only if they are destroyed in a manner that will prevent their function and future use a a firearm. The resulting severed receiver portions would not be subject to the provisions of 26 U.S.C. 5844 or 18 U.S.C. 922(o); however, these articles would be subject the provisions of the Arms Export Control Act, 18 U.S.C. 925, 22 U.S.C. 2778, and implementing regulation at 27 CFR Part 47. It is important to note that these machinegun receivers must be destroyed and cannot be imported whether they are serviceable or unserviceable.

An ATF-approved method of destruction for a Sten type machinegun receiver requires three diagonal torch cuts that sever or pass through the following areas: (1) the threaded portion of the receiver/chamber area, (2) the return spring cap socket, and (3) the sear slot in the lower side of the receiver. All cutting must be done with a cutting torch having a tip of sufficient size to displace at leas ¼ inch of material at each location. Each cut must completely sever the receiver in the designated area and must be done with a diagonal torch cut. Using a bandsaw or a cut-off wheel to destroy the receiver does not ensure destruction of the weapon.

his method of destruction is illustrated in the diagram below.

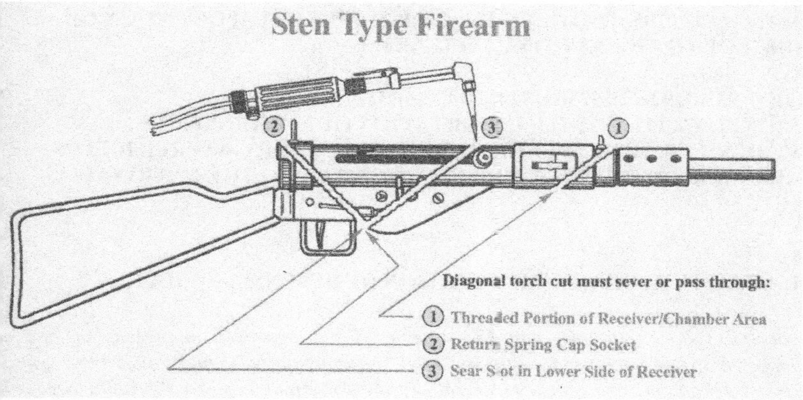

lternative methods of destruction may also be acceptable. These alternative methods must be quivalent in degree to the approved method of destruction. Receivers that are not sufficiently modified annot be approved for importation. To ensure compliance with the law, it is recommended that the nporter submit in writing the alternative method of destruction to the ATF Firearms Technology ranch (FTB) for review and approval prior to importation.

Ield, an ATF-approved method of destruction for a Sten type machinegun receiver will result in the evered portions of the receiver being importable for unrestricted commercial sale. The severed articles ould not be subject to the provisions of 26 U.S.C. 5844 or 18 U.S.C. 922(o), but would continue to be ubject to the provisions of the Arms Export Control Act, 22 U.S.C. 2778. Alternative methods of estruction may also be acceptable. It is recommended that such methods be reviewed and approved by e ATF Firearms Technology Branch prior to the weapon's importation.

ate signed: January 24, 2003

26 U.S.C. 5812, 5841, 5844, 5861, 5872
27 CFR 479.11, 479.26, 479.105, 479.111, 479.112, 479.114 – 479.119: IMPORTATION OF FIREARMS SUBJECT TO THE NATIONAL FIREARMS ACT.

18 U.S.C. 921(a)(3), 922(i), 922(o), 923(e), 924(d), 925(d)(3)
27 CFR 478.11, 478.22, 478.111 – 478.113: IMPORTATION OF MACHINEGUNS, DESTRUCTIVE DEVICES, SHORT –BARREL SHOTGUNS, SHORT-BARREL RIFLES, FIREARMS SILENCERS, AND OTHER FIREARMS SUBJECT TO THE NATIONAL FIREARMS ACT.

22 U.S.C. 2778
27 CFR 447.11, 447.21: TEMPORARY IMPORTATION OF DEFENSE ARTICLES

The Bureau of Alcohol, Tobacco, Firearms and Explosives (ATF) has approved an alternate method or procedure for importers to use when temporarily importing firearms subject to the National Firearms Act, the Gun Control Act and the Arms Export Control Act for inspection, testing, calibration, repair, o incorporation into another defense article.

ATF Rul. 2004-2

The Bureau of Alcohol, Tobacco, Firearms and Explosives (ATF) has received numerous inquiries fro importers who wish to temporarily import firearms subject to the Gun Control Act of 1968 (GCA), 18 U.S.C. Chapter 44, and the National Firearms Act (NFA), 26 U.S.C. Chapter 53, for inspection, testing calibration, repair, or incorporation into another defense article. Importers advise ATF that they generally obtain a temporary import license, DSP-61, from the Department of State authorizing the importation or comply with one of the regulatory exemptions from licensing in 22 CFR 123.4. They as whether such a license or exemption is sufficient to satisfy the requirements of the GCA and NFA.

Statutory Background

1. *The National Firearms Act*

The NFA imposes restrictions on certain firearms, including registration requirements, transfer approv requirements, and import restrictions. 26 U.S.C. 5812, 5841, 5844. The term "firearm" is defined in 2 U.S.C. 5845(a) to include machineguns, short-barrel shotguns, short-barrel rifles, silencers, destructive devices, and "any other weapons." Section 5844 of the NFA provides that no firearm may be importec into the United States unless the importer establishes that the firearm to be imported is –

> (1) Being imported or brought in for the use of the United States or any department, independent establishment, or agency thereof or any State or possession or any political subdivision thereof; or
>
> (2) Being imported or brought in for scientific or research purposes; or

(3) Being imported or brought in solely for testing or use as a model by a registered manufacturer or solely for use as a sample by a registered importer or registered dealer.

Regulations implementing the NFA in 27 CFR Part 479 require importers to obtain an ATF Form 6, Application and Permit for Importation of Firearms, Ammunition and Implements of War, prior to importing NFA firearms into the United States. 27 CFR 479.111. In addition, the regulations require importers to register the firearms they import by filing with the Director an accurate notice on Form 2, Notice of Firearms Manufactured or Imported, executed under the penalties of perjury, showing the importation of the firearm. 27 CFR 479.112. When an NFA firearm is to be exported from the United States, the exporter must file with the Director an application on form 9, Application and Permit for Exportation of Firearms, to obtain authorization to export the firearm. 27 CFR 479.114-119.

Regulations in 27 CFR Part 479 indicate that NFA firearms may be imported for scientific or research purposes or for testing or use as a model by a registered manufacturer or as a sample by a registered importer or a registered dealer. 27 CFR 479.111(a). However, section 479.105(c), implementing section 922(o) of the GCA, clarifies that machineguns manufactured on or after May 19, 1986, may be imported only with a purchase order for transfer to a governmental entity, or as a dealer's sales sample pursuant to section 479.105(d).

The regulations in Part 479 give the Director the authority to approve an alternate method or procedure in lieu of a method or procedure specifically prescribed in the regulations when it is found that:

(1) The alternate method or procedure is within the purpose of, and consistent with the effect intended by, the specifically prescribed method or procedure and that the alternate method or procedure is substantially equivalent to that specifically prescribed method or procedure; and

(2) Good cause is shown for the use of the alternate method or procedure.

(3) The alternate method or procedure will not be contrary to any provision of law and will not result in an increase in cost to the Government or hinder the effective administration of the GCA or regulations issued thereunder.

27 CFR 479.26

2. *The Gun Control Act*

Import provisions of the GCA, 18 U.S.C. 922(i) and 925(d)(3), generally prohibit the importation of firearms subject to the NFA, except for the use of governmental entities. 18 U.S.C. 925(a)(1). The term "firearm" is defined in section 921(a)(3) to include any weapon which will or is designed to or may be readily converted to expel a projectile by the action of an explosive; the frame or receiver of any such weapon; any firearm silencer; and any destructive device. In addition, section 922(o) of the GCA prohibits the transfer or possession of a machinegun manufactured on or after May 19, 1986, except for the official use of governmental entities.

Regulations implementing the GCA in 27 CFR part 478 require that persons importing firearms into the United States obtain an approved ATF Form 6, Application and Permit for Importation of Firearms, Ammunition and Implements of War, prior to bringing the firearm into the United States. 27 CFR 478.111-114. Regulations in Part 478 provide that the Director may approve an alternate method or procedure in lieu of a method or procedure specifically prescribed by the GCA and regulations when it is found that:

> (1) The alternate method or procedure is within the purpose of, and consistent with the effect intended by, the specifically prescribed method or procedure and that the alternate method or procedure is substantially equivalent to that specifically prescribed method or procedure; and
>
> (2) Good cause is shown for the use of the alternate method or procedure.
>
> (2) The alternate method or procedure will not be contrary to any provision of law and will not result in an increase in cost to the Government or hinder the effective administration of the GCA or regulations issued thereunder.
>
> 27 CFR 478.22

3. *The Arms Export Control Act*

The Arms Export Control Act (AECA), 22 U.S.C. 2778, gives the President the authority to control the export and import of defense articles and defense services in furtherance of world peace and the security and foreign policy of the United States. Authority to administer the permanent import provisions of the AECA was delegated to Attorney General, while the authority to administer the export and temporary import provisions of the AECA was delegated to the Secretary of State, Executive Order 11958 of January 18, 1977, as amended by Executive Order 13333 of January 23, 2003, 3 CFR Executive Order 13284.

The term "defense article" is defined in 27 CFR 447.11 as any item designated in sections 447.21 or 447.22. Section 447.21, the U.S. Munitions Import List, includes a number of defense articles that are also subject to the GCA and NFA. Category I, "Firearms," includes nonautomatic and semiautomatic firearms to caliber .50 inclusive, combat shotguns, shotguns with barrels less than 18 inches in length, and firearms silencers and suppressors. All Category I firearms are subject to the GCA. "Combat shotguns" include the USAS-12 shotgun and the Striker-12/Streetsweeper shotgun, which have been classified as destructive devices under the GCA and NFA. In addition, all shotguns with barrels of less than 18 inches in length are subject to both the GCA and NFA. All rifles with barrels of less than 16 inches in length are subject to both the GCA and NFA, and silencers are subject to both the GCA and NFA.

Category II, "Artillery Projectors," includes guns over .50, howitzers, mortars, and recoilless rifles. Firearms over .50 caliber have a bore of more than one-half inch in diameter and are "destructive devices" as defined in the GCA and NFA.

ategory IV, "launch Vehicles, Guided Missiles, Ballistic Missiles, Rockets, torpedoes, Bombs and Mines," includes rockets, bombs, grenades, torpedoes, and land and naval mines. All these articles are destructive devices" as defined in the GCA and NFA.

egulations of the Department of State implementing the AECA generally require a temporary import cense, DSP-61, for the temporary import and subsequent export of unclassified defense articles, unless therwise exempted. 22 CFR 123.3. Regulations in 22 CFR 123.4 provide an exemption from licensing the item temporarily imported:

(1) Is serviced (e.g., inspection, testing, calibration or repair, including overhaul, reconditioning and one-to-one replacement of any defective items, parts or components, but excluding any modification, enhancement, upgrade or other form of alteration or improvement that changes the basic performance of the item), and is subsequently returned to the country from which it was imported. Shipment may be made by the U.S. importer or a foreign government representative of the country from which the goods were imported; or

(2) Is to be enhanced, upgraded or incorporated into another item whish has already been authorized by the Office of Defense Trade Controls for permanent export; or

(3) Is imported for the purpose of exhibition, demonstration or marketing in the United States and is subsequently returned to the country from which it was imported; or

(4) Has been rejected for permanent import by the Department of the Treasury [after January 24, 2003, the Department of Justice] and is being returned to the country from which it was shipped; or

(5) Is approved for such import under the U.S. Foreign Military Sales (FMS) program pursuant to an executed U.S. Department of Defense Letter of Offer and Acceptance (LOA).

illful violations of the AECA are punishable by imprisonment for not more than 10 years, a fine of not ore than $1,000,000, or both. 22 U.S.C. 2778©. Articles imported in violation of the AECA are also bject to seizure and forfeiture. 18 U.S.C. 545.

iscussion

temporary import license authorizing the temporary importation and subsequent export of a defense ticle by the Department of State satisfies all legal requirements under the AECA. Importers may also mply with AECA requirements if the importation meets one of the exemptions in 22 CFR 123.4. owever, if the defense article is subject to the GCA and NFA, the importer must also comply with the quirements of those statutes. Neither the GCA nor NFA make a distinction between temporary portation and permanent importation, as is the case under the AECA. Regulations implementing the CA and NFA make it clear that an "importation" occurs when firearms are brought into the territory of e United States. 27 CFR 478.11 and 479.11. Accordingly, any bringing of firearms into the territory of e United States is subject to the import provisions of the GCA and NFA. Issuance of a temporary port license by the Department of State, or exemption from licensing under regulations in 22 CFR rt 123, will not excuse compliance with the GCA and NFA.

The statutes and regulations outlined above do not address the importation of machineguns manufactured after May 19, 1986, for scientific or research purposes or for testing, repair, or for use as model by a manufacturer or importer. Nor do the regulations address the importation of post-86 machineguns for repair, inspection, calibration, or incorporation into another defense article.

For other" defense articles" that are subject to the requirements of the GCA and NFA, such as silencers destructive devices, and short-barrel weapons, ATF has the authority to approve the importation of such firearms for scientific or research purposes or for testing or use as a model or sample by a registered importer or a registered dealer. However, such importation must comply with all applicable provisions of the NFA, including filing of a Form2, Notice of Firearms Manufactured or Imported, to effect registration. If such articles are subsequently exported, a Form 9, Application and Permit for Permanent Exportation of Firearms, must also be approved prior to exportation.

As with post-86 machineguns, neither the law nor regulation specifically address the importation of firearms subject to the NFA for purposes of repair, inspection, calibration, or for incorporation into another defense article.

ATF recognizes that inspection, repair, calibration, incorporation into another defense article, and reconditioning of machineguns, destructive devices, and other NFA firearms is often necessary for National defense. These defense articles are frequently sold to allies of the United States for their legitimate defense needs. Accordingly, ATF believes it is appropriate to recognize an alternate method that allows importers to temporarily import these firearms, subject to requirements to ensure the security of these defense articles while they are in the United States and accountability of the persons who impo them.

Pursuant to 27 CFR 478.22 and 479.26, ATF hereby authorizes an alternate method or procedure for importers of defense articles to use for temporary importation of such articles for inspection, calibratio repair, or incorporation into another defense article when such articles are subject to the requirements of the NFA and GCA. The procedure requires that importers-

(1) Be qualified under the GCA and NFA to import the type of firearms sought for importation

(2) Obtain a temporary import license, DSP-61, from the Department of State in accordance wi 22CFR 123.3 OR qualify for a temporary import license exemption pursuant to 22 CFR 123.4;

(3) Within 15 days of the release of the firearms from Customs custody, file an ATF Form2, Notice of Firearms Manufactured or Imported, showing the importation of the firearms. The DSP-61 must be attached to the Form 2. If the importation is subject to a licensing exempti under 22 VFR 123.4, the importer must submit with the ATF Form 2 a statement, under penalty of perjury, attesting to the exemption and stating that the article will be exported within four years of its importation into the United States;

(4) Maintain the defense articles in a secure place and manner to ensure that the articles are not diverted to criminal or terrorist use; and

(5) Export the articles within 4 years of importation into the United States.

mporters who follow the procedures outlined above will be in compliance with the provisions of the GCA, NFA, and AECA administered and enforced by ATF. All other provisions of the law must be ollowed.

ATF finds that the procedure outlined above meets the legal requirements for an alternate method or rocedure because there is a good cause to authorize the importation of defense articles for repair, nspection, calibration, or incorporation into another defense article. Because such defense articles are ften provided to allies of the United States, it is imperative that the original manufacturers have a awful method of importing such articles for repair and routing maintenance. The alternate method or rocedure is consistent with the effect intended by the procedure set forth in the GCA and NFA, because he firearms must be registered and stored securely. Finally, the alternate method is consistent with the equirements of the GCA and NFA and will not result in any additional costs to ATF or the Department f State.

"Transfers" of NFA Weapons After Importation

ATF recognizes that temporarily imported NFA firearms are sometimes "transferred" from the importer o a contractor within the United States for inspection, testing, calibration, repair, or incorporation into nother defense article. ATF has approved a procedure for authorizing the transportation or delivery of emporarily imported NFA firearms to licensed contractors for repair or manipulation, as noted above.

Conveyance of an NFA weapon to a licensee for purposes of inspection, testing, calibration, repair, or ncorporation into another defense article is generally not considered to be a "transfer" under 26 U.S.C. 845(j). ATF has taken the position that temporary custody by a licensee is not a transfer for purposes of he NFA since no sale, lease, or other disposal is intended by the owner. However, in order to document he transaction as a temporary conveyance and make clear that an actual "transfer" of a firearm has not aken place, ATF strongly recommends that the importer submit a Form 5, Application for Tax Exempt Transfer and Registration of Firearm, for approval prior to conveying a firearm for repair or nanipulation. In the alternative, the importer should convey the weapon with a letter to the contractor, tating: (1) the weapon is being temporarily conveyed for inspection, testing, calibration, repair, or ncorporation into another defense article; and (2) the approximate time period the weapon is to be in the ontractor's possession. The transferee must be properly licensed to engage in an NFA firearms usiness.

Held, pursuant to 27 CFR 478.22 and 479.26, the Bureau of Alcohol, Tobacco, Firearms and Explosives as approved an alternate method or procedure for importers to use when temporarily importing firearms ubject to the Gun Control Act, National Firearms Act, and the Arms Export Control Act for inspection, esting, calibration, repair, or incorporation into another defense article. This procedure applies to all efense articles that are also subject to the NFA and GCA. The procedure requires that importers-

(1) Be qualified under the GCA and NFA to import the type of firearms sought for importation;

(2) Obtain a temporary import license, DSP-61, from the Department of State in accordance with 22CFR 123.3 OR qualify for a temporary import license exemption pursuant to 22 CFR 123.4;

(3) Within 15 days of the release of the firearms from Customs custody, file an ATF Form2, Notice of Firearms Manufactured or Imported, showing the importation of the firearms. The DSP-61 must be attached to the Form 2. If the importation is subject to a licensing exemption under 22 VFR 123.4, the importer must submit with the ATF Form 2 a statement, under penalty of perjury, attesting to the exemption and stating that the article will be exported within four years of its importation into the United States;

(4)Maintain the defense articles in a secure place and manner to ensure that the articles are not diverted to criminal or terrorist use; and

(5) Export the articles within 4 years of importation into the United States.

Held further, temporary conveyance of NFA weapons from the importer to a contractor within the United States for purposes of inspection, testing, calibration, repair, or incorporation into another defense article may be accomplished through advance approval of ATF Form 5, Application for Tax Exempt Transfer and Registration of Firearm, or with a letter from the importer to the contractor stating (1) the weapon is being temporarily conveyed for inspection, testing, calibration, repair, or incorporation into another defense article; and (2) the approximate time period the weapon is to be in the contractor's possession. The transferee must be properly licensed to engage in an NFA firearms business.

Date signed: April 7, 2004

26 U.S.C. 5845(B): DEFINITIONS (MACHINEGUN)

27 CFR 479.11: MEANING OF TERMS

The 7.62mm Aircraft Machine Gun, identified in the U.S. military inventory as the "m-134" (Army), "GAU-2B/A" (Air Force), and "GAU-17/A" (Navy), is a machinegun as defined by 26 U.S.C. 5845(b), Rev. Rul. 55-528 modified.

ATF Rul. 2004-5

The Bureau of Alcohol, Tobacco, firearms and Explosives (ATF) has examined the 7.62mm Aircraft Machine Gun, commonly referred to as a "Minigun." The Minigun is a 36 pound, six barrel, electrically powered machinegun. It is in the U.S. military inventory and identified as the "M-134" (Army), "GAU 2B/A" (Air force), and "GAU 17/A" (Navy). It is a lightweight and extremely reliable weapon, capable of discharging up to 6,000 rounds per minute. It has been used on helicopters, fixed-wing aircraft, and wheeled vehicles. It is highly adaptable, being used with pintle mounts, turrets, pods, and internal installations.

The Minigun has six barrels and bolts which are mounted on a rotor. The firing sequence begins with the manual operation of a trigger. On an aircraft, the trigger is commonly found on the control column, or joy stick. Operation of the trigger causes an electric motor to turn the rotor. As the rotor turns, a stud on each bolt travels along an elliptical groove on the inside of the housing, which causes the bolts to move forward and rearward on tracks n the rotor. A triggering cam, or sear shoulder, trips the firing pin when the bolt has traveled forward through the full length of the bolt track. One complete revolution of the rotor discharges cartridges from all six barrels. The housing that surrounds the rotor, bolts and firing mechanism constitutes the frame or receiver of the firearm.

The National Firearms Act defines "machinegun" as "any weapon which shoots, is designed to shoot, or can be readily restored to shoot, automatically more than one shot, without manual reloading, by a single function of the trigger." 26 U.S.C. 5845(b). The term also includes "the frame or receiver of any such weapon, any part designed and intended solely and exclusively, or combination of parts designed and intended, for use in converting a weapon into a machinegun, and any combination of parts from which a machinegun can be assembled if such parts are in the possession or under the control of a person." Id; see 18 U.S.C. 921(a)(23); 27 CFR 478.11, 479.11.

ATF and its predecessor agency, the Internal Revenue Service (IRS), have historically held that the original, crank-operated Gatling Gun, and replicas thereof, are not automatic firearms or machineguns as defined. See Rev. Rul. 55-528, 1955-2 C.B. 482. The original Gatling gun is a rapid-firing, hand-operated weapon. The rate of fire is regulated by the rapidity of the hand cranking movement, manually controlled by the operator. It is not a "machinegun" as that term is defined in 26 U.S.C. 5845(b) because it is not a weapon that fires automatically.

The Minigun is not a Gatling Gun. It was not produced under the 1862 – 1893 patents of the original Gatling Gun. While using a basic design concept of the Gatling Gun, the Minigun does not incorporate any of Gatling's original components and its fed mechanisms are entirely different. Critically, the Minigun shoots more than one shot, without manual reloading, by a single function of the trigger, as prescribed by 26 U.S.C> 5845(b). See United States v. Fleischli, 305 F. 3d 643, 655-656 (7th Cir. 2002). See also Staples v. United States, 511 U.S. 600, 603 (1994) (automatic refers to a weapon that "once its trigger is depressed, the weapon will automatically continue to fire until its trigger is released or the ammunition is exhausted"); GEORGE C. NONTE, JR., FIREARMS ENCYCLOPEDIA 13 (Harper &

Rowe 1973) (the term "automatic" is defined to include "any firearm in which a single pull and continuous pressure upon the trigger (or other firing device) will produce rapid discharge of successive shots so long as ammunition remains in the magazine or feed device in other words, a machinegun"); WEBSTER'S II NEW RIVERSIDE-UNIVERSITY DICTIONARY (1988) (defining automatically as "acting or operating in a manner essentially independent of external influence or control"); JOHN B. QUICK, PH.D., DICTIONARY OF WEAPONS AND MILITARY TERMS 40 (McGraw-hill 1973) (defining automatic fire as "continuous fire from an automatic gun, lasting until pressure on the trigger is released").

The term "trigger" is generally held to be the part of a firearm that is used to initiate the firing sequence. See United States v. Fleischli, 305 F.3d at 655-56 (and cases cited therein); see also ASSOCIATION OF FIREARMS AND TOOLMARK EXAMINERS (ATFE) GLOSSARY 185 (1st ed. 1980) ("that part of a firearm mechanism which is moved manually to cause the firearm to discharge"); WEBSTER'S II NEW RIVERSIDE-UNIVERSITY DICTIONARY (1988) ("lever pressed by the finger in discharging a firearm").

Held, the 7.62mm Minigun is designed to shoot automatically more than one shot, without manual reloading, by a single function of the trigger. Consequently, the 7.62mm Minigun is a machinegun as defined in section 5845(b) of the National Firearms Act. See United States v. Fleischli, 305 F. 3d at 655-56. Similarly, the housing that surrounds the rotor is the frame or receiver of the Minigun, and thus is also a machinegun. Id .; *see 18 U.S.C. 921(a)(23); 27 CFR 478.11, 479.11.*

To the extent this ruling is inconsistent with Revenue Ruling 55-528 issued by the IRS, Revenue Ruling 55-528, 1955-2 C.B. 482 is hereby modified.

Date signed: August 18, 2004

Carl J. Truscott

Director

04/03/07 TUE 13:29 FAX 3046164501 BATF 002

U.S. Department of Justice

Bureau of Alcohol, Tobacco,
Firearms and Explosives

January 31, 2006

www.atf.gov

901040:GS
5320

Federal Firearms Licensees and National Firearms Act
Special (Occupational) Taxpayers

SUBJECT: Filing National Firearms Act Transaction Forms by Facsimile Transmission

This is to update information regarding the filing of certain National Firearms Act (NFA) forms by facsimile transmission as the functions of the NFA Branch have been relocated to Martinsburg, West Virginia from Washington, DC. Please note that the contact information for the NFA Branch is:

Bureau of Alcohol, Tobacco, Firearms and Explosives
NFA Branch
244 Needy Road
Martinsburg, WV 25401
Telephone: (304) 616-4500
Facsimile: (304) 616-4501

The Bureau of Alcohol, Tobacco, Firearms and Explosives (ATF) held in ATF Ruling 89-1, that certain forms required by ATF may be filed by facsimile transmission. This ruling and ATF Industry Circular 89-6 set forth the requirements and procedures.

The NFA Branch had determined that for forms filed under the NFA, the procedure would be available for ATF Form 2, Form 3, Form 9, and for Form 5 in situations where the application is not accompanied by fingerprint cards. The submission of NFA forms may only be done by Federal firearms licensees who have paid the special (occupational) tax to import, manufacture, or deal in NFA firearms.

If you choose to utilize the procedure, each person who will be filing forms on behalf of the licensee must first submit to us (in original) an affidavit declaring, under penalties of perjury, that certain conditions will be met. These conditions are detailed in the attached sample affidavit. The affidavit needs only to be submitted once. If you have previously filed an affidavit with us, there is no need to file another.

AFFIDAVIT FOR FACSIMILE TRANSMITTAL OF NFA NOTICE OR TRANSFER APPLICATION FORMS

TO: Bureau of ATF
NFA Branch
244 Needy Road, Martinsburg, WV 25401

AFFIDAVIT

I, ______________________, doing business as ______________

______________________ hereby state:

1. In addition to those applications sent via U.S. mail, I will be filing ATF Forms (2, 3, 5 and/or 9) by facsimile transmission;
2. The signature appearing on the facsimile copies will be mine and is intended as the original signature;
3. I will be submitting the facsimile copies as originals for purposes of approval under the provisions of Title 27, Code of Federal Regulations, Part 479, and that the facsimile copies are to be treated as originals; and
4. I will personally transmit the copies via facsimile machine.

I declare under penalty of perjury under the laws of the United States that the above information is true and correct.

______________________ Signature ______________ Date

Name ______________________

Title ______________________

Trade name ______________________

Address ______________________

Telephone ______________________

FFL number ______________________

EIN ______________________

ATF Ruling 89-1

The Bureau of Alcohol, Tobacco and Firearms has been asked whether certain forms and other documents could optionally be transmitted to the Bureau by means of facsimile machines. These include various application forms, comments on notices of proposed rulemaking, reports, and other documents submitted to the Bureau.

Recent technology advances have made transmitting documents by facsimile machine practical and the use of facsimile machines has become widespread. The Bureau has acquired several facsimile machines and found them to be acceptable for transmitting many types of documents.

The statutes administered by the Bureau (including the Internal Revenue Code, the Federal Alcohol Administration Act, the Gun Control Act, Title XI of the Organized Crime Control Act of 1970, and the Arms Export Control Act) generally specify the information to be included on documents that are submitted to the Bureau. However, the form of the documents and the manner of filing are generally left to the Secretary of the Treasury to prescribe by regulation. The regulations issued pursuant to these statutes generally require that forms and other documents be submitted in original and signed by the applicant, licensee, or permittee.

The Bureau has concluded that if the person transmitting the form or document intends that it be treated as an original and it is filed via a facsimile machine at the person's direction, then the facsimile copy received by the Bureau would be for all intents and purposes the original.

In order to verify the authenticity of a signature on forms and other documents required to be signed under penalties of perjury, the Bureau will require that an affidavit executed by the person signing the facsimile documents be submitted to the Bureau prior to acceptance of any facsimile transmitted copies of such forms or documents.

The affidavit should declare, under penalties of perjury, that the affiant will be filing the appropriate forms or documents by facsimile machine, that the signature appearing on the facsimile copies will be that of the affiant, and is intended as the original signature.

The affidavit should also state that the affiant will be submitting the facsimile copies as originals for approval or other purposes under the appropriate regulations and that the affiant intends that the facsimile copies be treated as originals.

Finally, the affidavit should also state that the person signing the form or other document will personally transmit the copy via facsimile machine, or, if a designated agent is to do the transmittal, the name of the agent who will do the actual transmittal.

The affidavit would only have to be submitted one time to each Bureau office to which facsimile transmitted copies are to be sent, and would not have to be resubmitted unless a different official would be signing or transmitting the copies.

The Bureau concludes that only documents that are legible and of suitable size (8 1/2" X 11") may be submitted by facsimile transmission. The Bureau also concludes that facsimile transmitted copies would not be acceptable with respect to documents which are accompanied by such articles as samples, fingerprint cards, photographs, checks, and labels for alcoholic beverage containers.

The acceptance of facsimile transmitted copies at some Bureau offices may be restricted due to limited availability of facsimile machines, telephone lines or other reasons. If an ATF form may be submitted by facsimile transmission, the instructions on the form will so specify, and will also indicate the locations and telephone numbers where the transmissions may be sent. Instructions for facsimile transmitted copies of comments on a notice of proposed rulemaking will be provided in the notice itself.

Accordingly, the Bureau has determined that facsimile transmitted copies of forms and other documents may be accepted under certain conditions.

Held, forms and other documents submitted to the Bureau may be accepted when transmitted by facsimile machines under the following conditions:

1. The Bureau office to where the forms or other documents are to be submitted is suitably equipped with facsimile machines and telephone lines to receive the facsimile transmitted copies.

2. The forms or other documents are legible and of suitable size (8 1/2" X 11") for facsimile transmission.

3. The forms or other documents are transmitted in accordance with instructions provided on the form, notice of proposed rulemaking, or by other competent authority.

4. The forms or other documents are not accompanied by other articles such as samples, fingerprint cards, photographs, checks, and labels for alcoholic beverage containers.

5. Prior to accepting any facsimile transmitted copies of forms and other documents required to be signed under penalties of perjury, the Bureau must have received an acceptable affidavit executed by the person whose signature appears on the forms or documents. The affidavit must declare, under penalties of perjury, that

a. the affiant will be filing the appropriate forms or other documents by facsimile machine,

b. that the signature appearing on the facsimile copies will be that of the affiant, and is intended as the original signature,

c. the affiant will be submitting the facsimile copies as originals for purposes of approval or other purpose under the appropriate regulations and that the affiant intends that the facsimile copies be treated as originals, and

d. the person signing the forms or other documents will personally transmit the copies via facsimile machine, or, if a designated agent is to do the transmittal, the name of the agent who will do the actual transmittal.

6. The affidavit is required to be submitted only one time to each Bureau office to which facsimile transmitted copies are to be sent, and need not be resubmitted unless a different official will be signing or transmitting the forms.

Industry Circular

Number: 89-6

Date: September 27, 1989

Department of the Treasury

Bureau of Alcohol, Tobacco and Firearms

Washington, DC 20226

FACSIMILE TRANSMITTED COPIES OF FORMS AND OTHER DOCUMENTS

All Alcohol, Tobacco, Firearms and Explosives Licensees, Permittees, and Others Concerned:

PURPOSE. The purpose of this circular is to advise industry members that an ATF ruling will be published in a future issue of the Alcohol, Tobacco and Firearms Quarterly Bulletin. The ruling will read substantially as follows:

The Bureau of Alcohol, Tobacco and Firearms has been asked whether certain forms and other documents could optionally be transmitted to the Bureau by means of facsimile machines. These include various application forms, comments on notices of proposed rulemaking, reports, and other documents submitted to the Bureau.

Recent technological advances have made transmitting documents by facsimile machine practical and the use of facsimile machines has become widespread. The Bureau has acquired several facsimile machines and found them to be acceptable for transmitting many types of documents.

The statutes administered by the Bureau (including the Internal Revenue Code of 1986, the Federal Alcohol Administration Act, the Gun Control Act of 1968, Title XI of the Organized Crime Control Act of 1970, and the Arms Export Control Act) generally specify the information to be included on documents that are submitted to the Bureau. However, the form of the documents and the manner of filing are generally left to the Secretary of the Treasury to prescribe by regulation. The regulations issued pursuant to these statutes generally require that forms and other documents be submitted in original and signed by the applicant, licensee, or permittee.

The Bureau has concluded that if the person transmitting the form or document intends that it be treated as an original and it is filed via a facsimile machine at the person's direction, then the facsimile copy received by the Bureau would be for all intents and purposes the original.

In order to verify the authenticity of a signature on forms and other documents required to be signed under penalties of perjury, the Bureau will require that an affidavit executed by the person signing the forms or documents be submitted to the Bureau prior to acceptance of any facsimile transmitted copies of such forms or documents.

The affidavit should declare, under penalties of perjury, that the affiant will be filing the appropriate forms or documents by facsimile machine, that the signature appearing on the facsimile copies will be that of the affiant, and that the signature on the facsimile is intended as the original signature.

The affidavit should also state that the affiant will be submitting the facsimile copies as originals for approval or other purposes under the appropriate regulations and that the affiant intends that the facsimile copies be treated as originals.

Finally, the affidavit should also state that the person signing the form or other document will personally transmit the copy via facsimile machine, or, if a designated agent is to do the transmittal, the name of the agent who will do the actual transmittal will be stated.

The affidavit would only have to be submitted one time to each Bureau office to which facsimile transmitted copies are to be sent, and would not have to be resubmitted unless a different official would be signing or transmitting the copies.

The Bureau concludes that only documents that are legible and of suitable size (8 1/2" X 11" or as specifically authorized by the receiving office) may be submitted by facsimile transmission. The Bureau also concludes that facsimile transmitted copies would not be acceptable with respect to fingerprint cards, bonds, photographs, checks, or labels for alcoholic beverage containers or the forms or documents required to be submitted with such items.

The acceptance of facsimile transmitted copies at some Bureau offices may be restricted due to limited availability of facsimile machines, telephone lines or other reasons. Instructions on the form or instructions otherwise provided will state whether a facsimile transmitted copy of a form or other document may be used.

The instructions will also indicate the locations and telephone numbers where the transmissions may be sent. Instructions for facsimile transmitted copies of comments on a notice of proposed rulemaking will be provided in the notice itself.

Accordingly, the Bureau has determined that facsimile transmitted copies of forms and other documents may be accepted under certain conditions.

Held: Forms and other documents submitted to the Bureau may be accepted when transmitted by facsimile machines under the following conditions:

1. The Bureau office to which the forms or other documents are to be submitted is suitably equipped with facsimile machines and telephone lines to receive the facsimile transmitted copies.
2. The forms or other documents are legible and of suitable size (8 1/2" X 11" or as specifically authorized by the receiving office) for facsimile transmission.
3. The forms or other documents are transmitted in accordance with instructions on the form, notice of proposed rulemaking, or otherwise provided.
4. The forms or other documents are not fingerprint cards, bonds, photographs, checks, or labels for alcoholic beverage containers and are not the forms or documents required to be submitted with such items.
5. Prior to accepting any facsimile transmitted copies of forms and other documents required to be signed under penalties of perjury, the Bureau must have received an affidavit executed by the person whose signature appears on the forms or documents. The affidavit must declare, under penalties of perjury, that:

   a. the affiant will be filing the appropriate forms or other documents by facsimile machine,
   b. the signature appearing on the facsimile copies will be that of the affiant, and is intended as the original signature,
   c. the affiant will be submitting the facsimile copies as originals for purposes of approval or other purpose under the appropriate regulations and that the affiant intends that the facsimile copies be treated as originals, and
   d. the person signing the forms or other documents will personally transmit the copies via facsimile machine, or, if a designated agent is to do the transmittal, the name of the agent who will do the actual transmittal.

6. The affidavit is required to be submitted only one time to each Bureau office to which facsimile transmitted copies are to be sent, and need not be resubmitted unless a different official will be

signing or transmitting the forms.

INQUIRIES. Inquiries concerning this circular should refer to its number and be addressed to the Associate Director (Compliance Operations), Bureau of Alcohol, Tobacco and Firearms, 1200 Pennsylvania Avenue, NW, Washington, DC 20226.

Daniel R. Black

Acting Director

Frequently Asked Questions - Silencers

Q1: What part of a silencer must be marked?

A: The silencer must be marked in accordance with 27 C.F.R. §§ 478.92 and 479.102. The regulations require that the markings be conspicuous and legible, meaning that the markings may be placed on any external part, such as the outer tube or end cap.

ATF strongly recommends that manufacturers place all required markings on the outer tube of the silencer, as this is the accepted industry standard. Moreover, this practice eliminates the need to remark in the event an end cap bearing the markings is damaged and requires replacement.

Q2: May a Federal firearms licensee repair a silencer by replacing worn or damaged components?

A: A person who is licensed under the Gun Control Act (GCA) to manufacture firearms and who has paid the special (occupational) tax to manufacture National Firearms Act (NFA) firearms may replace a component part or parts of a silencer. Repairs may not be done if they result in removal, obliteration, or alteration of the serial number, as this would violate 18 U.S.C. § 922(k). If a silencer part bearing the serial number, other than the outer tube, must be replaced, the new part must be marked with the same serial number as the replacement part.

The term "repair" does not include replacement of the outer tube of the silencer. The outer tube is the largest single part of the silencer, the main structural component of the silencer, and is the part to which all other component parts are attached. The replacement of the outer tube is so significant an event that it amounts to the "making" of a new silencer. As such, the new silencer must be marked, registered and transferred in accordance with the NFA and GCA.

In the event that identical replacement parts for a silencer are not available, new and different component parts may be used as long as the silencer retains the same dimensions and caliber. In addition, the repair may result in a minimal reduction in the length of the outer tube due to rethreading, but repair may not increase the length of the outer tube. Increasing the length of the outer tube significantly affects the performance of the silencer and results in the "making" of a new silencer. As stated above, a new silencer must be marked, registered and transferred in accordance with the NFA and GCA. Reducing the length of the tube by a minimal amount in order to repair a silencer is often necessary to replace damaged end caps, as the tube must be rethreaded. Such minimal reduction of the length of the tube uses all of the original parts, does not significantly affect performance of the silencer and may be done as part of a repair process without making a new silencer.

Persons other than qualified manufacturers may repair silencers, but replacement parts are "silencers" as defined in 18 U.S.C. § 921(a)(24) that must be registered and transferred in accordance with the NFA and GCA.

Q3: May the outer tube of a registered silencer be repaired due to damage? If so, may the repair be done by someone other than the original manufacturer?

A damaged outer tube may be repaired by any Federal firearms licensee qualified to perform gunsmithing or by the registered owner. The repair may not alter the dimensions or caliber of the silencer, except that the length of the outer tube may be reduced, as set forth above. The repair may not e performed if it results in the removal, obliteration, or alteration of the serial number, as this would iolate 18 U.S.C. § 922(k). In that case, the silencer may be returned to the registered owner in its riginal, damaged condition or destroyed. A replacement silencer must be registered and transferred to ne registrant of the damaged silencer in the same manner as a new silencer, subject to the registration nd transfer procedures of the NFA and GCA.

Q4: If the outer tube is destroyed or damaged beyond repair, may it be replaced?

A: Unless the outer tube is replaced by the manufacturer prior to its removal from the manufacturing remises for purposes of sale or distribution (see **Q6**), the replacement of the outer tube amounts to the naking of a new silencer. For the registered owner to fabricate a new outer tube, he or she must submit n ATF Form 1, Application to Make and Register a Firearm, pay the making tax of $200, and receive ATF approval. The application to make should indicate that the new tube is being fabricated for use in eplacing a damaged outer tube on a registered silencer, and the application should indicate the make, nodel and serial number of the registered silencer. It would be helpful for the applicant to include a opy of the approved registration for the silencer. Assembly of the newly fabricated tube with the other arts of the registered silencer does not require an additional application to make nor payment of another naking tax, as the one Form 1 will provide permission to fabricate the new tube and to assemble it with ne old silencer parts. The replacement tube must be marked in accordance with 27 C.F.R. § 479.102. The registrant may use the same serial number that appeared on the damaged tube.

f the registered owner wishes to acquire a replacement tube from a person other than a qualified nanufacturer, the replacement tube must be registered as a new silencer by the other person and ransferred to the registered owner in accordance with the NFA and GCA. The other person must ubmit an ATF Form 1, pay the $200 making tax, and receive ATF approval to make the replacement ube. The replacement tube must be marked in accordance with 27 C.F.R. §§ 478.92 and 479.102. The ther person would then transfer the replacement tube to the owner of the damaged silencer, subject to ne transfer tax, in accordance with the NFA and GCA. The new tube may be then be assembled with ne other parts. The original damaged silencer should be reported to the NFA Branch as destroyed.

Alternatively, a qualified manufacturer may replace the tube, report the manufacture on ATF Form 2, Notice of Firearms Manufactured or Imported, and transfer the replacement tube to the owner in ccordance with the NFA and GCA. The transfer must comply with the $200 transfer tax and all other rovisions of the NFA, as it would be a new silencer. The replacement tube must also be marked in ccordance with 27 C.F.R. §§ 478.92 and 479.102. The required markings include an individual serial umber and the name, city, and State of the manufacturer who replaced the tube. The replacement tube nay not be marked with the name, city, and State of the original manufacturer of the silencer, as this ould be a false marking. Although the new tube is a new silencer for purposes of the NFA, it would be replacement firearm of the same type as the original silencer, and it may be returned directly to the egistrant in interstate commerce in accordance with 18 U.S.C. § 922(a)(2). The original damaged ilencer should be reported to the NFA Branch as destroyed.

Q5: May a repair change the dimensions or caliber of a silencer?

A: If alterations to a silencer would increase the overall length or change the diameter or caliber of a ilencer, this is the making of a new silencer, as opposed to a repair. The new silencer must be egistered and transferred in accordance with the NFA and the GCA. Alterations to a registered silencer hat result in a minimal reduction in the overall length for purposes of rethreading are permissible as repairs. However, the reduction in length may not result in the removal, obliteration, or alteration of the existing serial number, as this would violate 18 U.S.C. § 922(k). If such a repair is necessary, the damaged silencer should be destroyed or returned to the registrant. If it is destroyed, destruction should be reported to the NFA Branch. Any replacement silencer must be registered and transferred in accordance with the NFA and the GCA. See Q2 and Q3 for further information on repairs.

Q6: If a silencer is found to be defective due to the manufacturing process, may it be replaced?

A: A silencer may be replaced only under the following circumstances:

(1) A manufacturer of silencers licensed as a manufacturer under the GCA who has paid special (occupational) tax under the NFA;

(2) prior to the time the silencer has left the manufacturer's premises;

(3) determines that a silencer of its own manufacture is defective.

If all the above criteria are satisfied, the manufacturer may destroy the defective silencer and replace it with another silencer. If the silencer has already been registered, the replacement silencer may be marked with the same serial number and markings as the original silencer. If the destruction is prior to registration on Form 2, the replacement silencer may be marked with the same serial number or another serial number. See also **Q2, Q3,** and **Q4**.

APPENDIX C

ATF FORMS

ATF Form 1 (5320.1)
Application to Make and Register a Firearm

DEPARTMENT OF THE TREASURY
BUREAU OF ALCOHOL, TOBACCO AND FIREARMS

APPLICATION TO MAKE AND REGISTER A FIREARM

(Detach this sheet before completing form)

INSTRUCTIONS

1. DEFINITIONS.

a. **NATIONAL FIREARMS ACT (NFA).** Title 26, United States Code, Chapter 53. The implementing regulations are found in Title 27, Code of Federal Regulations, Part 179.

b. **GUN CONTROL ACT (GCA).** Title 18, United States Code, Chapter 44. The implementing regulations are found in Title 27, Code of Federal Regulations, Part 178.

c. **FIREARM.** The term "firearm" means: (1) a shotgun having a barrel or barrels of less than 18 inches in length; (2) a weapon made from a shotgun if such weapon as modified has an overall length of less than 26 inches or a barrel or barrels of less than 18 inches in length; (3) a rifle having a barrel or barrels of less than 16 inches in length; (4) a weapon made from a rifle if such weapon as modified has an overall length of less than 26 inches or a barrel or barrels of less than 16 inches in length; (5) any other weapon, as defined in 18 U.S.C. 5845(e); (6) a machinegun; (7) a muffler or a silencer for any firearm whether or not such firearm is included within this definition; and (8) a destructive device.

d. **PERSON.** The term "person" means a partnership, company, association, trust, estate, or corporation as well as a natural person.

e. **EMPLOYER IDENTIFICATION NUMBER (EIN).** Required of taxpayer filing special (occupational) tax returns under 27 CFR § 179.35.

f. **SPECIAL (OCCUPATIONAL) TAX.** Required by the NFA to be paid by a Federal firearms licensee engaged in the business of manufacturing (Class 2), importing (Class 1), or dealing (Class 3) in NFA firearms.

g. **FEDERAL FIREARMS LICENSE.** A license issued under the provisions of the GCA to manufacture, import or deal in firearms.

h. **ATF OFFICER.** An officer or employee of the Bureau of Alcohol, Tobacco and Firearms (ATF) authorized to perform any function relating to the administration of the NFA.

i. **MAKE.** The term "make", and the various derivatives of such word, shall include manufacturing *(other than by one qualified to engage in such business under the NFA)*, putting together, altering, any combination of these, or otherwise producing a firearm.

j. **MISDEMEANOR CRIME OF DOMESTIC VIOLENCE.** A crime that is a misdemeanor under Federal or State law and has, as an element, the use or attempted use of physical force, or the threatened use of a deadly weapon, committed by a current or former spouse, parent, or guardian of the victim, by a person with whom the victim shares a child in common, by a person who is cohabiting with or has cohabited with the victim as a spouse, parent, or guardian. The term includes all misdemeanors that involve the use or attempted use of physical force *(e.g., simple assault, assault and battery)*, if the offense is committed by one of the defined parties. The person is NOT considered to have been convicted of such crime unless the person was represented by a lawyer or gave up the right to a lawyer, and, if the person was entitled to a jury trial, was tried by a jury or gave up the right to a jury trial.

k. **REACTIVATION.** The restoration of a registered unserviceable NFA firearm to a functional condition. This action incurs the making tax liability.

l. **UNSERVICEABLE FIREARM.** One which is incapable of discharging a shot by means of an explosive and incapable of being readily restored to firing condition. An acceptable method of rendering most firearms unserviceable is to fusion weld the chamber closed and fusion weld the barrel solidly to the frame.

2. PREPARATION OF APPLICATION AND PAYMENT OF TAX.

a. **AUTHORITY.** As provided by 26 U.S.C. § 5822, any person *(other than a qualified manufacturer of firearms (see paragraph b))* seeking to make a firearm must complete, in duplicate, a separate application on this form for each firearm. The applicant maker must furnish all the information called for on this application form.

b. **REGISTRATION BY QUALIFIED MANUFACTURER.** A person who has a Federal firearms license to manufacture firearms *(Type 07 or 10)* <u>and</u> who has paid special (occupational) tax to manufacture NFA firearms is exempt from the making tax and filing of the ATF F 1 application. Such qualified manufacturer must report and register each NFA firearm manufactured by filing ATF F 2, Notice of Firearms Manufactured or Imported, as required by 27 CFR § 179.102.

c. **PAYMENT OF/EXEMPTION FROM PAYMENT OF TAX.** As provided in 26 U.S.C. § 5821, there is a $200.00 tax on each firearm made, except as provided in 26 U.S.C. §§ 5852 and 5853, when an NFA firearm may be made without payment of the tax when made by, or on behalf of the United States or any State or political subdivision thereof. Documentation that the firearm is being made for a government entity, such as a United States government contract or a State or local government agency purchase order, must accompany the application. The reactivation of a registered unserviceable firearm is subject to the making tax.

d. **PHOTOGRAPHS AND FINGERPRINTS.** An individual maker must (1) attach to each copy in item 12 of the ATF F 1, a 2 inch x 2 inch photograph of his/her frontal view taken within 6 months prior to the date of the application and (2) submit two properly completed FBI Forms FD-258 *(Fingerprint Card with blue lines)* with the application. The fingerprints must be clear for accurate classification and taken by someone properly equipped to take them.

e. **SIGNATURES.** All signatures required on ATF F 1 must be original in ink on both copies.

f. **REMITTANCE.** If the application is subject to the making tax, a check or money order, made payable to the Bureau of Alcohol, Tobacco and Firearms, in the amount of $200.00 must be submitted with the ATF F 1. <u>Do not send cash</u>.

g. **PHOTOCOPIES OR COMPUTER GENERATED VERSIONS.** The ATF F 1 may be photocopied or a computer-generated version *(as long as it is in the same format and contains all required information)* may be used. This form may also be downloaded from the ATF internet website at www.atf.treas.gov.

h. **SERIAL NUMBERS AND OTHER MARKINGS.** If an existing firearm is being modified into an NFA firearm, enter the existing serial number of the firearm into item 4g and the name and address of the original manufacturer into item 4a. DO NOT ALTER OR MODIFY THE EXISTING SERIAL NUMBER. If the NFA firearm is being made from parts, your name and address are to be entered into 4a and a serial number you create is to be entered into item 4g.

i. **SUBMISSION.** All requested information must be entered in blue or black ink and must be legible. Illegible entries will be returned for correction. Send both copies of the ATF F 1 and attachments to the address located in the upper right hand corner on the face side of the ATF F 1. The return of the application or your sending it to any other address will only delay the processing.

3. **APPROVAL OF APPLICATION.** Upon approval of an application, the NFA Branch will affix the NFA tax stamp *(if any)* to the application, cancel it, and return the approved copy to the maker. The approval of the application effectuates registration of the firearm to the maker; however, the firearm must not be made until the application has been approved.

4. **WITHDRAWAL OF APPLICATION.** The application may be withdrawn prior to approval by submission of a written request from the maker. The NFA Branch will arrange for a refund of any tax paid.

5. **CANCELLATION OF APPROVED APPLICATION.** An approved application may be cancelled <u>only</u> if the firearms had not been made or modified. The maker must return the approved application with a written request for cancellation, citing the need and that the making of the firearm did not take place. The NFA Branch will arrange for a refund of any tax paid.

6. **DISAPPROVAL OF APPLICATION.** If the application is disapproved, the NFA Branch will note the reason for disapproval on the application and return one copy to the maker. The NFA Branch will arrange for a refund of any tax paid.

7. **REASONS FOR DISAPPROVAL.** 26 U.S.C. § 5822 provides that applications shall be denied if the making or possession of the firearm would place the maker in violation of law.

a. **STATE OR LOCAL LAW.** If State or local law prohibits the making or possession of the firearm being made, the application will be disapproved.

b. **MACHINEGUNS AND SEMIAUTOMATIC ASSAULT WEAPONS.**

(1) 18 U.S.C. § 922(o) provides that machineguns may be made only for government use or export. An application will be denied unless the making meets these criteria.
(2) 18 U.S.C. § 922(v) generally prohibits the manufacture of a semiautomatic assault weapon. The making of an NFA firearm does not provide relief from this prohibition.

c. **PERSONS PROHIBITED FROM MAKING A FIREARM.** The application will be disapproved if the maker is a person prohibited from possessing a firearm by 18 U.S.C. § 922(g), which provides that is shall be unlawful for any person-

(1) who has been convicted in any court of a crime punishable by imprisonment for a term exceeding one year;
(2) who is a fugitive from justice;
(3) who is an unlawful user of or addicted to any controlled substance;
(4) who has been adjudicated as a mental defective or who has been committed to a mental institution;
(5) who, being an alien, is illegally or unlawfully in the United States or, except as provided in 18 U.S.C. § 922(y)(2), has been admitted to the United States under a nonimmigrant visa *(as that term is defined in 8 U.S.C. § 1101(a)(26))*;
(6) who has been discharged from the Armed Forces under dishonorable conditions;
(7) who, having been a citizen of the United States, has renounced his citizenship;
(8) who is subject to a court order that-

(A) was issued after a hearing of which such person received actual notice; and at which such person had an opportunity to participate;

ATF F 1 (5320.1) (3-2001)

INSTRUCTIONS *(Continued)*

(B) restrains such person from harassing, stalking, or threatening an intimate partner of such person or child of such intimate partner or person, or engaging in other conduct that would place an intimate partner in reasonable fear of bodily injury to the partner or child; and

(C) (i) includes a finding that such person represents a credible threat to the physical safety of such intimate partner or child; or (ii) by its terms explicitly prohibits the use, attempted use, or threatened use of physical force against such intimate partner or child that would reasonably be expected to cause bodily injury, or

(9) who has been convicted in any court of a misdemeanor crime of domestic violence, to ship or transport in interstate or foreign commerce, or possess in or affecting commerce, or possess in or affecting commerce, any firearm or ammunition; or to receive any firearm or ammunition which has been shipped or transported in interstate or foreign commerce.

8. **STATUS INQUIRIES AND QUESTIONS.** The NFA Branch telephone number is (202) 927-8330. Any inquiry relating to the status of an applicable to make an NFA firearm or about procedures in general should be directed to the NFA. Please be aware that the information relating to the registration of firearms is defined as "return" or "return information" by 26 U.S.C. § 6103 and disclosure is generally prohibited to anyone other than the registrant. Information relating to the NFA and other firearms laws is also available at the ATF Internet website **www.atf.treas.gov**.

9. **PENALTIES.** Any person who violates or fails to comply with any of the requirements of the NFA shall, upon conviction, be fined not more than $10,000 or be imprisoned for not more than 10 years, or both. Any firearm involved in a violation of the NFA shall be subject to seizure and forfeiture. It is unlawful for any person to make or cause the making of a false entry on any application or record required by the NFA knowing such entry to be false.

10. **COMPLIANCE WITH THE GUN CONTROL ACT.** All provisions of the GCA must also be complied with.

PRIVACY ACT INFORMATION

1. **AUTHORITY.** Solicitation of this information is made pursuant to the National Firearms Act *(26 U.S.C. §§ 5821 and 5822)*. Disclosure of this information by the applicant is mandatory for any person *(other than a manufacturer qualified under the National Firearms Act)* making a firearm as defined in the National Firearms Act.

2. **PURPOSE.** To verify payment of the tax imposed by 26 U.S.C. § 5821; to determine that the making would not be in violation of law; and to effect registration of the firearm.

3. **ROUTINE USES.** The information will be used by ATF to make the determinations set forth in paragraph 2. In addition, to effect registration of the firearm, information as to the identification of the firearm, date of registration, and the identification and address of person entitled to possess the firearm will be entered into the National Firearms Registration and Transfer Record. No information obtained from a application, registration, or records required to be submitted by a natural person in order to comply with any provision of the National Firearms Act or regulations issued thereunder, shall, except in connection with prosecution or other action for furnishing false information, be used, directly or indirectly, as evidence against that person in any criminal proceedng with respect to a violation of law occurring prior to or concurrently with the filing of the application. The information from this application may only be disclosed to Federal authorities for purpose of prosecution for violation of the National Firearms Act.

4. **EFFECTS OF NOT SUPPLYING INFORMATION REQUESTED.** Failure to supply complete information will delay processing and may cause denial of the application.

PAPERWORK REDUCTION ACT NOTICE

This form is in accordance with the Paperwork Reduction Act of 1995. The information you provide is used to establish that a transferee's receipt and possession of the firearm would be in conformance with Federal, State, and local law. The data is used as proof of lawful registration of a firearm to the manufacturer. The furnishing of this information is mandatory *(26 U.S.C. § 5822)*.

The estimated average burden associated with this collection of information 4 hours per respondent or recordkeeper, depending on individual circumstances. Comments concerning the accuracy of this burden estimate and suggestions for reducing this burden should be addressed to Reports Management Officer, Document Services Branch, Bureau of Alcohol, Tobacco and Firearms, Washington, DC 20226.

An agency may not conduct or sponsor, and a person is not required to respond to, a collection of information unless it displays a currently valid OMB control number.

ATF F 1 (5320.1) (3-2001)

OMB No. 1512-0024 (02/29/2004)

DEPARTMENT OF THE TREASURY
BUREAU OF ALCOHOL, TOBACCO AND FIREARMS

APPLICATION TO MAKE AND REGISTER A FIREARM

(Submit in duplicate. See instructions attached.)

ATF CONTROL NUMBER

TO: National Firearms Act Branch, Bureau of Alcohol, Tobacco and Firearms, Washington, DC 20226

The undersigned hereby makes application, as required by Sections 5821 and 5822 of the National Firearms Act, Title 26 U.S.C., Chapter 53, to make and register the firearm described below.

1. TYPE OF APPLICATION *(Check one)*

☐ a. TAX PAID. Submit with your application a check or money order for $200 made payable to the Department of the Treasury. Upon approval of the application, this office will acquire, affix, and cancel the required "National Firearms Act" stamp for you.

☐ b. TAX EXEMPT because firearm is being made on behalf of the United States, or any department, independent establishment, or agency thereof.

☐ c. TAX EXEMPT because firearm is being made by or on behalf of any State or possession of the United States, or any political subdivison thereof, or any official police organization of such a government entity engaged in criminal investigations.

2. APPLICATION IS MADE BY: ☐ INDIVIDUAL ☐ CORPORATION OR OTHER BUSINESS ENTITY ☐ GOVERNMENT ENTITY

3a. TRADE NAME *(If any)*

3b. APPLICANT'S NAME AND MAILING ADDRESS *(Type or print below and between the dots)*

3c. IF P.O. BOX IS SHOWN ABOVE, STREET ADDRESS MUST BE GIVEN HERE

3d. COUNTY

3e. TELEPHONE AREA CODE AND NUMBER

IMPORTANT: COMPLETE THE REVERSE SIDE. INDIVIDUALS *(INCLUDING FEDERALLY LICENSED COLLECTORS)* MUST ALSO SUBMIT, IN DUPLICATE, FBI FORM FD-258, FINGERPRINT CARD.

4. DESCRIPTION OF FIREARM *(Complete items a through i)*

a. NAME AND LOCATION OF ORIGINAL MANUFACTURER OF FIREARM *(RECEIVER) (If prototype, furnish plans and specifications) (See instruction 2h)*

b. TYPE OF FIREARM TO BE MADE *(See instruction 1c)*

c. CALIBER, GAUGE OR SIZE *(Specify)*

d. MODEL

LENGTH *(inches)*: e. OF BARREL; f. OVERALL

g. SERIAL NUMBER *(See instruction 2h.)*

h. ADDITIONAL DESCRIPTION *(Include all numbers and other identifying data which will appear on the firearm)*

i. STATE WHY YOU INTEND TO MAKE FIREARM *(Use additional sheet if necessary)*

5. APPLICANT'S FEDERAL FIREARMS LICENSE *(If any)*
(Give complete 15-digit number)

6. SPECIAL (OCCUPATIONAL) TAX STATUS

a. EMPLOYER IDENTIFICATION NUMBER *(If applicable)*

b. CLASS

IMPORTANT: GIVE FULL DETAILS ON SEPARATE SHEET FOR ALL "YES" ANSWERS IN ITEMS 7 AND 8

7. ARE YOU:	YES	NO
a. Charged by information or under indictment in any court for a crime for which the judge could imprison you for more than one year. An information is a formal accusation of a crime made by a prosecuting attorney.		
b. A **fugitive** from justice?		
c. An alien who is **illegally** in the United States?		
d. Under 21 years of age?		
e. An unlawful user of or addicted to, marijuana, or any depressant, stimulant, or narcotic drug, or any other controlled substance?		
f. Subject to a court order restraining you from harassing, stalking or threatening an intimate partner or child of such partner? *(See instruction 7c)*		

8. HAVE YOU:	YES	NO
a. Been convicted in any court of a crime for which the judge could have imprisoned him/her for more than one year, even if the judge actually gave him/her a shorter sentence?		
b. Been discharged from the armed forces under **dishonorable** conditions?		
c. Been adjudicated mentally defective or been committed to a mental institution?		
d. Renounced his or her United States citizenship?		
e. Been convicted in any court of a misdemeanor crime of domestic violence? *(See instruction 1m)*		

UNDER PENALTIES OF PERJURY, I DECLARE that I have examined this application including accompanying documents, and to the best of my knowledge and belief it is true, accurate and complete and the making and possession of the firearm described above would not constitute a violation of Chapter 44, Title 18, U.S.C., Chapter 53, Title 26, U.S.C., or any provisions of State or local law.

9. SIGNATURE OF APPLICANT

10. NAME AND TITLE OF AUTHORIZED OFFICIAL OF FIRM OR CORPORATION *(If applicable)*

11. DATE

THE SPACE BELOW IS FOR THE USE OF THE BUREAU OF ALCOHOL, TOBACCO AND FIREARMS

By authority of the Director, Bureau of Alcohol, Tobacco and Firearms, this application has been examined and the applicant's making and registration of the firearm described above is:

☐ APPROVED *(With the following conditions, if any)*

☐ DISAPPROVED *(For the following reasons)*

AUTHORIZED ATF OFFICIAL

DATE

ATF F 1 (5320.1) (3-2001) PREVIOUS EDITIONS ARE OBSOLETE

ADDITIONAL REQUIREMENTS

12. PHOTOGRAPH

AFFIX
RECENT PHOTOGRAPH HERE
(Approximately 2" x 2")
(See instruction 2d)

13. **LAW ENFORCEMENT CERTIFICATION** *(See IMPORTANT note below)*

I certify that I am the chief law enforcement officer of the organization named below having jurisdiction in the area of residence of

____
(Name of Maker)

I have no information indicating that the maker will use the firearm or device described on this application for other than lawful purposes. I have no information that POSSESSION OF THE FIREARM DESCRIBED IN ITEM 4 ON THE FRONT OF THIS FORM WOULD PLACE THE MAKER IN VIOLATION OF STATE OR LOCAL LAW.

____
(Signature and Title of Chief Law Enforcement Officer - see IMPORTANT note below)

BY *(See IMPORTANT NOTE BELOW)* ____
(Signature and Title of Delegated Person)

____
(Organization)

____
(Street Address)

________________________________ ____________
(City, State, and ZIP Code) *(Date)*

IMPORTANT: The chief law enforcement officer is considered to be the Chief of Police for the maker's city or town of residence, the Sheriff for the maker's county of residence; the Head of the State Police for the maker's State of residence; a State or local district attorney or prosecutor having jurisdiction in the maker's area of residence; or another person whose certification is acceptable to the Director, Bureau of Alcohol, Tobacco and Firearms. If someone has specific delegated authority to sign on behalf of the Chief of Police, Sheriff, etc., this fact must be noted by printing the Chief's, Sheriff's, or other authorized official's name and title, followed by the word "by" and the full signature and title of the delegated person. Item 13 must be completed for an individual maker. The certification must be dated no more than one year prior to the date of receipt of the application

IMPORTANT INFORMATION FOR CURRENTLY REGISTERED FIREARMS

If this registration document evidences the current registration of the firearm described on it, please note the following information.

ESTATE PROCEDURES: For procedures regarding the transfer of firearms in an estate resulting from the death of the registrant identified in item 3, the executor should contact the NFA Branch, Bureau of ATF, Washington, DC 20226.

CHANGE OF ADDRESS: The registrant shall notify the NFA Branch, Bureau of Alcohol, Tobacco and Firearms, Washington, DC 20226, in writing, of any change to the address in item 3a.

INTERSTATE MOVEMENT: If the firearm identified in item 4 is a machinegun, short-barreled rifle, short-barreled shotgun, or destructive device, the registrant may be required by 18 U.S.C. § 922(a)(4) to obtain permission from ATF prior to any transportation in interstate or foreign commerce.

CHANGE OF DESCRIPTION: The registrant shall notify the NFA Branch, Bureau of Alcohol, Tobacco and Firearms, Washington, DC 20226, in writing, of any change to the description of the firearms in item 4.

RESTRICTIONS ON POSSESSION: Any restriction *(see approval block on face of form)* on the possession of the firearm identified in item 4 continues with the further transfer of the firearm.

PERSONS PROHIBITED FROM POSSESSING FIREARMS: If the registrant becomes prohibited by 18 U.S.C. § 922 from possessing a firearm, the registrant shall notify the NFA Branch, Bureau of Alcohol, Tobacco and Firearms, Washington, DC 20226, in writing, immediately upon becoming pro-hibited for guidance on the disposal of the firearm.

PROOF OF REGISTRATION: This approved application is the registrant's proof of registration and it shall be made available to any ATF officer upon request.

ATF F 1 (5320.1) (3-2001)

ATF Form 2 (5320.2)
Notice of Firearms Manufactured or Imported

INSTRUCTIONS

1. Preparation of Notice of Firearms Manufactured or Imported.

a. This form is required to effect the registration of all firearms imported, manufactured, remanufactured or reactivated by qualified Federal firearms licensees who have paid the special (occupational) tax to import or manufacture firearms.

b. Reactivation of an NFA firearm - Any NFA firearm (including the frame or receiver of such firearm) must be registered to the possessor in order to be lawfully possessed. A Form 2 to register a reactivated NFA firearm will not be accepted if the unserviceable firearm is not registered to the applicant. The firearm in that event, would be considered contraband and would be subject to seizure and forfeiture provisions of the law.

c. Remanufacture of a firearm - Any NFA firearm has been properly destroyed, and thereby removed from the purview of the NFA, which are subsequently restored. (This could also include the converting of semi-automatic firearms to full automatic firearms). Public Law 99-308, §922(o) restrictions apply.

d. Manufacture of NFA firearms - Other types of manufacturing operations which result in the manufacture or making of an NFA firearm. (This includes producing firearms from scratch).

e. A separate Form 2 must be submitted for the four categories of manufacture, remanufacture, reactivation and importation of firearms.

f. If the importation involves more than one import permit, a separate Form 2 must be filed to report those firearms imported under each permit.

g. Serial numbers - Sections 178.92 and 179.102 of the regulations require that an individual serial number, *not duplicating any serial number placed by the manufacturer or importer on any other firearm*, must be placed on the firearm. However, please do not alter existing serial numbers.

h. When large numbers of firearms of the same type and model are reported, lists of serial numbers (double-spaced) may be attached to this form. The number of firearms in item 3b should agree with total number of firearms reported. Lists must be referenced to this form by date and manufacturer or importer in such a way that if pages become separated they can be matched up to the Form 2.

i. Photostatic or carbon signatures are not acceptable. Although typed forms are preferred, pen and ink may be used; forms completed in pencil will not be accepted. Facsimile copies will be accepted, if an original affidavit is on file with us.

j. If any questions arise concerning the preparation of this form, please contact the nearest ATF Office or the National Firearms Act Branch at (202) 927-8330.

2. Where to File Form - Submit completed forms to the National Firearms Act Branch, Bureau of Alcohol, Tobacco and Firearms, Room 5440, 650 Massachusetts Ave., N.W., Washington, DC 20226.

3. When to File Forms -

a. All firearms manufactured, remanufactured or reactivated during a single day must be filed no later than the close of the next business day. (27 CFR 179.112).

4. Disposition of Form - The manufacturer or importer must prepare the form, in duplicate, file the original with the National Firearms Act Branch and keep the copy with the firearms records required to be retained at the premises covered by the required special (occupational) tax stamp.

5. Receipt of Form by the Bureau of Alcohol, Tobacco and Firearms-

a. The receipt of this form properly prepared and executed by a manufacturer will register the firearms listed on the form to the manufacturer or importer, with the exception noted in item 1f. of these instructions.

b. Timely receipt by ATF of a properly prepared and executed form and timely receipt by the Imports Branch of a copy of ATF Form 6a (required by 27 CFR 179.112) covering the firearm(s) reported on the form by the importer, will register the listed firearms to the importer.

Paperwork Reduction Act Notice

This request is in accordance with the Paperwork Reduction Act of 1980. The information you provide as a qualified licensed firearms manufacturer or importer is to register, as required by law, firearms within the jurisdiction of the National Firearms Act which have been lawfully manufactured or imported. The data is used to determine applicant's eligibility to register the firearms described. The furnishing of the information is mandatory (26 U.S.C. 5841c).

The estimated average burden associated with this collection of information is 30 minutes per respondent or recordkeeper depending on individual circumstances. Comments concerning the accuracy of this burden estimate and suggestions for reducing this burden should be addressed to Reports Management Officer, Information Programs Branch, Bureau of Alcohol, Tobacco and Firearms, Washington, D.C. 20226, and the Office of Management and Budget, Paperwork Reduction Project (1512-0025), Washington, D.C. 20503.

*U.S. GPO: 1993-342-646/64479 ATF F 2 (5320.2) (8-93)

Form Approved: OMB No. 1512-0025 (02/28/95)

DEPARTMENT OF THE TREASURY
BUREAU OF ALCOHOL, TOBACCO AND FIREARMS

NOTICE OF FIREARMS MANUFACTURED OR IMPORTED

(Complete in duplicate - See Instructions on reverse)

TO: THE DIRECTOR, BUREAU OF ALCOHOL, TOBACCO AND FIREARMS, WASHINGTON, DC 20226

The undersigned hereby serves notice of the manufacture, reactivation, or importation of firearms as required by section 5841 of the National Firearms Act, Title 26, U.S.C. Chapter 53.

2. PRINT NAME AND TITLE OF PERSON AUTHORIZED TO SIGN FOR A BUSINESS OR FIRM

3. NAME AND ADDRESS *(Include trade name)*

☐ SOLE PROPRIETOR ☐ PARTNERSHIP ☐ CORPORATION

4. WHERE ARE FIREARMS KEPT?

4a. TELEPHONE NUMBER *(Include area code)*

1. TYPE OF NOTICE

a. FIREARMS ON THIS NOTICE ARE: *(Check one)*

☐ MANUFACTURED

☐ REMANUFACTURED

☐ REACTIVATED

☐ IMPORTED *(If imported complete items c and d)*

b. NUMBER OF FIREARMS COVERED BY THIS NOTICE *(See Instruction 1.c.)*

FOR IMPORTED FIREARMS ONLY

c. IMPORTATION PERMIT NUMBER *(See Instruction 1.d. on reverse)*

d. PERMIT EXPIRATION DATE

5. DESCRIPTION OF FIREARMS *(Complete all items)*

DATE OF MANUFACTURE OR REACTIVATION *(If imported, give date released from Customs custody & manufacturer's name) (If remanufactured or reactivated, indicate name of original manufacuter)* a	TYPE OF FIREARM *(Shortbarreled rifle, machine gun, destructive device, etc.)* b	CALIBER GAUGE OR SIZE c	MODEL d	LENGTH *(in.)* OF BAR-REL e	OVER-ALL f	SERIAL NUMBER *(See Instructions 1.e. and 1.g.)* g

ADDITIONAL DESCRIPTION ON THE FIREARM(S) DESCRIBED ABOVE; I.E. IF CONVERTED BY KIT. (Attach additional sheets if necessary)

6. FEDERAL FIREARMS LICENSE			7. SPECIAL (OCCUPATIONAL) TAX STAMP	
LICENSE NUMBER	TYPE OF BUSINESS	EXPIRATION DATE	EMPLOYER IDENTIFICATION NUMBER	CLASS

UNDER PENALTIES OF PERJURY, I DECLARE that I have examined this notice of firearms manufactured, remanufactured, reactivated or imported and, to the best of my knowledge and belief, it is true, correct and complete.

8. SIGNATURE OF MANUFACTURER OR IMPORTER *(Or authorized official shown in item 2)*

9. DATE

ATF F2 (5320.2) (8-93)

ATF Form 3 (5320.3)

Application for Tax-Exempt Transfer of Firearm and Registration to Special (Occupational) Taxpayer

INSTRUCTIONS

This form must be completed, in duplicate, and used only between those who have paid the appropriate fees to maintain a current Federal firearms license which is issued every three years and valid special *(occupational)* tax stamp which must be renewed every July 1st. The FFL and SOT stamp qualifies a person to import, manufacture or deal in National Firearms Act (NFA) firearms. The applicant must furnish all the information called for on the form. The transferor's and transferee's names must be shown exactly as they appear on the federal firearms license.

1. **PRINT** or type firearm **buyer's *full name*** including middle name and ***address.*** P.O. Box number, rural route or highway and box number and ***trade name.***

2a. **PRINT** or type firearm **owner's *full name*** including middle and complete ***address,*** former and current ***trade name.***

2b. The **form** will not be mailed to this address. It will be **returned** to the address in item 2a.

2c. The firearm **seller's *telephone number.***

3a - g. **PRINT** or type **original manufacturer's** or importer's ***name*** and/or tradename and address and ensure that all of the information referencing the firearm is complete and accurate. See General Information below for more details.

3h. **PRINT** or type additional information referencing the firearm, such as a complete description including serial number of the Title I firearm into which a registered auto sear may be placed.

4. **PRINT** or type firearm **buyer's *VALID*** 15 digit **FFL #.**

5a. **PRINT** or type firearm **buyer's *VALID*** 9 digit **EIN #.**

5b. Is the firearm **buyer** a **Class** 1, 2, or 3?

6. **PRINT** or type firearm **owner's *VALID*** 15 digit **FFL #.**

7a. **PRINT** or type firearm **owner's *VALID*** 9 digit **EIN #.**

7b. Is the firearm **owner** a **Class** 1, 2, or 3?

8. Firearm **owner's original *signature.***

9. **PRINT** or type firearm **owner's *name*** and ***title.***

10. **DATE** firearm **owner *prepared.***

GENERAL INFORMATION

1. **Ensure** that the **state** in which the firearm buyer resides **deos not prohibit NFA firearms.** See "Firearms State Laws and Published Ordinances", ATF P 5300.5.
2. Prior to completing and **mailing** the Form 3 to the NFA Branch, BATF, Wash., DC 20226, please **make a physical examination of the firearm** to **ensure** that the information in **items 3a - g** is **complete and accurate** and ***matches the information permanently placed on the firearm.***
3. The firearm **owner can not deliver** the firearm to the firearm buyer until the approved form is received from the Bureau of ATF.
4. If the firearm transaction cannot be completed immediately, please void the transfer.
5. Always **initial** and **date** any ***corrections*** you make on the form.
6. **RECORDKEEPING REQUIREMENTS:** Pursuant to Title 18, U.S.C., Chapter 44, the transferor and the transferee, as firearm licensees, shall record their disposition and acquisition of the described firearm as required by subpart H of Title 27 CFR, Part 178.

PAPERWORK REDUCTION ACT NOTICE

This form is in accordance with the Paperwork Reduction Act of 1995. The information you provide will be used to apply to transfer firearms tax exempt from one Federal firearms licensee and special *(occupational)* taxpayer qualified to deal in NFA firearms to another qualified special taxpayer. The data is used to verify lawful transfer and registration of firearms. The information being furnished is mandatory (26 U.S.C. 5812).

The estimated average burden associated with this collection of information is 30 minutes per respondent or recordkeeper, depending on individual circumstances. Comments concerning the accuracy of this burden estimate and suggestions for reducing this burden should be addressed to the Reports Management Officer, Document Services Branch, Bureau of Alcohol, Tobacco and Firearms, Washington, DC 20226.

An agency may not conduct or sponsor, and a person is not required to respond to, a collection of information unless it displays a currently valid OMB control number.

PRIVACY ACT INFORMATION

1. **AUTHORITY.** Solicitation of this information is made pursuant to the National Firearms Act (26 U.S.C. §5812 and 5851 and 5852). Disclosure of this information by the applicant is mandatory for any person *(other than a manufacturer qualified under the National Firearms Act)* making a firearm as defined in the National Firearms Act.
2. **PURPOSE.** To verify the tax exemption imposed by 26 U.S.C. § 5851; to determine that the transfer would not be in violation of law; and to effect registration of the firearm.
3. **ROUTINE USES.** The information will be used by ATF to make the deteminations set forth in paragraph 2. In addition, to effect registration of the firearm, information as to the identification of the firearm, date of registration, and the identification and address of person entitled to possess the firearm will be entered into the National Firearms Registration and Transfer Record. No information obtained from an application, registration, or records required to be submitted by a natural person in order to comply with any provision of the National Firearms Act or regulations issued thereunder, shall, except in connection with prosecution or other action for furnishing false information, be used, directly or indirectly, as evidence against that person in any criminal proceeding with respect to a violation of law occuring prior to or concurrently with the filing of the application. The information from this application may only be disclosed to Federal authorities for purpose of prosecution for violation of the National Firearms Act.
4. **EFFECTS OF NOT SUPPLYING INFORMATION REQUESTED.** Failure to supply complete information will delay processing and may cause denial of the application.

ATF F 3 (5320.3) (2-98)

OMB No. 1512-0026 (09/30/98)

DEPARTMENT OF THE TREASURY
BUREAU OF ALCOHOL, TOBACCO AND FIREARMS
APPLICATION FOR TAX-EXEMPT TRANSFER OF FIREARM AND REGISTRATION TO SPECIAL (OCCUPATIONAL) TAXPAYER National Firearms Act (Title 26, U.S.C. Chapter 53)

1. TRANSFEREE'S NAME AND ADDRESS *(Show Trade Name, if any, and give number, street, city, State and ZIP Code, P.O. Box alone is not sufficient.)*

☐ SOLE PROPRIETOR ☐ PARTNERSHIP ☐ CORPORATION

To be submitted in duplicate by transferor of firearm

TO: National Firearms Act Branch
Bureau of Alcohol, Tobacco and Firearms
Washington, D.C. 20226

2a. TRANSFEROR'S NAME AND MAILING ADDRESS *(Show Trade Name, if any)*

☐ SOLE PROPRIETOR ☐ PARTNERSHIP ☐ CORPORATION

2b. NUMBER, STREET, CITY, STATE AND ZIP CODE OF LICENSED PREMISES IF DIFFERENT FROM ITEM 2a.

2c. TELEPHONE NO. *(Include Area Code)*

The above-named and undersigned transferor and special (occupational) taxpayer hereby makes application as required by Section 5812 of the National Firearms Act to transfer, without payment of tax, and register the firearm described below to the special *(occupational)* taxpayer identified as the transferee in this application.

3. DESCRIPTION OF FIREARM *(Complete items a through h, if applicable)*

a. NAME AND ADDRESS OF MANUFACTURER AND/OR IMPORTER OF FIREARM	b. TYPE OF FIREARM *(Shortbarreled rifle, machine gun, destructive device, any other weapon, etc.)*	c. CALIBER, GAUGE OR SIZE *(Specify)*	d. MODEL LENGTH *(Inches)* e. OF BARREL: f. OVERALL: g. SERIAL NUMBER

h. ADDITIONAL DESCRIPTION OR DATA APPEARING ON FIREARM *(Attach additional sheet if necessary)*

4. TRANSFEREE'S FEDERAL FIREARMS LICENSE *(If any)*
(Give complete 15-digit number)

First 6 digits	2 digits	2 digits	5 digits

5. TRANSFEREE'S SPECIAL (OCCUPATIONAL) TAX STATUS

a. EMPLOYER IDENTIFICATION NUMBER	b. CLASS

6. TRANSFEROR'S FEDERAL FIREARMS LICENSE *(If any)*
(Give complete 15-digit number)

First 6 digits	2 digits	2 digits	5 digits

7. TRANSFEROR'S SPECIAL (OCCUPATIONAL) TAX STATUS

a. EMPLOYER IDENTIFICATION NUMBER	b. CLASS

I believe I am entitled to exemption from payment of the transfer tax imposed by Section 5811, National Firearms Act (NFA), on the firearm described above because the transferee named herein is qualified under the NFA to manufacture, import or deal in that type of firearm.

UNDER PENALTIES OF PERJURY, I DECLARE that I have examined this application, and to the best of my knowledge and belief it is true, correct and complete.

8. ORIGINAL SIGNATURE OF TRANSFEROR *(Or authorized official)*	9. NAME AND TITLE OF AUTHORIZED OFFICIAL *(Print or type)*	10. DATE

THE SPACE BELOW IS FOR THE USE OF THE BUREAU OF ALCOHOL, TOBACCO AND FIREARMS

BY AUTHORITY OF THE DIRECTOR, THIS APPLICATION HAS BEEN EXAMINED, AND THE TRANSFER AND REGISTRATION OF THE FIREARM DESCRIBED HEREIN AND THE INTERSTATE MOVEMENT OF THAT FIREARM, WHEN APPLICABLE, TO THE TRANSFEREE ARE:

☐ APPROVED *(with the following conditions, if any)* ☐ DISAPPROVED *(For the following reasons)*

AUTHORIZED ATF OFFICIAL	DATE

ATF F 3 (5320.3) (2-98) PREVIOUS EDITION IS USEABLE

TF Form 4 (5320.4)
pplication for Tax paid Transfer and Registration of Firearm

Form Approved: OMB 1512-0027(01/31/00)

DEPARTMENT OF THE TREASURY — BUREAU OF ALCOHOL, TOBACCO AND FIREARMS

APPLICATION FOR TAX PAID TRANSFER AND REGISTRATION OF FIREARM

DETACH THIS SHEET BEFORE COMPLETING FORM
INSTRUCTIONS

1. DEFINITIONS

a. FIREARM. The term "firearm" means: (1) a shotgun having a barrel or barrels of less than 18 inches in length; (2) a weapon made from a shotgun if such weapon as modified has an overall length of less than 26 inches or a barrel or barrels of less than 18 inches in length; (3) a rifle having a barrel or barrels of less than 16 inches in length; (4) a weapon made from a rifle if such weapon as modified has an overall length of less than 26 inches or a barrel or barrels of less than 16 inches in length; (5) any other weapon, as defined in b. below; (6) a machinegun; (7) a muffler or a silencer for any firearm whether or not such firearm is included within this definition; and (8) a destructive device. The term "firearm" shall not include an antique firearm or any device (other than a machinegun or destructive device) which, although designed as a weapon, the Director, Bureau of Alcohol, Tobacco and Firearms, or authorized delegate, finds by reason of the date of its manufacture, value, design and other characteristics is primarily a collector's item and is not likely to be used as a weapon.

b. ANY OTHER WEAPON. The term "any other weapon" means any weapon or device capable of being concealed on the person and from which a shot can be discharged through the energy of an explosive; a pistol or revolver having a barrel with a smooth bore designed or redesigned to fire a fixed shotgun shell; weapons with combination shotgun and rifle barrels 12 inches or more, less than 18 inches in length, from which only a single discharge can be made from either barrel without manual reloading; and shall include any such weapon which may be readily restored to fire. Such term shall not include a pistol or a revolver having a rifled bore, or rifled bores, or weapons designed, made, or intended to be fired from the shoulder and not capable of firing fixed ammunition.

c. TRANSFEROR. The term "transferor" means the registered owner of a firearm who is applying to transfer it.

d. TRANSFEREE. The term "transferee" means the person desiring to acquire the firearm.

e. PERSON. The term "person" means any individual, company, corporation, association, firm, partnership, joint stock company, trust or society other than a special (occupational) taxpayer qualified to deal in NFA firearms.

2. PREPARATION OF APPLICATION FOR TAX PAID TRANSFER AND REGISTRATION OF A FIREARM.

a. Reference §§5811 and 5812, National Firearms Act, United States Code, Chapter 53), persons seeking to transfer a firearm tax paid must complete, in duplicate, a separate application on this form for each firearm. The applicant transferor must furnish all the information called for on this application form.

b. If the transferor of a destructive device, machinegun, shortbarreled shotgun or short-barreled rifle is a licensee under 18 U.S.C., Chapter 44, and the transferee is anyone other than a qualified licensee under the National Firearms Act [special (occupational) taxpayer], the transferee must complete item 2 (Applicant Certification) on the reverse side of ATF Form 4 (5320.4). Item 3 (Law Enforcement Certification) must be completed for the transfer of any NFA firearm to an individual other than a special (occupational) taxpayer.

c. If the transferee is an individual, including a collector licensed under 18 U.S.C., Chapter 44, in addition to satisfying the requirements of 2b, above, a recent 2" x 2" photograph must be affixed in item 1 on the reverse of ATF Form 4 (5320.4) and completed FBI Form FD-258, Fingerprint Card, must be submitted in duplicate. The fingerprints should be taken by a person qualified to do so, and must be clear, unsmudged and classifiable

d. All signatures required on this form must be entered in ink on both copies. Photocopies or other facsimile signatures are not acceptable. It is preferred that the form be prepared by the use of a typewriter, using carbon paper to make an exact duplicate. Pen and ink may be used, but under no circumstances will a form filled in by use of a lead pencil be accepted. All changes made on this form must be initialed and dated.

e. If the transferee is acquiring the firearm as other than an individual, the tradename and business address should be shown in item 2 on the face of ATF Form 4 (5320.4). Firearms may not be acquired as a part of the business inventory of a firearms licensee who is not a special (occupational) taxpayer. In such case, the home address of the transferee should be shown.

f. If the firearm is being transferred from an estate, item 3a. should show the transferor as: Executor's name, title (Executor (or executrix, administrator, administratrix) of the Estate of (name)), and the executor's address. Item 3c, should reflect the last address of the decedent and date of death.

g. If any questions arise concerning the preparation of this form, please contact the NFA Branch, Bureau of Alcohol, Tobacco and Firearms, Washington, DC 20226 or the nearest Alcohol, Tobacco and Firearms office.

3. DISPOSITION OF APPLICATION FOR TRANSFER AND REGISTRATION OF FIREARM. The transferor will forward both copies of the form to the National Firearms Act Branch, Bureau of Alcohol, Tobacco and Firearms, P.O. Box 73201 Chicago, IL 60673-7201. This office will deposit the tax and forward the application to the Bureau of Alcohol, Tobacco and Firearms, National Firearms Act Branch, Washington, DC 20226 for processing. If approved, the original of the approved form will be returned to the transferor for delivery to the transferee with the firearm, and this office will retain the duplicate. Approval of this form will effect registration of the firearm to the transferee. The transferor shall not under any circumstances deliver the firearm to the transferee until the approved form is received. This form must be retained by the transferee and be available for inspection by Government officers until such time as the firearm may later be transferred after approval by this office. [If the application is disapproved, the original with any accompanying check or money order, will be returned to the transferor with the reason for disapproval.]

4. APPROVED TRANSFER NOT PHYSICALLY ACCOMPLISHED. The transferor should mark the front of the form "VOID," sign and date the form and return the voided form to the Director. This will assure that the firearm will remain registered in the name of the transferor who submitted the application. Upon receipt of the voided form, instructions for obtaining a tax refund will be provided.

5. LATER TRANSFERS OF THE FIREARM. If the firearm is to be transferred later by the new owner, a new application form covering the proposed transfer must be filed with the Director.

TF Form 4 (5320.4) (7-88)

INSTRUCTIONS (Continued)

6. RATE OF TAX ON THE TRANSFER OF A FIREARM. The tax on the transfer of a firearm is $200, except that the transfer tax is $5 on any firearm classified as "any other weapon" as defined in instruction 1b.

7. PERSONS PROHIBITED FROM RECEIVING FIREARMS. Under 18 U.S.C., Chapter 44 and Title VII of Public Law 90-351, as amended, (18 U.S.C., Appendix), the following persons are prohibited from receiving a firearm, including an unserviceable firearm which has a frame or receiver: (1) fugitives from justice (any crime); (2) persons under indictment for, or who have been convicted of, a crime punishable by imprisonment, for a term exceeding 1 year; (3) narcotic addicts or unlawful drug users; (4) persons adjudicated mental defectives or mentally incompetent, or who have been committed to any mental institution; (5) veterans discharged under dishonorable conditions; (6) persons who have renounced their United States citizenship; (7) aliens illegally or unlawfully in the United States; and (8) where the transferor is a firearms licensee, persons under 21 years of age in the case of any firearm other than a shotgun or a rifle and under 18 years of age in the case of a shotgun or a rifle. In addition, section 5812 of the National Firearms Act requires that an application to transfer a firearm be denied if the transfer, receipt or possession of the firearm would place the transferee in violation of law. The term "law" in this statute includes Federal laws as well as State statutes and local ordinances applicable to the locality where the transferee resides.

8. INQUIRIES ABOUT STATUS OF A TRANSFER APPLICATION. The transfer application form will be reviewed and returned to the transferor promptly if additional information or corrections are required. However, because of the necessity for an FBI record check on an individual transferee's fingerprints, approximately 60 days must be allowed for processing such applications. Under the provisions of Section 6103(a), Title 26, United States Code, disclosure of any "return" or "return information" is generally prohibited to anyone except the person filing the return. Therefore, information about the status of a transfer application may be given only to the transferor. Such information cannot be divulged to the transferee.

9. PENALTIES. Any person who violates or fails to comply with any of the requirements of the National Firearms Act shall, upon conviction, be fined not more than $10,000 or be imprisoned for not more than 10 years, or both, in the discretion of the court. Moreover, any firearm involved in any violation of the provisions of the National Firearms Act or any regulations issued thereunder shall be subject to seizure and forfeiture. It is unlawful for any person to make or cause the making of a false entry on any application or record required by the National Firearms Act knowing such entry to be false.

NOTE: All provisions of Title I of the Gun Control Act must also be complied with, including the recordkeeping requirements for licenses and retention of ATF Form 4473 (5300.9) by licensees for dispositions to nonlicensees.

SPECIAL NOTE: A direct interstate transfer will be approved only to a licensee who is also a special (occupational) taxpayer qualified to deal in the particular type NFA firearm; or a government entity; or a lawful heir; or, in the case of a firearm which has been designated as a "curio or relic," to a licensed collector.

PAPERWORK REDUCTION ACT NOTICE

This information request is in accordance with the Paperwork Reduction Act of 1980. The information you provide is used in applying to transfer serviceable firearms taxpaid to persons other than special taxpayers qualified to deal in NFA firearms or government entities. Data is used to identify transferor, transferee, and firearm, and to ensure legality for transfer under Federal, State and local laws. The furnishing of this information is mandatory. (26 U.S.C. 5812)

The following information is provided pursuant to Section 3 of the Privacy Act of 1974 (5 U.S.C. §552(e)(3)):

1. *AUTHORITY.* Solicitation of this information is made pursuant to the National Firearms Act (26 U.S.C. §5812). Disclosure of this information by the applicant is mandatory for transfer of an NFA firearm, unless the transfer is otherwise exempt from tax.

2. *PURPOSE.* To insure payment of the tax imposed by 26 U.S.C. §5811; to insure that the transfer would not be in violation of law; and to effect registration of the firearm.

3. *ROUTINE USES.* The information will be used by ATF to make the determinations set forth in paragraph 2. In addition, to effect registration of the firearm, information as to the identification of the firearm, date of registration, and the identification and address of person entitled to possess the firearm will be entered into the National Firearms Registration and Transfer Record. No information obtained from an application, registration, or records required to be submitted by a natural person in order to comply with any provision of the National Firearms Act or regulations issued thereunder, shall, except in connection with prosecution or other action for furnishing false information, be used, directly or indirectly, as evidence against that person in any criminal proceeding with respect to a violation of law occurring prior to or concurrently with the filing of the application. The information from this application may only be disclosed to Federal authorities for purposes of prosecution for violation of the National Firearms Act.

4. *EFFECTS OF NOT SUPPLYING INFORMATION REQUESTED.* Failure to supply complete information will delay processing and may cause denial of the application.

ATF Form 4 (5320.4) (7-86)

DEPARTMENT OF THE TREASURY — BUREAU OF ALCOHOL, TOBACCO AND FIREARMS

APPLICATION FOR TAX PAID TRANSFER AND REGISTRATION OF FIREARM

SEE INSTRUCTIONS ATTACHED.
TO BE SUBMITTED IN DUPLICATE
TO: National Firearms Act Branch
Bureau of Alcohol, Tobacco and Firearms
P.O. Box 73201
Chicago, IL 60673-7201

2a. TRANSFEREE'S NAME AND ADDRESS (If transferee is a Special (Occupational) Taxpayer who is acquiring firearm for personal use, rather than as part of his business inventory, show personal name below and check here: ☐)

1. TYPE OF TRANSFER *(Check one)* *(See instructions 1 and 6)*

☐ $5 ☐ $200

2b. TRADE NAME *(See instruction 2e)*

2c. COUNTY

3a. TRANSFEROR'S NAME AND MAILING ADDRESS (If the firearm is registered under your trade name, enter your trade name. EXECUTORS: See instruction 2f.)

Submit with your application a check or money order for the appropriate amount made payable to the Department of the Treasury. Upon approval of this application, this office will acquire, affix and cancel the required "National Firearms Act" stamp for you. *(See Instruction 6)*

3b. NUMBER, STREET, CITY, STATE AND ZIP CODE IF DIFFERENT FROM ITEM 3a.

3c. IF APPLICABLE: DECEDENT'S NAME, ADDRESS, AND DATE OF DEATH

The above-named and undersigned tranferor hereby makes application as required by Section 5812 of the National Firearms Act to transfer and register the firearm described below to the transferee.

4. DESCRIPTION OF FIREARM *(Complete items a through h)*

a. NAME AND ADDRESS OF MANUFACTURER AND/OR IMPORTER OF FIREARM

b. TYPE OF FIREARM *(Short-barreled rifle, machine gun, destructive device, any other weapon, etc.)*

c. CALIBER, GAUGE OR SIZE *(Specify)*

d. MODEL

LENGTH *(Inches)* e. OF BARREL: f. OVERALL:

g. SERIAL NUMBER

h. ADDITIONAL DESCRIPTION OR DATA APPEARING ON FIREARM *(Attach additional sheet if necessary)*

5. TRANSFEREE'S FEDERAL FIREARMS LICENSE *(If any)* *(Give complete 15-digit number)*

First 6 digits	2 digits	2 digits	5 digits

6. TRANSFEREE'S SPECIAL (OCCUPATIONAL) TAX STATUS

a. ATF IDENTIFICATION NUMBER

b. CLASS

7. TRANSFEROR'S FEDERAL FIREARMS LICENSE *(If any)* *(Give complete 15-digit number)*

First 6 digits	2 digits	2 digits	5 digits

8. TRANSFEROR'S SPECIAL (OCCUPATIONAL) TAX STATUS

a. ATF IDENTIFICATION NUMBER

b. CLASS

UNDER PENALTIES OF PERJURY, I DECLARE that I have examined this application and to the best of my knowledge and belief it is true, correct and complete, and that the transfer of the described firearm to the transferee and receipt and possession of it by the transferee are not prohibited by the provisions of Chapter 44, Title 18, United States Code; Chapter 53, Title 26, United States Code; or Title VII of the Omnibus Crime Control and Safe Streets Act, as amended; or any provisions of State or local law.

9. SIGNATURE OF TRANSFEROR *(Or authorized official)*

10. NAME AND TITLE OF AUTHORIZED OFFICIAL *(Print or type)*

11. DATE

THE SPACE BELOW IS FOR THE USE OF THE BUREAU OF ALCOHOL, TOBACCO AND FIREARMS

BY AUTHORITY OF THE DIRECTOR, THIS APPLICATION HAS BEEN EXAMINED, AND THE TRANSFER AND REGISTRATION OF THE FIREARM DESCRIBED HEREIN AND THE INTERSTATE MOVEMENT OF THAT FIREARM, WHEN APPLICABLE, TO THE TRANSFEREE ARE:

STAMP NUMBER

☐ APPROVED *(With the following conditions, if any)*

☐ DISAPPROVED *(For the following reasons)*

SIGNATURE OF DIRECTOR, BUREAU OF ALCOHOL, TOBACCO AND FIREARMS

DATE

ATF Form 4 (5320.4) (7-88)

CERTIFICATIONS

If the transferor of a destructive device, machinegun, short-barreled shotgun or short-barreled rifle is a Federal firearms licensee, and the transferee is anyone other than a licensee qualified to deal in the firearm to be transferred, the transferee must sign the Applicant Certification (item 2 below) in the presence of the law enforcement officer signing item 3 below. The Law Enforcement Certification (item 3 below) must be completed for the transfer of any registered firearm to an individual other than a licensee qualified to deal in the firearm to be transferred. In addition, the individual transferee must affix a recent photograph (taken within the past year) in item 1 and submit, in duplicate (to the transferor) two completed copies of FBI Form FD-258, Fingerprint Card. (See Important note below.)

1. PHOTOGRAPH

AFFIX
RECENT PHOTOGRAPH HERE
(Approximately 2" x 2")

2. APPLICANT CERTIFICATION

I, ______________________________ (Name of Transferee), have a reasonable necessity to possess the device or weapon described on this application for the following reason(s) ______________________________

____

and my possession of the device or weapon would be consistent with public safety (18 U.S.C. 922(b) (4) and 27 CFR 178.98).

UNDER PENALTIES OF PERJURY, I declare that I have examined this application, and to the best of my knowledge and belief it is true, correct and complete, and that receipt and possession of the firearm described on this form will not place me in violation of the provisions of Chapter 44, Title 18, U.S.C.; Chapter 53, Title 26, U.S.C.; or Title VII of the Omnibus Crime Control and Safe Streets Act, as amended, or any provisions of State or local law.

______________________________ (Signature of Transferee or official authorized to sign for firm)

______________ Date

3. LAW ENFORCEMENT CERTIFICATION (See IMPORTANT note below)

I certify that I am the chief law enforcement officer of the organization named below having jurisdiction in the area of residence of ______________________________ (Name of Transferee). I have no information indicating that the transferee will use the firearm or device described on this application for other than lawful purposes. I have no information that the receipt and/or possession of the firearm described in item 4 of this form would place the transferee in violation of State or local law.

______________________________ (Signature and Title of Chief Law Enforcement Officer — See IMPORTANT note below)

______________ (Date)

______________________________ (Organization and Street Address)

______________ (County) ______________ (Phone — include area code)

IMPORTANT: The chief law enforcement officer is considered to be the Chief of Police for the transferee's city or town of residence, the Sheriff for the transferee's county of residence; the Head of the State Police for the transferee's State of residence; a State or local district attorney or prosecutor having jurisdiction in the transferee's area of residence; or another person whose certification is acceptable to the Director, Bureau of Alcohol, Tobacco and Firearms. If someone has specific delegated authority to sign on behalf of the Chief of Police, Sheriff, etc., this fact must be noted by printing the Chief's, Sheriff's, or other authorized official's name and title, followed by the word "by" and the full signature and title of the delegated person.

ATF Form 4 (5320.4) (7-88)

ATF Form 5 (5320.5)
Application for Tax Exempt Transfer and Registration of a Firearm

OMB No. 1512-0028 (01/31/00)

DEPARTMENT OF THE TREASURY — BUREAU OF ALCOHOL, TOBACCO AND FIREARMS

APPLICATION FOR TAX EXEMPT TRANSFER AND REGISTRATION OF A FIREARM

DETACH THIS SHEET BEFORE COMPLETING
INSTRUCTIONS

1. DEFINITIONS.

a. FIREARM. The term "firearm" means: (1) a shotgun having a barrel or barrels of less than 18 inches in length; (2) a weapon made from a shotgun if such weapon as modified has an overall length of less than 26 inches or a barrel or barrels of less than 18 inches in length; (3) a rifle having a barrel or barrels of less than 16 inches in length; (4) a weapon made from a rifle if such weapon as modified has an overall length of less than 26 inches or a barrel or barrels of less than 16 inches in length; (5) any other weapon, as defined in b. below; (6) a machinegun; (7) a muffler or a silencer or any firearm whether or not such firearm is included within this definition; and (8) a destructive device. The term "firearm" shall not include an antique firearm or any device (other than a machinegun or destructive device) which, although designed as a weapon, the Director, Bureau of Alcohol, Tobacco and Firearms, or authorized delegate, finds by reason of the date of its manufacture, value, design and other characteristics is primarily a collector's item and is not likely to be used as a weapon.

b. ANY OTHER WEAPON. The term "any other weapon" means any weapon or device capable of being concealed on the person and from which a shot can be discharged through the energy of an explosive; a pistol or revolver having a barrel with a smooth bore designed or redesigned to fire a fixed shotgun shell; weapons with combination shotgun and rifle barrels 12 inches or more, less than 18 inches in length, from which only a single discharge can be made from either barrel without manual reloading; and shall include any such weapon which may be readily restored to fire. Such term shall not include a pistol or a revolver having a rifled bore, or rifled bores, or weapons designed, made, or intended to be fired from the shoulder and not capable of firing fixed ammunition.

c. TRANSFEROR. The term "transferor" means the registered owner of a firearm who is applying to transfer it.

d. TRANSFEREE. The term "transferee" means the person desiring to acquire the firearm.

e. PERSON. The term "person" means any individual, company, corporation, association, firm, partnership, joint stock company, trust or society which is not a special (occupational) taxpayer.

2. PREPARATION OF APPLICATION FOR TAX-EXEMPT TRANSFER AND REGISTRATION OF A FIREARM.

a. As set forth in section 5812, National Firearms Act (26 U.S.C., Chapter 53), and except as otherwise provided, persons seeking to transfer a firearm without payment of tax must complete, in duplicate, a separate application on this form for each firearm. The applicant transferor must furnish all the information called for on this application form.

b. If the transferor of a destructive device, machinegun, shortbarreled shotgun or short-barreled rifle is a licensee under 18 U.S.C., Chapter 44, and the transferee is anyone other than a qualified licensee under the National Firearms Act [special (occupational) taxpayer], the transferee must complete item 2 (Applicant Certification) on the reverse side of ATF Form 5 (7560.5).

Item 3 (Law Enforcement Certification) must be completed for the transfer of any NFA firearm to an individual other than a special (occupational) taxpayer.

c. If the transferee is an individual (including a licensed collector), in addition to satisfying the requirements of 2b. above, a recent 2" x 2" photograph must be affixed in item 1 on the reverse of ATF Form 5 and completed FBI Form FD-258, Fingerprint Card, in duplicate, must be submitted. The fingerprints should be taken by a person qualified to do so, and must be clear, unsmudged and classifiable.

d. It is preferred that the form be prepared by the use of a typewriter, using carbon paper to make an exact duplicate. Pen and ink may be used, but under no circumstances will a form filled in by use of a lead pencil be accepted. All signatures on both copies MUST be original. Photocopies or other facsimiles are not acceptable. All changes made on the face of the form by the transferor must be initialed and dated.

e. If the transferee is acquiring the firearm as other than an individual, the trade name should be shown in item 2b. on the face of ATF Form 5 and the business address should appear in items 2a. and 2c. A firearms licensee who is not a special (occupational) taxpayer may not acquire firearms as part of the business inventory and item 2b. should be left blank. Items 2a. and 2c. should, in such case, reflect the home address.

f. If the firearm is being transferred from an estate, item 3a. should show the transferor as: Executor's name, title (Executor (or executrix, administrator, administratrix) of the Estate of (name)), and the executor's address. Item 3b. should reflect the last address of the decedent and date of death.

g. If any questions arise concerning the preparation of this form, please contact the NFA Branch, Bureau of Alcohol, Tobacco and Firearms, Washington, DC 20226 or the nearest Alcohol, Tobacco and Firearms office.

3. DISPOSITION OF APPLICATION FOR TRANSFER AND REGISTRATION OF FIREARM. The transferor will forward both copies of the form to the NFA Branch, Bureau of Alcohol, Tobacco and Firearms, Washington, DC 20226. If approved, the original of the approved form will be returned to the transferor for delivery to the transferee with the firearm, and this office will retain the duplicate. Approval of this form will effect registration of the firearm to the transferee. The transferor shall not under any circumstances deliver the firearm to the transferee until the approved form is received. This form must be retained by the transferee and be available for inspection by Government officers until such time as the firearm may later be transferred after approval by this office. If the application is disapproved, the original will be returned to the transferor with the reason for disapproval.

4. APPROVED TRANSFER NOT PHYSICALLY ACCOMPLISHED. The transferor should mark the front of the form "VOID," sign and date the form and return the voided form to the Director. This will assure that the firearm will remain registered in the name of the transferor who submitted the application.

ATF FORM 5 (5320.5) (4-89)   PREVIOUS EDITIONS ARE OBSOLETE

INSTRUCTIONS (Continued)

5. LATER TRANSFERS OF THE FIREARM. If the firearm is to be transferred later by the new owner, a new application form covering the proposed transfer must be filed with the Director.

6. TRANSFERS EXEMPT FROM TAX. A registered firearm may be transferred without payment of tax under any of the following conditions:

a. If it is an unserviceable firearm being transferred as a curio or ornament. (NOTE: an unserviceable firearm is defined as one which is incapable of discharging a shot by means of an explosive and incapable of being readily restored to a firing condition. An acceptable method of rendering most firearms unserviceable is to fusion weld the chamber closed and fusion weld the barrel solidly to the frame. Certain unusual firearms require other methods to render the firearm unserviceable. Contact ATF for instructions when in doubt.)

b. To or from the United States, or any department, independent establishment or agency thereof.

c. To or from any State or possession of the United States or any political subdivision thereof, or any official police organization of such a government entity engaged in criminal investigations.

d. To a lawful heir as defined in the laws of the State of residence of the decedent.

e. The National Firearms Act also provides that registered firearms may be transferred tax exempt between firearms dealers who have also paid special (occupational) tax. In such instances, however, the application must be submitted on ATF Form 3.

7. PERSONS PROHIBITED FROM RECEIVING FIREARMS. Under 18 U.S.C., Chapter 44 and Title VII of Public Law 90-351, as amended, (18 U.S.C., Appendix), the following persons are prohibited from receiving a firearm, including an unserviceable firearm which has a frame or receiver: (1) fugitives from justice (any crime); (2) persons under indictment for, or who have been convicted of, a crime punishable for a term exceeding 1 year; (3) narcotic addicts or unlawful drug users; (4) persons adjudicated mental defectives or mentally incompetent, or who have been committed to any mental institution; (5) veterans discharged under dishonorable conditions; (6) persons who have renounced their United States citizenship; (7) aliens illegally or unlawfully in the United States; and (8) where the transferor is a firearms licensee, persons under 21 years of age in the case of any firearm other than a shotgun or a rifle and under 18 years of age in the case of a shotgun or a rifle. In addition, section 5812 of the National Firearms Act requires that an application to transfer a firearm be denied if the transfer, receipt or possession of the firearm would place the transferee in violation of law. The term "law" in this statute includes Federal laws as well as State statutes and local ordinances applicable to the locality where the transferee resides.

8. PENALTIES. Any person who violates or fails to comply with any of the requirements of the National Firearms Act shall, upon conviction, be fined not more than $10,000 or be imprisoned for not more than 10 years, or both, in the discretion of the court. Moreover, any firearm involved in any violation of the provisions of the National Firearms Act or any regulations issued thereunder shall be subject to seizure and forfeiture. It is unlawful for any person to make or cause the making of a false entry on any application or record required by the National Firearms Act knowing such entry to be false.

NOTE: All provisions of Title I of the Gun Control Act must also be complied with, including the recordkeeping requirements for licenses.

SPECIAL NOTE: A direct interstate transfer will be approved only to a licensee who is also a special (occupational) taxpayer qualified to deal in the particular type NFA firearm; or a government entity; or a lawful heir; or, in the case of a firearm which has been designated as a "curio or relic," to a licensed collector.

PAPERWORK REDUCTION ACT NOTICE

This form meets the clearance requirements of Section 3507, PL 96-511, 12/11/80. The information you provide is used to apply the tax-exempt transfer of an unserviceable firearm to anyone other than a Federal firearms licensee who has paid the required special (occupational) tax to deal in NFA firearms. The data is used to ensure legality of transfer under Federal, State and local law. The furnishing of this information is mandatory (26 USC 5812).

The estimated average burden associated with this collection of information is 4 hours per respondent or recordkeeper, depending on individual circumstances. Comments concerning the accuracy of this burden estimate and suggestions for reducing this burden should be addressed to Reports Management Officer, Document Services Branch, Bureau of Alcohol, Tobacco and Firearms, Washington, DC 20226.

An agency may not conduct or sponsor, and a person is not required to respond to, a collection of information unless it displays a currently valid OMB control number.

PRIVACY ACT INFORMATION

The following information is provided pursuant to Section 3 of the Privacy Act of 1974 (5 U.S.C. §552(e)(3)):

1. *AUTHORITY.* Solicitation of this information is made pursuant to the National Firearms Act (26 U.S.C. §5812). Disclosure of this information by the applicant is mandatory for any transfer without payment of transfer tax as provided in 26 U.S.C. §5852.

2. *PURPOSE.* To verify that the proposed transfer is exempt from transfer tax; to verify that the transfer would not be in violation of law; and to effect registration of the firearm.

3. *ROUTINE USES.* The information will be used by ATF to make the determinations set forth in paragraph 2. In addition, to effect registration of the firearm, information as to the identification of the firearm, date of registration, and the identification and address of person entitled to possess the firearm will be entered into the National Firearms Registration and Transfer Record. No information obtained from an application, registration, or records required to be submitted by a natural person in order to comply with any provision of the National Firearms Act or regulations issued thereunder, shall, except in connection with prosecution or other action for furnishing information, be used, directly or indirectly, as evidence against that person in any criminal proceeding with respect to a violation of law occurring prior to or concurrently with the filing of the application. The information from this application may only be disclosed to Federal authorities for purposes of prosecution for violation of the National Firearms Act.

4. *EFFECTS OF NOT SUPPLYING INFORMATION REQUESTED.* Failure to supply complete information will delay processing and may cause denial of the application.

PREVIOUS EDITIONS ARE OBSOLETE

ATF FORM 5 (5320.5) (4-89)

Form Approved: OMB No. 1512-0028 (09/30/95)

DEPARTMENT OF THE TREASURY — BUREAU OF ALCOHOL, TOBACCO AND FIREARMS

APPLICATION FOR TAX EXEMPT TRANSFER AND REGISTRATION OF A FIREARM

SEE INSTRUCTIONS ATTACHED. TO BE SUBMITTED IN DUPLICATE TO: National Firearms Act Branch, Bureau of Alcohol, Tobacco and Firearms, Washington, DC 20226

2a. TRANSFEREE'S NAME AND ADDRESS (If transferee is a Special (Occupational) Taxpayer who is acquiring firearm for personal use, rather than as part of his business inventory, show personal name below and check here: ☐)

1. TYPE OF TRANSFER: TAX EXEMPT

I believe that I am entitled to exemption from the payment of the transfer tax imposed by Section 5811 (26 U.S.C. Chapter 53) on the firearm described herein for the following reason *See Instruction 6)*:

- ☐ FIREARM IS UNSERVICEABLE AND IS BEING TRANSFERRED AS A CURIO OR ORNAMENT (6a)
- ☐ FIREARM IS BEING TRANSFERRED TO OR FROM A GOVERNMENT ENTITY (6b & 6c)
- ☐ FIREARM IS BEING TRANSFERRED TO A LAWFUL HEIR (6d)
- ☐ OTHER (Specify)

2b. TRADE NAME *(See instruction 2e)*

2c. COUNTY

3a. TRANSFEROR'S NAME AND MAILING ADDRESS *(If firearm is registered under your trade name, enter your trade name. EXECUTORS: See instruction 2f.)*

3c. NUMBER, STREET, CITY, STATE AND ZIP CODE OF RESIDENCE (OR FIREARMS BUSINESS PREMISES) IF DIFFERENT FROM ITEM 3a.

3b. IF APPLICABLE: DECEDENT'S NAME, ADDRESS, AND DATE OF DEATH

The above-named and undersigned tranferor hereby makes application as required by Section 5812 of the National Firearms Act to transfer and register the firearm described below to the transferee.

4. DESCRIPTION OF FIREARM *(Complete items a through i)*

a. NAME AND ADDRESS OF MANUFACTURER AND/OR IMPORTER OF FIREARM	b. TYPE OF FIREARM *(Short-barreled rifle, machine gun, destructive device, any other weapon, etc.)*	c. CALIBER, GAUGE OR SIZE *(Specify)*	d. MODEL
			LENGTH *(Inches)* e. OF BARREL: f. OVERALL:
			g. SERIAL NUMBER

h. ADDITIONAL DESCRIPTION OR DATA APPEARING ON FIREARM *(Attach additional sheet if necessary)*

i. IS THE FIREARM UNSERVICEABLE AS DEFINED IN INSTRUCTION 6a? ☐ YES ☐ NO *(If "No," describe any other method by which firearm has been rendered unserviceable. Use additional sheets.)*

5. TRANSFEREE'S FEDERAL FIREARMS LICENSE *(If any)*
(Give complete 15-digit number)

First 6 digits	2 digits	2 digits	5 digits

6. TRANSFEREE'S SPECIAL (OCCUPATIONAL) TAX STATUS

a. ATF IDENTIFICATION NO.	b. CLASS

7. TRANSFEROR'S FEDERAL FIREARMS LICENSE *(If any)*
(Give complete 15-digit number)

First 6 digits	2 digits	2 digits	5 digits

8. TRANSFEROR'S SPECIAL (OCCUPATIONAL) TAX STATUS

a. ATF IDENTIFICATION NO.	b. CLASS

UNDER PENALTIES OF PERJURY, I DECLARE that I have examined this application, and to the best of my knowledge and belief it is true, correct and complete, and that the transfer of the described firearm to the transferee and receipt and possession of it by the transferee are not prohibited by the provisions of Chapter 44, Title 18, United States Code; Chapter 53, Title 26, United States Code; or Title VII of the Omnibus Crime Control and Safe Streets Act, as amended; or any provisions of State or local law.

9. SIGNATURE OF TRANSFEROR *(Or authorized official)*	10. NAME AND TITLE OF AUTHORIZED OFFICIAL *(Print or type)*	11. DATE

THE SPACE BELOW IS FOR THE USE OF THE BUREAU OF ALCOHOL, TOBACCO AND FIREARMS

BY AUTHORITY OF THE DIRECTOR, THIS APPLICATION HAS BEEN EXAMINED, AND THE TRANSFER AND REGISTRATION OF THE FIREARM DESCRIBED HEREIN AND THE INTERSTATE MOVEMENT OF THAT FIREARM, WHEN APPLICABLE, TO THE TRANSFEREE ARE:

☐ APPROVED *(With the following conditions, if any)*		☐ DISAPPROVED *(For the following reasons)*	
EXAMINER	DATE	AUTHORIZED ATF OFFICIAL	DATE

ATF FORM 5 (5320.5) (4-89) PREVIOUS EDITIONS ARE OBSOLETE

CERTIFICATIONS

If the transferor of a destructive device, machinegun, short-barreled shotgun or short-barreled rifle is a Federal firearms licensee, and the transferee is anyone other than a licensee qualified to deal in the firearm to be transferred, the transferee must sign the Applicant Certification (item 2 below) in the presence of the law enforcement officer signing item 3 below. The Law Enforcement Certification (item 3 below) must be completed for the transfer of any registered firearm to an individual other than a licensee qualified to deal in the firearm to be transferred. In addition, the individual transferee must affix a recent photograph (taken within the past year) in item 1 and submit, in duplicate (to the transferor) two completed copies of FBI Form FD-258, Fingerprint Card. (See Important note below.)

1. PHOTOGRAPH

AFFIX
RECENT PHOTOGRAPH HERE
(Approximately 2" x 2")

2. APPLICANT CERTIFICATION

I, ______________________________ , have a reasonable necessity to possess the device or
(Name of Transferee)
weapon described on this application for the following reason(s) ______________________________

____

and my possession of the device or weapon would be consistent with public safety (18 U.S.C. 922(b) (4) and 27 CFR 178.98).

UNDER PENALTIES OF PERJURY, I declare that I have examined this application, and to the best of my knowledge and belief it is true, correct and complete, and that receipt and possession of the firearm described on this form will not place me in violation of the provisions of Chapter 44, Title 18, U.S.C.; Chapter 53, Title 26, U.S.C.; or Title VII of the Omnibus Crime Control and Safe Streets Act, as amended, or any provisions of State or local law.

______________________________ ______________
(Signature of Transferee or official authorized to sign for firm) Date

3. LAW ENFORCEMENT CERTIFICATION (See IMPORTANT note below)

I certify that I am the chief law enforcement officer of the organization named below having jurisdiction in the area of residence of ______________________________ . I have no information indicating that the transferee will use the fire-
(Name of Transferee)
arm or device described on this application for other than lawful purposes. I have no information that the receipt and/or possession of the firearm described in item 4 of this form would place the transferee in violation of State or local law.

____
(Signature and Title of Chief Law Enforcement Officer — See IMPORTANT note below)

______________________________ ______________
(Date)

____
(Organization and Street Address)

____
(County) (Phone — include area code)

IMPORTANT: The chief law enforcement officer is considered to be the Chief of Police for the transferee's city or town of residence, the Sheriff for the transferee's county of residence; the Head of the State Police for the transferee's State of residence; a State or local district attorney or prosecutor having jurisdiction in the transferee's area of residence; or another person whose certification is acceptable to the Director, Bureau of Alcohol, Tobacco and Firearms. If someone has specific delegated authority to sign on behalf of the Chief of Police, Sheriff, etc., this fact must be noted by printing the Chief's, Sheriff's, or other authorized official's name and title, followed by the word "by" and the full signature and title of the delegated person.

PREVIOUS EDITIONS ARE OBSOLETE ATF FORM 5 (5320.5) (4-89)

TF Form 9 (5320.9)
pplication and Permit for Permanent Exportation of Firearms

NSTRUCTIONS (*See 27 CFR Part 479)*

Any person desiring to permanently export a firearm without payment of the transfer tax must submit ATF Form 9, in to the Director, Bureau of Alcohol, Tobacco and Firearms, Washington, D.C. 20226, for a permit providing for deferment of the transfer tax. No shipment may be made until that permit is received from the Director.

a. ATF Form 9 must be submitted until a State Department License for the exportation has been received by the applicant.

b. Part 1 of ATF Form 9 must be completed by the applicant and submitted to ATF with a certified copy of a written order, contract of sale, or other evidence showing firearm is to be shipped to a foreign destination. The transferor's written certification attesting to that fact must appear on the documentation. If the applicant wishes to transfer the firearm(s) tax free to another person who will export the firearm, the applicant must submit Form 9 with documentation that the transfer is a part of the exportation process.

c. The form must be completed by typewriter or pen and ink; penciled entries are unacceptable. The signature in item 10 must be entered in ink on all four copies. Photostatic, facsimile or carbon copy signatures are not acceptable. Where the exporter is a Federal Firearms Licensee, only those individuals Form 7, Application for License, or an individual whom the licensee has certified to ATF as a responsible official, may sign as the authorized official in item 10, Part 1.

d. Applications approved by ATF will serve as the permit to export the firearm(s) described on the application.

e. In the event exportation is not effected, all copies of the form must be *immediately* returned to ATF for cancellation. (27 CFR 179.116).

When the firearms are to be exported by other than parcel post, two copies each of the permit and the Shipper's Export Declaration (Commerce Form 7525-V), and a copy of the State Department License, must be submitted to the District Director of U.S. Customs Service at the port of exportation, an must precede or accompany the shipment in order to permit appropriate inspection prior to lading. The Customs official, after execution Part 3, will retain one copy of the permit and return one copy to the Director, Bureau of Alcohol, Tobacco and Firearms, NFA Branch, Washington, DC 20226, as provided in Customs Manual Supplement No. 3284.02 dated March 12, 1979.

When the firearms are to be exported by parcel post, one copy of the permit must be presented to the postmaster at the office receiving the parcel. The postmaster will execute Part 4 of ATF Form 9 and return the permit to the exporter.

Proof of exportation must be furnished by the exporter to ATF within a 6-month period from date of issuance of the permit by ATF. Satisfactory evidence of exportation would be:

a. The certificate of exportation executed by a Customs Official.

b. The certificate of mailing by parcel post executed by a postmaster.

c. A certificate of landing executed by a Customs official of the foreign country to which the firearm is exported.

d. A sworn statement of the foreign consignee covering the receipt of the firearm.

e. A return receipt, or photostatic copy, signed by the addressee or an authorized agent of the addressee, if the shipment of a firearm was made by insured or registered parcel post.

TE: *ATF receipt of the required documentation that the firearm(s) has/have been exported will relieve the transferor from the transfer tax liability. IF SATISFACTORY EVIDENCE IS NOT FURNISHED WITHIN THE STATED PERIOD, THE TRANSFER TAX WILL BE ASSESSED.*

ACTION BY U.S. CUSTOMS SERVICE

n receipt of an approved ATF Form 9, in duplicate, the Customs official may order such inspection as deemed necessary prior to lading of the merchandise. If satisfied that the shipment is per and the information contained in the permit to export is in agreement with the information shown in the shipper's export declaration, the Customs official will, after the merchandise has n exported, execute Part 3 of ATF Form 9. One copy will be retained with the shipper's export declaration. Customs will forward the remaining copy to the Bureau of Alcohol, Tobacco and arms, NFA Branch, Washington, DC 20226.

PRIVACY ACT INFORMATION

following information is provided pursuant to Section 3 of the Privacy Act of 1974 (5 U.S.C. §552a(e)(3)):

AUTHORITY. Solicitation of this information is made pursuant to the National Firearms Act (26 U.S.C. §5854). Disclosure of this information is mandatory for the exportation of a firearm exempt from transfer tax.

PURPOSE. To determine whether the proposed transfer qualifies as an exportation exempt from tax.

ROUTINE USES. The information will be used by ATF to make the determination set forth in para. 2. In addition, ATF will use the information to annotate the National Firearms Registration and Transfer Record. No information obtained from an application, registration, or records required to be submitted by a natural person in order to comply with any provision of the National Firearms Act, or regulations issued thereunder, shall, except in connection with prosecution, or other action, for furnishing false information, be used, directly or indirectly, as evidence against that person in any criminal proceeding with respect to a violation of law occurring prior to or concurrently with the filing of the application. The information from this application may only be disclosed to Federal authorities as provided in Section 6103, 26 USC (as amended by the Tax Reform Act of 1979).

EFFECTS OF NOT SUPPLYING INFORMATION REQUESTED. Failure to supply complete information will delay processing and may cause denial of the application.

PAPERWORK REDUCTION ACT NOTICE

form meets the clearance requirements of Section 3507, PL 96-511, 12/11/80. The information you provide is used to verify exportation of a firearm and justify removal of the firearm from National Firearms Registration and Transfer Record (NFRTR). The information is required to show satisfactory proof that a firearm may be exported without payment of the transfer tax in anner as prescribed by the Secretary of the Treasury. The furnishing of this information is mandatory (26 USC 5854).

estimated average burden associated with this collection of information is 3.4 hour(s) per respondent or recordkeeper, depending on individual circumstances. Comments concerning the uracy of this burden estimate and suggestion for reducing this burden should be addressed to Reports Management Officer, Information Programs Branch, Bureau of Alcohol, Tobacco and arms, Washington, D.C. 20226, and the Office of Management and Budget, Paperwork Reduction Project (1512-0020), Washington, D.C. 20503.

F 9 (5320.9) (7/84)

Form Approved OMB No. 1512-0020 (05/31/93)

DEPARTMENT OF THE TREASURY - BUREAU OF ALCOHOL, TOBACCO AND FIREARMS

APPLICATION AND PERMIT FOR PERMANENT EXPORTATION OF FIREARMS

(CHAPTER 53, TITLE 26, UNITED STATES CODE)
(SUBMIT IN QUADRUPLICATE. SEE INSTRUCTIONS ON REVERSE.)

TO: DIRECTOR, BUREAU OF ALCOHOL, TOBACCO AND FIREARMS, WASHINGTON, DC 20226

PART 1 - APPLICATION. The undersigned transferor hereby makes application to export the firearm(s) described herein. The application is supported by the attached certified copy of written order or contract of sale of such firearm(s) to consignee.

1. NAME AND ADDRESS OF FOREIGN CONSIGNEE	2. INTENDED PORT OF EXPORTATION (Including air freight)	3. NUMBER OF FIREARMS INCLUDED IN THIS APPLICATION

4. DESCRIPTION OF FIREARM(S) (If additional space is needed, continue on a separate sheet using the format below.)

LINE NO.	TYPE (Machine gun, destructive device, short barreled shotgun or rifle, etc.) (a)	CALIBER, GAUGE OR SIZE (b)	MODEL (c)	LENGTH OF BARREL(S) (d)	OVERALL LENGTH (e)	SERIAL NUMBER (f)
1						
2						
3						
4						
5						

5. NAME OF EXPORTER (And trade name if any)	6. ADDRESS (Number, street, city, county, state, ZIP Code)

7. FEDERAL FIREARMS LICENSE (if any) (Give complete 15-digit number)	8. ATF IDENTIFICATION NUMBER (If any)	9. STATE DEPARTMENT LICENSE NUMBER FOR THIS SPECIFIC SHIPMENT

Under penalties of perjury, I certify that I am the lawful possessor of the firearm(s) described on this form and any accompanying sheets, and that I have examined this application and, to the best of may knowledge and belief, it is true, correct and complete.

10. SIGNATURE AND TITLE (Owner or authorized official) (See instruction 1c.)	11. DATE OF APPLICATION

PART 2 - PERMIT (This portion to be completed by Bureau of Alcohol, Tobacco and Firearms)

12. THIS APPLICATION IS: ☐APPROVED ☐DISAPPROVED: (Reason)	13. ASSESSMENT OF TAX INCURRED BY THIS TRANSFER WILL BE MADE UNLESS PROPER EVIDENCE OF EXPORTATION IS RECEIVED ON OR BEFORE:

14. EXAMINER	15. DATE	16. AUTHORIZED ATF OFFICIAL	17. DATE

PART 3 - CERTIFICATION BY CUSTOMS

I hereby certify that the described merchandise, covered by a shipper's export declaration on file in this office, was laden and cleared as described below:

LADEN IN (Name and/or type of carrier)	DATE CLEARED	FOREIGN DESTINATION

SIGNATURE OF CUSTOMS OFFICIAL	PORT OF	DATE

PART 4 - CERTIFICATION OF MAILING BY PARCEL POST/EXPORTATION

I certify that there has (have) been posted at this office today, parcel(s) addressed as indicated in Item 1, Part 1 of this application, declared to be firearms by the transferor named in Item 5, Part 1, or the transferor's authorized agent, who has waived the right to withdraw same from the mails.

POST OFFICE NAME OR STAMP	SIGNATURE OF POSTMASTER, BY	DATE POSTED

ATF F 9 (5320.9) (7-84)

ATF Form 10 (5320.10)

Application for Registration of Firearms Acquired by Certain Government Entities

SPECIAL NOTICE

1. If a firearm is registered to the person from whom you obtained it, that person may submit ATF Form 5 for the tax exempt transfer of the firearm to you, in lieu of your submission of ATF Form 10. If the transfer is approved, the restriction on further transfer of the firearm which applies to firearms registered on ATF Form 10 would not apply.

2. Title 27 CFR 179.104 provides that any state, any political subdivision thereof, or any official police organization of such a governmental entity engaged in criminal investigations which acquires for official use a National Firearms Act weapon not registered to it, such as by abandonment or by forfeiture, will register such firearm with the Director by filing ATF Form 10, Application for Registration of Firearms Acquired by Certain Governmental Entities, and that such registration shall become a part of the National Firearms Registration and Transfer Record. Registration of such firearms has been required since passage of the original National Firearms Act in 1934. On April 15, 1971, the Supreme Court decided in the matter of *U.S. vs. Freed, et al.*, and noted that, "only possessors who lawfully make, manufacture, or import firearms can and must register them." However, in order to assist law enforcement agencies, while curtailing the flow of "gangster type" weapons into interstate commerce, the cited regulation was promulgated to permit the limited registration of firearms by governmental entities for official use only. When registration of a firearm on this form by a governmental entity is approved, the Bureau will approve subsequent transfer of such firearm to another qualified governmental entity only, for official use. Otherwise, such firearm must be destroyed or abandoned to ATF.

INSTRUCTIONS

1. Preparation of this form.
   a. This form must be submitted in duplicate. Photostatic copies of this form are not acceptable.
   b. Only one firearm may be listed on each form.
   c. It is preferred that the original and carbon copy be typed, although pen and ink entries are acceptable.
   d. The signature on each copy must be in ink. Pencilled, photostatic, or facsimile signatures are not acceptable.
   e. Serial Number - If the firearm being registered does not bear a serial number, please contact the nearest Alcohol, Tobacco and Firearms office to have an ATF serial number assigned and placed on the frame or receiver of the firearm prior to the submission of the form.

2. Disposition of this form - The applicant will forward the original and one copy of the form to the Director, Bureau of Alcohol, Tobacco and Firearms, Washington, DC 20226, Attention: Technical Services Division. The Director will return the original form, showing approval or disapproval, to the applicant. Approval authorized by the Director will effect the registration of the firearm to the governmental entity. The approved form must be retained with the permanent records of the entity. Subsequent transfer of the firearm will be approved to another qualified governmental entity only.

3. Firearms Held for Use as Evidence - Firearms being held for use as evidence in a criminal proceeding need not be registered if they are to be destroyed or abandoned to ATF when no longer needed as evidence.

DEFINITIONS

The following types of firearms, whether serviceable or unserviceable, fall within the purview of the National Firearms Act and must be registered to the possessor to be lawfully possessed:

1. Short-barreled shotgun - Shotguns with barrels less than 18 inches long or any weapon made from a shotgun having an overall length of less than 26 inches or a barrel less than 18 inches in length.

2. Short-barreled rifle - Rifles with barrels less than 16 inches long. This includes a pistol with a shoulder stock unless it has been specifically exempted; or any weapon made from a rifle having an overall length of less than 26 inches or a barrel less than 16 inches in length.

3. Any other weapon - Any other weapon, except a conventional pistol or revolver having a rifled bore, capable of firing a shot and being concealed on the person. Examples include: "pen" guns ostensibly designed to expel tear gas but which fire fixed ammunition, H & R Handy Guns; Ithaca Auto-Burglar guns; cane guns; and gadget-type firearms.

4. Machinegun - Any weapon which shoots, is designed to shoot, or can be readily restored to shoot, automatically more than one shot, without manual reloading, by a single function of the trigger. The term also includes the frame or receiver of any such weapon, any combination of parts designed and intended for use in converting a weapon into a machinegun, and any combination of parts from which a machinegun can be assembled if such parts are in the possession or under the control of a person.

5. Destructive device - Destructive devices include explosive, incendiary (including so called "molotov cocktails") or poison gas bombs, grenades, rockets, missiles, mines, and similar devices. Included in this category are anti-tank guns, bazookas and mortars and other military type weapons with a bore of more than one-half inch diameter, other than a sporting shotgun or shotgun ammunition.

6. Muffler or silencer - A muffler or silencer for any firearm whether or not such a firearm is included within this listing.

PAPERWORK REDUCTION ACT NOTICE

This form meets the clearance requirements of the Paperwork Reduction Act of 1995. The information you provide is used to properly identify the registrant and the firearms to be registered. The form when approved, registers the firearm to the law enforcement agency. The furnishing of this information is mandatory (26 U.S.C. 5853a).

The estimated average burden associated with this collection of information is 30 minutes per respondent or recordkeeper, depending on individual circumstances. Comments concerning the accuracy of this burden estimate and suggestions for reducing this burden should be addressed to Reports Management Officer, Document Services Branch, Bureau of Alcohol, Tobacco and Firearms, Washington, DC 20226.

An agency may not conduct or sponsor, and a person is not required to respond to, a collection of information unless it displays a currently valid OMB control number.

ATF F 10 (5320.10) (10-83)

OMB No. 1512-0029 (01/31/05)

DEPARTMENT OF THE TREASURY
BUREAU OF ALCOHOL, TOBACCO AND FIREARMS

APPLICATION FOR REGISTRATION OF FIREARMS ACQUIRED BY CERTAIN GOVERNMENTAL ENTITIES

(Submit in duplicate)

PLEASE READ INSTRUCTIONS ON REVERSE CAREFULLY BEFORE COMPLETING THIS FORM

TO: The Director, Bureau of Alcohol, Tobacco and Firearms, Washington, DC 20226

The undersigned hereby makes application to register to the governmental entity identified in this application the firearm described below.

27 CFR 179.104 Provides for limited registration of otherwise unregistrable firearms by certain governmental entities, for official use only. If this application is approved, it is with the condition that the firearm is for "OFFICIAL USE ONLY."

1. NAME AND COMPLETE ADDRESS *(Including ZIP code)* OF DEPARTMENT OR AGENCY MAKING APPLICATION TO REGISTER FIREARM

2. TELEPHONE NUMBER *(Include Area Code)*

3. DESCRIPTION OF FIREARM *(Complete items (a) through (h).)*

NAME AND ADDRESS OF MANUFACTURER AND/OR IMPORTER OF FIREARM a	TYPE OF FIREARM *(See Definitions)* b	*(Check One)* ☐CALIBER ☐GAUGE ☐SIZE c	MODEL d	LENGTH *(Inches)* OF BARREL e	 OVERALL f	SERIAL NUMBER *(See Instruction 1)* g

h. ADDITIONAL DESCRIPTION *(Include all numbers on firearm, and location of each number. If firearm is unserviceable, describe how it was made unserviceable.)*

4. LOCATION WHERE FIREARM IS USUALLY KEPT

5. FROM WHOM WAS FIREARM RECEIVED? *(Optional response - This information may assist ATF in removing records relating to previous registrations of this firearm.)*

I CERTIFY THAT the above described firearm is for OFFICIAL USE ONLY and that I have examined this application and, to the best of my knowledge and belief, it is true, correct and complete.

6. SIGNATURE OF AUTHORIZED OFFICIAL *(See Instruction 1)*	7. PRINT NAME AND TITLE OF AUTHORIZED OFFICIAL	8. DATE

SPACE BELOW IS FOR USE BY THE BUREAU OF ATF - PLEASE DO NOT WRITE BELOW

By authority of the Director, Bureau of Alcohol, Tobacco and Firearms, this application has been examined and the registration of the described firearm to the government entity described above is:

☐ APPROVED FOR OFFICIAL USE

☐ DISAPPROVED

REMARKS

AUTHORIZED ATF OFFICIAL

DATE

ATF F 10 (5320.10) (10-83)

TF Form 5320.20

pplication To Transport Interstate or to Temporarily Export Certain NFA Firearms

INSTRUCTIONS

. A written request and prior authorization from ATF to transport interstate or in foreign commerce any destructive device, machinegun, short-barreled rifle, or short-barreled shotgun is required under the provisions of Section 922(a)(4), Title 18, U.S.C., and Section 178.28, Title 27, CFR. A letter of request, in duplicate, containing all information required on this form, may be submitted in lieu of the form.

The registered owner of NFA firearm(s) shall complete two copies of ATF F 5320.20 and forward the forms to the Director, Bureau of Alcohol, Tobacco and Firearms, Washington, D.C. 20226 (Attention: NFA Branch).

All signatures on both copies of the form shall be in ink. A facsimile signature is not acceptable. All other entries on the form shall be printed in ink or typewritten.

d. The original of ATF F 5320.20 will be returned to the registered owner. Approval authorizes the registered owner to transport the designated firearm(s) only during the time period specified in item 3. THE AUTHORIZATION DOES NOT CARRY OR IMPORT RELIEF FROM ANY STATUTORY OR REGULATORY PROVISIONS RELATING TO FIREARMS OTHER THAN 27 CFR 178.28.

e. In the event item 2 is checked "yes" and the firearm(s) is not returned to the original location by the date specified, the registered owner shall submit a new application on ATF F 5320.20 to receive approval to return the firearm(s).

f. If a contract or common carrier is used to transport the firearm(s) a copy of ATF F 5320.20 shall be furnished to the carrier and shall be in the possession of the carrier for the duration of the transportation. This will meet the requirements of sections 922(e) and (f) of the Gun Control Act of 1968.

PRIVACY ACT INFORMATION

he following information is provided pursuant to Section 3 of the Privacy Act of 1974 5 U.S.C. 552a(e)(3)):

. AUTHORITY. Solicitation of this information is made pursuant to the Gun Control Act of 1968 (18 U.S.C. 922(a)(4)). Disclosure of this information by he applicant is mandatory if the applicant wishes to transport in interstate or foreign commerce any destructive device, machinegun, short-barreled hotgun, or short-barreled rifle.

. PURPOSE. To determine whether the proposed transaction of the listed items is reasonably necessary and consistent with the public safety and pplicable State and local law.

. ROUTINE USES. This information will be used by ATF to make the determination set forth in paragraph 2. No information obtained from an aplication, registration, or records required to be submitted by a natural person in order to comply with the provisions of the National Firearms Act or egulations issued thereunder, shall, except in connection with prosecurition or other action for furnishing false information be used, directly or inirectly, as evidence against that person in any criminal proceeding with the filing of the application.

. EFFECTS OF NOT SUPPLYING INFORMATION REQUESTED. Failure to supply complete information will delay processing and may cause denial f the application.

PAPERWORK REDUCTION ACT NOTICE

his form is in accordance with the Paperwork Reduction Act of 1995. Its purpose is to obtain the information necessary to provide authorization to a erson who is not a qualified firearms licensee who wishes to transport interstate or engage in foreign commerce temporarily, any destructive device, achine gun, short-barreled shotgun or short-barreled rifle. The information will be used to identify the registrant, the firearms to be transported and he destination of the firearms. The furnishing of this information is mandatory (18 USC 922a).

he estimated average burden associated with this collection of information is 30 minutes per respondent or recordkeeper, depending on individual ircumstances. Comments concerning the accuracy of this burden estimate and suggestions for reducing this burden should be addressed to Reports anagement Officer, Document Services Branch, Bureau of Alcohol, Tobacco and Firearms, Washington, D.C. 20226.

n agency may not conduct or sponsor, and a person is not required to respond to, a collection of information unless it displays a currently valid OMB control umber.

ATF F 5320.20 (9-84)

OMB No. 1512-0022 (01/31/00)

DEPARTMENT OF THE TREASURY — BUREAU OF ALCOHOL, TOBACCO AND FIREARMS

APPLICATION TO TRANSPORT INTERSTATE OR TO TEMPORARILY EXPORT CERTAIN NATIONAL FIREARMS ACT (NFA) FIREARMS

(See Reverse for Instructions and Privacy Act Information)

FOR ATF USE ONLY

NFA CONTROL NO.

PART I – APPLICATION (Submit in Duplicate) (Print or Type all Entries)

1. NAME AND ADDRESS OF REGISTERED OWNER *(Full Name, Number, Street, City, State, and ZIP Code)*

2. FIREARMS TO BE RETURNED TO ORIGINAL LOCATION? ☐ YES ☐ NO

3. Dates firearm(s) will be away from original location, if to be returned to that location, (or dates in transit, if to be permanently relocated).

1a. TELEPHONE NUMBER *(Include Area Code)*

FROM *(Month, Day, Year)* TO *(Month, Day, Year)*

4. FIREARM(S) TO BE TRANSPORTED *(Complete New Forms For Additional Firearms)*

MANUFACTURER	TYPE OF FIREARM (Machine gun, etc.)	SERVICE-ABLE (Yes or No)	CALIBER OR GAUGE	MODEL	BARREL LENGTH	OVER-ALL LENGTH	SERIAL NUMBER

5. REASON FOR TRANSPORTATION OF FIREARMS *(Example: Permanent change of Address)*

6. TRANSPORTING FROM *(Number, Street, City, County, State, and ZIP Code)*

7. TRANSPORTING TO *(Number, Street, City, County, State, and ZIP Code)*

8. MODE OF TRANSPORTATION *(Name and Address of carrier, if by common or contract carrier)*

Items 9 thru 11 shall be completed if firearm(s) is/are being temporarily exported.

9. STATE DEPARTMENT LICENSE NO.

10. PORT OF EXIT

11. PORT OF REENTRY

NOTE: If firearm(s) will not be returned or relocated on or before the date specified, submit a new ATF F 7560.8 *(See Instructions "e" and "f")*.

The undersigned certifies that the information on this form is to the best of my knowledge and belief, true and complete, that the transportation does not involve a transfer of title, that the transportation or possession of the listed firearm(s) is not inconsistent with the laws at the place of destination, that all such laws will be complied with, and that, where applicable, all temporary export license provisions under the Arms Export Control Act of 1976 have been complied with.

12. SIGNATURE OF OWNER

13. DATE

PART II – ACTION BY THE BUREAU OF ALCOHOL, TOBACCO AND FIREARMS

By authority of the Director, application to transport or temporarily export the above-listed firearm(s) to and from the location, and for the time period indicated, is:

14. ☐ APPROVED *(with the following conditions, if any)*:

15. ☐ DISAPPROVED *(for the following reason(s))*:

16. SIGNATURE OF AUTHORIZED ATF OFFICIAL

ATF F 5320.20 (9-84) PREVIOUS EDITIONS ARE OBSOLETE

.TF Form 4587 (5330.4)

.pplication to Register as an Importer of U.S. Munitions Import List Articles

INSTRUCTIONS

1. Persons engaged in the business of importing articles on the United States Munitions Import List are required to register pursuant to Section 38 of the Arms Export Control Act of 1976.

2. Complete this form in duplicate using a typewriter or ball point pen. The owner, a partner, a corporate officer or corporate general manager must sign all copies of the application in ink. Submit both copies to:

Bureau of Alcohol, Tobacco & Firearms
P.O. Box 73198
Chicago, IL 60673

3. Item 4 shows the fee schedule which is prescribed by 27 CFR Part 47. This registration must be accompanied by a check or money order made payable to the Bureau of Alcohol, Tobacco, and Firearms in the amount which is applicable to the number of years for which registration is requested.

4. Fees paid in advance for whole future years of a multiple year registration will be refunded upon request if the registrant ceases to engage in importing articles on the U.S. Munitions Import List. A request for a refund must be submitted to the Director, Bureau of Alcohol, Tobacco and Firearms, Washington, D.C. 20226, Attention: Firearms and Explosives Imports Branch, prior to the beginning of any year for which a refund is claimed.

5. If the application is approved, the Director will return the original to the applicant and retain a copy.

6. After registration, importation of U.S. Munitions Import List articles must be effected in accordance with the procedures set forth in 27 CFR Parts 47, 178 and 179, which provides for the use of Form 6 Part I (5330.3A), Application and Permit for Importation of Firearms, Ammunition and Implements of War, and Form 6A (5330.3C), Release and Receipt of Imported Firearms, Ammunition and Implements of War.

7. Federal Firearms License (Item 7) - is a license issued by the Bureau of Alcohol, Tobacco and Firearms pursuant to Chapter 44 of Title 18, U.S.C. and 27 CFR Part 178. Any person engaged in the business of importing firearms or ammunition as defined in 18 U.S.C. 921(a) must be licensed under the provisions of 27 CFR Part 178.

8. Special (occupational) tax stamp (Item 8) - is a stamp issued by the Internal Revenue Service pursuant to registration required by the National Firearms Act, Chapter 53, Internal Revenue Code of 1954 and 27 CFR Part 179. Any person engaged in the business of importing firearms which fall within the definition of 26 U.S.C. 5845(a) must also register and pay a special (occupational) tax pursuant to the provisions of 27 CFR Part 179.

PRIVACY ACT INFORMATION

The following information is provided pursuant to Section 3 of the Privacy Act of 1974 (5 U.S.C. §552a(e)(3).

1. AUTHORITY. Solicitation of this information is made pursuant to section 38 of the Arms Export Control Act of 1976, as amended (22 U.S.C. §2778). Disclosure of this information by the applicant is mandatory if the applicant desires to import U.S. Munitions Import List articles.

2. PURPOSE. To collect the appropriate fee and to effect registration as an importer under the Arms Export Control Act.

3. ROUTINE USES. This information is used by ATF to aid in its law enforcement and regulatory activities with respect to the regulation of firearms and ammunition. This information may also be disclosed to other Federal, State, foreign and local law enforcement and regulatory agency personnel to verify the information on the application and to aid in the performance of their duties with respect to the regulation of firearms and ammunition; and to the State Department in connection with its duties and responsibilities in the area of foreign affairs. The information may further be disclosed to the Justice Department, if it appears that the furnishing of false information may constitute a violation of Federal law.

4. EFFECTS OF NOT SUPPLYING INFORMATION REQUESTED. Failure to supply complete information will delay processing and may cause denial of the application.

The following information is provided pursuant to Section 7(b) of the Privacy Act of 1974:

Disclosure of the individual's social security number is voluntary. Solicitation of this information is pursuant to 22 U.S.C. §2778. The number may be used to verify the individual's identity.

PAPERWORK REDUCTION ACT NOTICE

This request is in accordance with the Paperwork Reduction Act of 1980. This information collection is mandatory pursuant to 22 U.S.C. 2778. The purpose of this information collection is to allow ATF (1) to determine if the registrant qualifies to engage in the business of importing a firearm or firearms, ammunition, and the implements of war, and (2) to facilitate the collection of registration fees.

The estimated average burden associated with this collection of information is 30 minutes per respondent or recordkeeper, depending on individual circumstances. Comments concerning the accuracy of this burden estimate and suggestions for reducing this burden should be addressed to Reports Management Officer, Document Services Branch, Bureau of Alcohol, Tobacco and Firearms, Washington, D.C. 20226, and the Office of Management and Budget, Paperwork Reduction Project (1512-0021), Washington, D.C. 20503.

ATF F 4587 (5330.4) (7-91)

Form Approved: OMB No. 1512-0021(09/30/97)

DEPARTMENT OF THE TREASURY BUREAU OF ALCOHOL, TOBACCO AND FIREARMS **APPLICATION TO REGISTER AS AN IMPORTER OF U.S. MUNITIONS IMPORT LIST ARTICLES** **(SECTION 38, ARMS EXPORT CONTROL ACT OF 1976)** *(See instruction on reverse)*	FOR ATF USE ONLY	
	REGISTRATION NUMBER	EXPIRATION DATE

To: Director, Bureau of Alcohol, Tobacco and Firearms, Attention: Firearms and Explosives Imports Branch, Washington, D.C. 20226.

The undersigned hereby makes application to register as an importer of U.S. Munitions Import List articles are required by Section 38, Arms Export Control Act of 1976.

1. NAME AND ADDRESS *(Includes ZIP Code)*	2. TELEPHONE NO. *(Area Code)*	3. CURRENT A.E.C.A. REGISTRATION NO.

4. REQUEST TO REGISTER FOR: *(Check one, enclose check or money order for payment of fee indicated. See Instruction 3.)*

☐ 1 YEAR - FEE IS $250.00 ☐ 2 YEARS - FEE IS $500.00 ☐ 3 YEARS - FEE IS $700.00 ☐ 4 YEARS - FEE IS $850.00 ☐ 5 YEARS - FEE IS $1,000.00

5. APPLICANT IS: *(Check one)*

☐ INDIVIDUAL ☐ PARTNERSHIP ☐ CORPORATION ☐ OTHER *(Specify)*__________

6. DATE AND PLACE OF INCORPORATION OR COMMENCEMENT OF BUSINESS

7. FEDERAL FIREARMS LICENSE *(if any)*			8. SPECIAL (OCCUPATIONAL) TAX STAMP *(if any)*	
LICENSE NUMBER	CLASS OF LICENSE	EXPIRATION DATE	STAMP NUMBER	CLASS

9. LIST BELOW THE INFORMATION REQUIRED FOR EACH INDIVIDUAL OWNER, PARTNER, AND PRINCIPAL CORPORATE OFFICER IN THE APPLICANT BUSINESS. IF A FEMALE, LIST GIVEN NAMES AND MAIDEN, IF MARRIES, e.g., "MARY ALICE (SMITH) JONES, JONES", NOT "MRS. JOHN JONES." (IF ADDITIONAL SPACE IS NEEDED USE A SEPARATE SHEET).

FULL NAME	POSITION AND SOC. SEC. NO.	HOME ADDRESS *(Include ZIP Code)*	PLACE OF BIRTH	DATE OF BIRTH

10. DESCRIBE SPECIFIC ACTIVITY APPLICANT IS ENGAGED IN, OR INTENDS TO ENGAGE IN, WHICH REQUIRES REGISTRATION UNDER THE ARMS EXPORT CONTROL ACT

11. UNITED STATES MUNITIONS IMPORT LIST ARTICLES USUALLY IMPORTED *(Specify categories)*

12. PURPOSE OF IMPORTING MUNITIONS IMPORT LIST ARTICLES	13. WHAT PERCENTAGE IS IMPORTED FOR U.S. GOVERNMENT

I declare under the penalties provided by law that this request for registration as an importer of U.S. Munitions Import List articles, including the documents submitted in support of it, has been examined by me and, to the best of my knowledge and belief, it is true, correct & complete.

14. SIGNATURE *(Sign all copies in ink. See instruction 2)*	15. TITLE	16. DATE

APPROVAL (FOR ATF USE ONLY)

17. SIGNATURE OF DIRECTOR, BUREAU OF ALCOHOL, TOBACCO AND FIREARMS	18. DATE

ATF F 4587(5330.4) (7-91) PREVIOUS EDITIONS ARE OBSOLETE

ATF Form 5630.7
Special Tax Registration and Return, National Firearms Act

INSTRUCTION SHEET
ATF F 5630.7, SPECIAL TAX REGISTRATION AND RETURN
FIREARMS

GENERAL INSTRUCTIONS

If you are engaged in one or more of the National Firearms Act (NFA) activities listed on this form *(see definition)*, you are required to file this form and pay special occupational tax before beginning business. This form is for NFA taxpayers only. Businesses engaged in alcohol and tobacco related activities subject to special tax should file ATF F 5630.5. You may file one return to cover several locations or several types of activity. However, you must submit a separate return for each tax period. The special occupational tax period runs from July 1 through June 30 and payment is due annually by July 1. If you do not pay on a timely basis interest will be charged and penalties may be incurred.

If you engage in a taxable activity at more than one location, attach to your return a sheet showing your name, trade name, address and employer identification number and the complete street addresses of all additional locations.

As evidence of tax payment, you will be issued a Special Tax Stamp, ATF F 5630.6A, for each location and/or business. You must have a Federal Firearms License (FFL) for the location, appropriate to the type of activity conducted. The type of business *(sole propriator, partnership, corporation)* must be the same for the taxable activity and the FFL. If a trade name is used, it must be the same on the tax stamp and the FFL.

The special tax rates listed on this form became effective January 1, 1988. If you were engaged in NFA firearms related activity prior to this date and did not pay special occupational tax, please contact the ATF National Revenue Center for assistance.

SECTION I - TAX IDENTIFYING INFORMATION

Complete Section I, Taxpayer Identifying Information, as specified on the form. Enter the tax period covered by the return in the space provided. Your return must contain a valid EMPLOYER IDENTIFICATION NUMBER (EIN). The EIN is a unique number for business entities issued by the Internal Revenue Service (IRS). **You must have an EIN whether you are an individual ownership, partnership, corporation, or agency of the government.** If you do not have an EIN, contact your local IRS office immediately to obtain one. While ATF may assign a temporary identification number *(beginning with XX)* to allow initial processing of a return which lacks an EIN, **a tax stamp will not be issued until you have submitted a valid EIN.** Do not delay submission of your return and payment past the due date pending receipt of your EIN. If you have not received a number by the time you file this return, write *"number applied for"* in the space for the number. Submit your EIN by separate correspondence after receipt from the IRS.

SECTION II - TAX COMPUTATION

To compete Section II, enter the number of locations in Col. (d) on the appropriate line(s) and multiply by the tax rate, Col. (c). Insert the tax due in Col. (e). Compute the taxes due for each class and enter the total amount due in the block *"Total Tax Due"*.

INSTRUCTIONS FOR REDUCED RATE TAXPAYERS

The reduced rates for certain tax classes, indicated with an asterisk (*) in Section II, apply only to those taxpayers whose total gross receipts for your most recent income tax year are less than $500,000 *(not just receipts relating to the activity subject to special occupational tax)*. However, if you are a member of a controlled group as defined in section 5061(e)(3) of the Internal Revenue Code, you are not eligible for this reduced rate unless the total gross receipts for the entire group are less than $500,000. If your business is beginning an activity subject to special tax for the first time, you may qualify for a reduced rate in your initial tax year if gross receipts for the business *(or the entire control group, if a member of a control group)* were under $500,000 the previous year. If you are eligible for the reduced rate, check item 12 in Section III and compute your tax using the reduced rate in Section II.

SECTION III - BUSINESS REGISTRATION

Please complete the ownership information in Section III. Supply the information specified in item 11 for each individual owner, partner or responsible person. For a corporation, partnership or association, a responsible person is anyone with the power to control the management policies or buying or selling practices pertaining to firearms. For a corporation, association or similar organization, it also means any person owning 10 percent or more of the outstanding stock in the business.

CHANGES IN OPERATIONS

For a change of address, location or trade name, an amended ATF F 5630.7 must be filed and approved before the change is made. Return your Special Tax Stamp, ATF F 5630.6A, along with the completed ATF F 5630.7 to: Bureau of ATF, 8002 Federal Office Building, 550 Main Street, Cincinnati, OH 45202-3263 and an amended ATF F 5630.6A will be issued. All taxpayers with such changes must return their FFL to the ATF Firearms and Explosives Licensing Center *(address listed on FFL)* for amendment.

If special taxpayers do not register these changes within the appropriate time frames, additional tax and interest will be charged and penalties may be incurred. For a change in ownership or control of an activity, consult the ATF Firearms and Explosives Licensing Center, at 404-679-5040, before beginning the activity. If the Federal firearms licensee discontinues business and retains NFA firearms, this retention may be in violation of law. The licensee should check with State or local authorities.

DEFINITION

IMPORTERS, MANUFACTURERS, and DEALERS of FIREARMS subject to the National Firearms Act *(tax class codes 61, 62, 63, 71, or 72)* are individuals or business entities who import, manufacture or deal in machineguns, short-barreled shotguns and rifles, destructive devices, etc. See 26 U.S.C. 5845 for additional information on the types of weapons subject to the National Firearms Act. *(NOTE: This tax is not required from those persons or entities who deal only in conventional, sporting type firearms.)*

MISCELLANEOUS INSTRUCTIONS

If you do not intend to pay the special tax for the next year, you must dispose of any machineguns manufactured or imported after May 19, 1986, prior to your special tax status lapsing. Title 18, United States Code, section 922(o) makes it unlawful to possess these machineguns unless you are properly qualified. As provided in Title 27, Code of Federal Regulations, Part 179,105(f), the disposition must be made to a government agency or qualified licensee or the weapon must be destroyed.

This form must be signed by the individual owner, a partner, or, in the case of a corporation, by an individual authorized to sign for the corporation.

Please sign and date the return, make check or money order payable to BUREAU OF ALCOHOL, TOBACCO AND FIREARMS, for the amount in the Total Tax Due block, and MAIL THE FORM ALONG WITH THE PAYMENT TO BUREAU OF ATF, Attention: NFA, P.O. Box 371970, Pittsburgh, PA 15250-7970.

IF YOU NEED FURTHER ASSISTANCE
CONTACT ATF NATIONAL REVENUE CENTER
AT
1-800-937-8864
OR
513-684-2979

*U.S. GPO: 1999-461-270/23706

ATF F 5630.7 (10-99)

OMB No. 1512-0472 (09/30/2002)

DEPARTMENT OF THE TREASURY
BUREAU OF ALCOHOL, TOBACCO AND FIREARMS

SPECIAL TAX REGISTRATION AND RETURN
NATIONAL FIREARMS ACT (NFA)

(Please Read Instructions on Back Carefully Before Completing This Form)

SECTION I - TAXPAYER IDENTIFYING INFORMATION

1. EMPLOYER IDENTIFICATION NUMBER *(Required - see instructions)*
2. BUSINESS TELEPHONE NUMBER ( )

FOR ATF USE ONLY

T

FF

FP

I

T

3. NAME *(Last, First, Middle)* or CORPORATION *(If Corporation)*
4. TRADE NAME
5. MAILING ADDRESS *(Street address or P.O. box number)*
6. CITY STATE ZIP CODE

ACTUAL LOCATION *(IF DIFFERENT THAN ABOVE)*

7. PHYSICAL ADDRESS OF PRINCIPAL PLACE OF BUSINESS *(Show street address)*
8. CITY STATE ZIP CODE
9. TAX PERIOD COVERING *(Only one tax period per form)*

FROM: ________ *(mm/dd/yy)*

TO: JUNE 30, ______ *(yy)*

SECTION II - TAX COMPUTATION

(a) TAX CLASS *(For Items Marked*, See Instructions)*	(b) TAX CLASS CODE	(c) TAX RATE ($) *(Annual)*	(d) NUMBER OF LOCATIONS	(e) TAX DUE
CLASS 1 - IMPORTER OF FIREARMS	61	$1,000		
CLASS 1 - IMPORTER OF FIREARMS *(REDUCED)**	71*	500		
CLASS 2 - MANUFACTURER OF FIREARMS	62	1,000		
CLASS 2 - MANUFACTURER OF FIREARMS *(REDUCED)**	72*	500		
CLASS 3 - DEALER IN FIREARMS	63	500		

MAKE YOUR CHECK OR MONEY ORDER PAYABLE TO "BUREAU OF ALCOHOL, TOBACCO AND FIREARMS", WRITE YOUR EMPLOYER IDENTIFICATION NUMBER ON THE CHECK AND SEND IT WITH THE RETURN TO BUREAU OF ATF, ATTENTION NFA, P.O. BOX 371970, PITTSBURGH, PA 15250-7970.

TOTAL TAX DUE

SECTION III - BUSINESS REGISTRATION

10. OWNERSHIP INFORMATION: *(Check One Box Only)* ☐ INDIVIDUAL OWNER ☐ PARTNERSHIP ☐ CORPORATION ☐ OTHER *(Specify)*______
11. OWNERSHIP RESPONSIBILITY *(See instructions on back; use a separate sheet of paper if additional space is needed.)*

FULL NAME	ADDRESS	POSITION

12. ☐ GROSS RECEIPTS less than $500,000 *(See instructions on back; use a separate sheet of paper if additional space is needed.)*
13. ☐ NEW BUSINESS — DATE BUSINESS COMMENCED *(mm, dd, yy)*
14. ☐ EXISTING BUSINESS WITH CHANGE IN:
    - ☐ (a) NAME/TRADE NAME *(Indicate)*
    - ☐ (b) ADDRESS *(Indicate)*
    - ☐ (c) OWNERSHIP *(Indicate)*
    - ☐ (d) EMPLOYER IDENTIFICATION NUMBER (OLD: NEW:)
    - ☐ (e) BUSINESS TELEPHONE NUMBER ( )
15. ☐ DISCONTINUED BUSINESS

SECTION IV - TAXPAYER CERTIFICATION

Under penalties of perjury, I declare that the statements in this return/registration are true and correct to the best of my knowledge and belief; that this return/registration applies only to the specified business and location or, where the return/registration is for more than one location, it applies only to the businesses at the locations specified on the attached list. **NOTE:** Violation of Title 26, United States Code 7206, is punishable upon conviction by a fine of not more than $100,000 *($500,000 in the case of a corporation)*, or imprisonment for not more 3 years, or both, together with the costs of prosecution.

16. SIGNATURE	17. TITLE	18. DATE

ATF F 5630.7 (10-99) PREVIOUS EDITIONS ARE OBSOLETE

ATF E-From 5013.3

Form6 Access Request

Instructions

You must complete this form in order to receive a user ID and password to obtain access to ATF's eForm 6 system. Each user must obtain an individual user ID and password which is not to be shared with anyone. Sharing your user ID and password can result in cancellation of your eForm 6 privileges.

Section A – You must check the appropriate box:

(1) Check "Add User" if you want access to eForm 6 for the first time.

(2) Check "Modify User" if you want to change any of the information you originally supplied in Section B or C. In all cases, supply your User ID (Section A, Item 2) and complete only those items in Section B and C that have changed.

For changes to Section C: (i) You must notify ATF's National Licensing Center in Atlanta, GA of any changes to your company name, address, or Federal firearms license numbers, or the Firearms and Explosives Imports Branch of any changes to your Arms Export Control Act registration information, before making any changes in eForm 6; and (ii) If you are deleting a Federal firearms license or Arms Export Control Act registration number, indicate in Section C, Item 15 which number(s) you want deleted from eForm 6.

(3) Check "Delete User" if you no longer want access to eForm 6 for yourself or another user. Please provide the User ID of the user to be deleted, if known (Section A, Item 2).

(4) Check "Reactivate User" if we cancelled your original User ID due to inactivity and you wish to begin using the eForm 6 system again. You must also complete the remainder of the form as instructed below and include your previous User ID.

Section B – You must enter the required information about the individual requesting access to eForm 6 in items 3-13. Also include your business telephone and FAX numbers. Each Federal firearms licensee or Arms Export Control Act registrant must submit one eForm 6 Access Request from a responsible person as indicated on item 14. This person will be able to review the User Profiles of all other users registered under that Federal firearms license or Arms Export Control Act registration number. This individual will also receive a confirmation email for each application submitted to ATF, whether via the eForm 6 or paper submission. A Responsible Person is defined as a sole proprietor, or in the case of a corporation, partnership or association, any individual possessing the power to direct or cause the direction of the management, policies and practices of the corporation, partnership or association as they relate to firearms, and in the case of a corporation, partnership, or association any person holding ten percent or more of the outstanding shares of stock issued by the applicant and the officers of that organization. These persons are listed on the ATF Form 7, Application for Federal Firearms License and ATF Form 4587, Application to Register as an Importer of U.S. Munitions Import List Articles.

Section C – You must enter the required information about the company for which you are requesting to file applications. This information must appear exactly as it does on the Federal Firearms License and/or Arms Export Control Act registration. Be sure to enter the correct number in item 16 (example: 1-23-456-08-5A-98765 or A-12-345-6789). You (the individual requesting access) must sign and date the form in items 17-18. If you are both a Federal firearms licensee and an Arms Export Control Act registrant, you must enter **both** numbers in item 16.

Section D – A person listed as responsible person on the ATF Form 7 or ATF Form 4587, must sign and print his or her name and title, and date the form in items 19-21.

You must send the original of this form to:

Firearms and Explosives Imports Branch
Bureau of Alcohol, Tobacco, Firearms and Explosives
650 Massachusetts Avenue, NW.
Washington, DC 20226

Your user ID and password will be sent to you separately for security reasons.

Privacy Act Information

We provide this information to comply with Section 3 of the Privacy Act of 1974 (5 U.S.C. 552a(e)(3)).

We require this information under the authority of 18 U.S.C. 925(d). You must disclose this information so we may identify the company on whose behalf applicant claims to act, to verify the scope of the applicant's authority to act, and to evaluate the applicant's qualifications for access to the system.

We use this informaion to approve, grant and control access to sensitive information systems. In addition, the information may be disclosed to other Federal, State and local law enforcement and regulatory agency personnel to verify information on the application and to aid in the performance of their duties. Disclosure may otherwise be made pursuant to the routine uses most recently published in the Federal Register for ATF's Regulatory Enforcement Records System (Treasury/ATF.008).

If you fail to supply complete information then there will be a delay in the processing of your application.

Disclosure of your Social Security Number is voluntary. Solicitation of this information is pursuant to section 925(d), Title 18 U.S.C. The Social Security Number may be used to verify the applicant's identity. If you fail to supply your Social Security Number, there will either be a delay in processing your application or you will not be granted access to the system.

Paperwork Reduction Act Notice

This request is in accordance with the Paperwork Reduction Act of 1995. We use this information to authenticate end users in the program to electronically file ATF Form 6 Part I (5330.3A). The information is used by the Government to verify the identity of the end users prior to issuing them passwords. The information we request is voluntary, however, if the requested information is not submitted, the users will not be granted a password and cannot participate in the electronic program.

The estimated average burden associated with this collection is 18 minutes per respondent or recordkeeper depending on the individual circumstances. Comments concerning the accuracy of this burden estimate and suggestions for reducing this burden should be directed to the Reports Management Officer, Documents Services Branch, Bureau of Alcohol, Tobacco, Firearms and Explosives, Washington, DC 20226.

ATF may not conduct or sponsor, and you are not required to respond to a collection of information unless it displays a currently valid OMB control number.

ATF E-Form 5013.3
Revised February 2004

U.S. Department of Justice
Bureau of Alcohol, Tobacco, Firearms and Explosives

OMB No. 1140-0087 (01/3

eForm 6 Access Request

Submit this form to the Firearms and Explosives Imports Branch, ATF, Washington, DC 20226	ATF Tracking Number

A - Action Requested

1. ☐ Add User ☐ Modify User Information ☐ Delete User ☐ Reactivate User	2. If Modifying or Deleting, Provide User IL *known*

B - User Information: *Please complete this section with all the required data to establish a user identification record*

3. First Name	4. Middle Initial	5. Last Name	6. Suffix (*i.e., Jr.,*
7. Social Security Number *(last 4 digits)*	8. Date of Birth		9. Mother's Maiden Name
10. Employee Title	11. Business E-mail Address		12. Business Phone Number
13. Business Fax Number	14. ☐ Check Here if User Will Be Responsible Person (*See instructions for defin*		

C - Company Information: *Provide information about the company for which you work*

15a. Name of Company as it Appears on Your Federal Firearms License and/or Arms Export Control Act Registration			
15b. Street Address of Company as it Appears on Your Federal Firearms License and/or Arms Export Control Act Registration			
15c. City	15d. State	15e. ZIP Code	15f. County
16. Federal Firearms License Number and/or Arms Export Control Act Registration Number			

Requester's Certification: I hereby attest that the entries on this form are true and correct and that the unique username and password or digital sig that the Bureau of Alcohol, Tobacco, Firearms and Explosives assigns to me are intended as my original signature and I intend that such submission treated as bearing an original signature for all intents and purposes when submitting firearm import applications electronically via the eForm 6 Syste have read and agree to be bound by the terms set out in the eForm 6 Notices and Agreement governing the use of the eForm 6 System.

17. Requester's Signature	18. Date

D - Approval Required: ***Signature of responsible person with signature authority required to grant access to eForm 6 System***

Responsible Person's Certification: I authorize the above-named user to complete and execute, on behalf of the company named in Item 15, firear import applications via the eForm 6 System. I attest that the company intends to be bound by the entries on any such applications and intends that su applications be treated as bearing an origianal signature for all intents and purposes. I have read and on behalf of the company agree to be bound by terms set out in the eForm 6 Notices and Agreement governing the use of the eForm 6 System. I hereby ratify and confirm all that the user shall law or cause to be done by virtue of this authorization.

19. Company Approval Signature By Responsible Person	20. Print Name and Title of Responsible Person	21. Date

ATF Use Only

User Verification Completed	Date	Comments
System Owner Approval	Date	Comments
ATF Operations Completed	Date	Comments
System Administrator Completed	Date	Comments
User Notification Completed	Date	Comments

ATF E-Form 5
Revised Febru

APPENDIX D

SAMPLE FORM LETTERS

FORM LETTER OF GOVERNMENT AGENCY PURCHASING IMPORTED FIREARMS

(Note: In this example, the department is purchasing firearms from a licensed dealer. If the order includes NFA firearms, the dealer must also be qualified under the NFA to deal in NFA firearms. The dealer will submit the letter to the importer for attachment to the import application.)
(*Letterhead of the Department*)
(*Date*)

Mr. John Doe
Doe's Police Supplies
(*Address*)

Dear Mr. Doe:

This is an order for (*quantity*) (*description of firearms, including type, manufacturer, model, and caliber*) at the price of (*amount*) each, plus shipping charges, for (*name of department, sheriff's office, or agency*).

The number of sworn full time officers in our (*department, office, or agency*) is (*number of officers*). These officers are authorized by law to make arrests and carry firearms in the performance of their official duties.

This order is being paid for with funds that our (*department, office, or agency*) is authorized to use under law for the acquisition of equipment (*Note: If funds of individual officers are being used to buy departmental firearms for the officers' official use, then add "including funds from individual officers who may be required to pay in whole or in part for their equipment to carry out their official duties.*"). These firearms will be the property of our (*department, office, or agency*) and are not being acquired for the purpose of resale or transfer, and they will be used to carry out its official responsibilities and duties.

The telephone contact number for this (*department, office, or agency*) is (telephone number).

(*If a purchase order has been issued by the department for the purchase of the firearms, then add the statement here that "Enclosed is our (department's, office's, or agency's) purchase order for these firearms.)*

Sincerely yours,

(*Name of department, office or agency*)

By: ________________________
(*Name and title of signing official*)

FROM IMPORTER
FORM LETTER OF DEALER ORDERING NFA SALES SAMPLE

(Note: In this example, a licensed dealer qualified to deal in NFA firearms is purchasing an NFA sales sample from a licensed importer qualified under the NFA to import such firearms. This letter, together with the letter of the law enforcement agency requesting a demonstration of the firearm, will be attached to the importer's application to import the firearm on ATF Form 6.)

(*Letterhead of the Dealer*)

(*Date*)

Mr. Richard Roe
Roe & Son, Importers
(*Address*)

Dear Mr. Roe:

This is an order for (*describe the firearm by type, manufacturer, model, and caliber)* at the price of (*amount*), plus shipping charges.

The firearm will be used by Doe's Police Supplies in a demonstration of the firearm to the (*name of the government department, office, or agency*). Attached is the letter of (*name of the department, office, or agency*), dated (*date*), requesting the demonstration.

The firearm is particularly suitable for use as a law enforcement weapon (*add here a statement showing why it is suitable for such use and cite examples of its use by other agencies if known*). The firearm is currently in production by (*name and location of foreign manufacturer*), so quantities of the firearm are readily available to fill any subsequent orders for the firearm from (*name of the department, office, or agency asking for the demonstration*).[233]

Sincerely yours,

Doe's Police Supplies

By:____________________
(*Name and title of person signing*)

---

[233] If the dealer has no information on the availability of the firearm to fill subsequent orders, the importer should include a statement in its letter transmitting the permit application to ATF showing availability for future sales. Also, if the firearm is not currently in production, other facts must be shown establishing that there are quantities available to fill future orders.

FORM LETTER FOR GOVERNMENT AGENCY TO REQUEST A DEMONSTRATION OF AN NFA FIREARM

(Note: In this example, the department is requesting a licensed dealer to obtain an imported NFA "sales sample" for demonstration to the department. The dealer must be qualified under the NFA to deal in NFA firearms. The dealer will submit the letter to the importer for attachment to the import application.)

(*Letterhead of the Department*)

(*Date*)

Mr. John Doe
Doe's Police Supplies
(*Address*)

Dear Mr. Doe:

The (*name of the department, office, or agency*) would like a demonstration of the (*describe the firearm by type, manufacturer, model and caliber*) for possible future purchase and use of our officers in the performance of their official duties.

The number of sworn full time officers in our (*department, office, or agency*) is (*number of officers*). These officers are authorized by law to make arrests and carry firearms in the performance of their official duties.

The firearm requested for demonstration is particularly suitable for use as a law enforcement weapon. (*add here a statement showing why it is suitable for such use and cite examples of its use by other agencies if known*).

The telephone contact number for this (*department, office, or agency*) is (*telephone number*).

Sincerely yours,

(*Name of department, office, or agency*)

By:______________________

(*Name and title of signing official*)

Made in the USA
Monee, IL
04 November 2026

67381315R00174